Springer Handbook
of Nanotechnology

Bharat Bhushan (Ed.)

3rd revised and extended edition

Springer Handbook of Nanotechnology

Since 2004 and with the 2nd edition in 2006, the Springer Handbook of Nanotechnology has established itself as the definitive reference in the nanoscience and nanotechnology area. It integrates the knowledge from nanofabrication, nanodevices, nanomechanics, nanotribology, materials science, and reliability engineering in just one volume. Beside the presentation of nanostructures, micro/nanofabrication, and micro/nanodevices, special emphasis is on scanning probe microscopy, nanotribology and nanomechanics, molecularly thick films, industrial applications and microdevice reliability, and on social aspects. In its 3rd edition, the book grew from 8 to 9 parts now including a part with chapters on biomimetics. More information is added to such fields as bionanotechnology, nanorobotics, and (bio) MEMS/NEMS, bio/nanotribology and bio/nanomechanics. The book is organized by an experienced editor with a universal knowledge and written by an international team of over 145 distinguished experts. It addresses mechanical and electrical engineers, materials scientists, physicists and chemists who work either in the nano area or in a field that is or will be influenced by this new key technology.

"The strong point is its focus on many of the practical aspects of nanotechnology... Anyone working in or learning about the field of nanotechnology would find this an excellent working handbook."

IEEE Electrical Insulation Magazine

"Outstandingly succeeds in its aim... It really is a magnificent volume and every scientific library and nanotechnology group should have a copy."

Materials World

"The integrity and authoritativeness... is guaranteed by an experienced editor and an international team of authors which have well summarized in their chapters information on fundamentals and applications."

Polymer News

List of Abbreviations
1 Introduction to Nanotechnology

Part A Nanostructures, Micro-/Nanofabrication and Materials

2 Nanomaterials Synthesis and Applications: Molecule-Based Devices
3 Introduction to Carbon Nanotubes
4 Nanowires
5 Template-Based Synthesis of Nanorod or Nanowire Arrays
6 Templated Self-Assembly of Particles
7 Three-Dimensional Nanostructure Fabrication by Focused Ion Beam Chemical Vapor Deposition
8 Introduction to Micro-/Nanofabrication
9 Nanoimprint Lithography-Patterning of Resists Using Molding
10 Stamping Techniques for Micro- and Nanofabrication
11 Material Aspects of Micro- and Nanoelectromechanical Systems

Part B MEMS/NEMS and BioMEMS/NEMS

12 MEMS/NEMS Devices and Applications
13 Next-Generation DNA Hybridization and Self-Assembly Nanofabrication Devices
14 Single-Walled Carbon Nanotube Sensor Concepts
15 Nanomechanical Cantilever Array Sensors
16 Biological Molecules in Therapeutic Nanodevices
17 G-Protein Coupled Receptors: Progress in Surface Display and Biosensor Technology
18 Microfluidic Devices and Their Applications to Lab-on-a-Chip
19 Centrifuge-Based Fluidic Platforms
20 Micro-/Nanodroplets in Microfluidic Devices

Part C Scanning-Probe Microscopy

21 Scanning Probe Microscopy-Principle of Operation, Instrumentation, and Probes
22 General and Special Probes in Scanning Microscopies
23 Noncontact Atomic Force Microscopy and Related Topics
24 Low-Temperature Scanning Probe Microscopy
25 Higher Harmonics and Time-Varying Forces in Dynamic Force Microscopy
26 Dynamic Modes of Atomic Force Microscopy
27 Molecular Recognition Force Microscopy: From Molecular Bonds to Complex Energy Landscapes

Part D Bio-/Nanotribology and Bio-/Nanomechanics

28 Nanotribology, Nanomechanics, and Materials Characterization
29 Surface Forces and Nanorheology of Molecularly Thin Films
30 Friction and Wear on the Atomic Scale
31 Computer Simulations of Nanometer-Scale Indentation and Friction
32 Force Measurements with Optical Tweezers
33 Scale Effect in Mechanical Properties and Tribology
34 Structural, Nanomechanical, and Nanotribological Characterization of Human Hair Using Atomic Force Microscopy and Nanoindentation
35 Cellular Nanomechanics
36 Optical Cell Manipulation
37 Mechanical Properties of Nanostructures

Part E Molecularly Thick Films for Lubrication

38 Nanotribology of Ultrathin and Hard Amorphous Carbon Films
39 Self-Assembled Monolayers for Nanotribology and Surface Protection
40 Nanoscale Boundary Lubrication Studies

Part F Biomimetics

41 Multifunctional Plant Surfaces and Smart Materials
42 Lotus Effect: Surfaces with Roughness-Induced Superhydrophobicity, Self-Cleaning, and Low Adhesion
43 Biological and Biologically Inspired Attachment Systems
44 Gecko Feet: Natural Hairy Attachment Systems for Smart Adhesion

Part G Industrial Applications

45 The *Millipede*-A Nanotechnology-Based AFM Data-Storage System
46 Nanorobotics

Part H Micro-/Nanodevice Reliability

47 MEMS/NEMS and BioMEMS/BioNEMS: Materials, Devices, and Biomimetics
48 Friction and Wear in Micro- and Nanomachines
49 Failure Mechanisms in MEMS/NEMS Devices
50 Mechanical Properties of Micromachined Structures
51 High-Volume Manufacturing and Field Stability of MEMS Products
52 Packaging and Reliability Issues in Micro-/Nanosystems

Part I Technological Convergence and Governing Nanotechnology

53 Governing Nanotechnology: Social, Ethical and Human Issues

Subject Index

使 用 说 明

1.《纳米技术手册》原版为一册，分为A~I部分。考虑到使用方便以及内容一致，影印版分为7册：第1册—Part A，第2册—Part B，第3册—Part C，第4册—Part D，第5册—Part E，第6册—Part F，第7册—Part G、H、I。

2.各册在页脚重新编排页码，该页码对应中文目录。保留了原书页眉及页码，其页码对应原书目录及主题索引。

3.各册均给出完整7册书的章目录。

4.作者及其联系方式、缩略语表各册均完整呈现。

5.主题索引安排在第7册。

6.目录等采用中英文对照形式给出，方便读者快速浏览。

材料科学与工程图书工作室

联系电话　0451-86412421
　　　　　0451-86414559

邮　　箱　yh_bj@yahoo.com.cn
　　　　　xuyaying81823@gmail.com
　　　　　zhxh6414559@yahoo.com.cn

Springer 手册精选系列

纳米技术手册

微纳机电系统和生物微纳机电系统

【第2册】

Springer
Handbook of
Nanotechnology

〔美〕Bharat Bhushan　主编

（第三版影印版）

哈尔滨工业大学出版社
HARBIN INSTITUTE OF TECHNOLOGY PRESS

黑版贸审字 08-2013-001号

Reprint from English language edition:
Springer Handbook of Nanotechnology
by Bharat Bhushan
Copyright © 2010 Springer Berlin Heidelberg
Springer Berlin Heidelberg is a part of Springer Science+Business Media
All Rights Reserved

This reprint has been authorized by Springer Science & Business Media for distribution in China Mainland only and not for export there from.

图书在版编目（CIP）数据

纳米技术手册：第3版. 2, 微纳机电系统和生物微纳机电系统 = Handbook of Nanotechnology.2,MEMS/NEMS and BioMEMS/NEMS：英文 /（美）布尚(Bhushan,B.)主编. — 哈尔滨：哈尔滨工业大学出版社, 2013.1
（Springer手册精选系列）
ISBN 978-7-5603-3948-1

Ⅰ.①纳… Ⅱ.①布… Ⅲ.①纳米技术 – 手册 – 英文 ②纳米技术 – 机电系统 – 手册 – 英文 Ⅳ.①TB303-62②TM7-62

中国版本图书馆CIP数据核字(2013)第007509号

责任编辑	杨　桦　许雅莹　张秀华
出版发行	哈尔滨工业大学出版社
社　　址	哈尔滨市南岗区复华四道街10号 邮编 150006
传　　真	0451-86414749
网　　址	http://hitpress.hit.edu.cn
印　　刷	哈尔滨市石桥印务有限公司
开　　本	787mm×960mm 1/16 印张 16
版　　次	2013年1月第1版　2013年1月第1次印刷
书　　号	ISBN 978-7-5603-3948-1
定　　价	48.00元

（如因印刷质量问题影响阅读，我社负责调换）

Foreword by Neal Lane

In a January 2000 speech at the California Institute of Technology, former President W.J. Clinton talked about the exciting promise of *nanotechnology* and the importance of expanding research in nanoscale science and engineering and, more broadly, in the physical sciences. Later that month, he announced in his State of the Union Address an ambitious US$ 497 million federal, multiagency national nanotechnology initiative (NNI) in the fiscal year 2001 budget; and he made the NNI a top science and technology priority within a budget that emphasized increased investment in US scientific research. With strong bipartisan support in Congress, most of this request was appropriated, and the NNI was born. Often, federal budget initiatives only last a year or so. It is most encouraging that the NNI has remained a high priority of the G.W. Bush Administration and Congress, reflecting enormous progress in the field and continued strong interest and support by industry.

Nanotechnology is the ability to manipulate individual atoms and molecules to produce nanostructured materials and submicron objects that have applications in the real world. Nanotechnology involves the production and application of physical, chemical and biological systems at scales ranging from individual atoms or molecules to about 100 nm, as well as the integration of the resulting nanostructures into larger systems. Nanotechnology is likely to have a profound impact on our economy and society in the early 21st century, perhaps comparable to that of information technology or cellular and molecular biology. Science and engineering research in nanotechnology promises breakthroughs in areas such as materials and manufacturing, electronics, medicine and healthcare, energy and the environment, biotechnology, information technology and national security. Clinical trials are already underway for nanomaterials that offer the promise of cures for certain cancers. It is widely felt that nanotechnology will be the next industrial revolution.

Nanometer-scale features are built up from their elemental constituents. Micro- and nanosystems components are fabricated using batch-processing techniques that are compatible with integrated circuits and range in size from micro- to nanometers. Micro- and nanosystems include micro/nanoelectro-mechanical systems (MEMS/NEMS), micromechatronics, optoelectronics, microfluidics and systems integration. These systems can sense, control, and activate on the micro/nanoscale and can function individually or in arrays to generate effects on the macroscale. Due to the enabling nature of these systems and the significant impact they can have on both the commercial and defense applications, industry as well as the federal government have taken special interest in seeing growth nurtured in this field. Micro- and nanosystems are the next logical step in the *silicon revolution*.

The discovery of novel materials, processes, and phenomena at the nanoscale and the development of new experimental and theoretical techniques for research provide fresh opportunities for the development of innovative nanosystems and nanostructured materials. There is an increasing need for a multidisciplinary, systems-oriented approach to manufacturing micro/nanodevices which function reliably. This can only be achieved through the cross-fertilization of ideas from different disciplines and the systematic flow of information and people among research groups.

Prof. Neal Lane
Malcolm Gillis University Professor,
Department of Physics and Astronomy,
Senior Fellow,
James A. Baker III Institute for Public Policy
Rice University
Houston, Texas

Served in the Clinton Administration as Assistant to the President for Science and Technology and Director of the White House Office of Science and Technology Policy (1998–2001) and, prior to that, as Director of the National Science Foundation (1993–1998). While at the White House, he was a key figure in the creation of the NNI.

Nanotechnology is a broad, highly interdisciplinary, and still evolving field. Covering even the most important aspects of nanotechnology in a single book that reaches readers ranging from students to active researchers in academia and industry is an enormous challenge. To prepare such a wide-ranging book on nanotechnology, Prof. Bhushan has harnessed his own knowledge and experience, gained in several industries and universities, and has assembled internationally recognized authorities from four continents to write chapters covering a wide array of nanotechnology topics, including the latest advances. The authors come from both academia and industry. The topics include major advances in many fields where nanoscale science and engineering is being pursued and illustrate how the field of nanotechnology has continued to emerge and blossom. Given the accelerating pace of discovery and applications in nanotechnology, it is a challenge to capture it all in one volume. As in earlier editions, professor Bhushan does an admirable job.

Professor Bharat Bhushan's comprehensive book is intended to serve both as a textbook for university courses as well as a reference for researchers. The first and second editions were timely additions to the literature on nanotechnology and stimulated further interest in this important new field, while serving as invaluable resources to members of the international scientific and industrial community. The increasing demand for up-to-date information on this fast moving field led to this third edition. It is increasingly important that scientists and engineers, whatever their specialty, have a solid grounding in the fundamentals and potential applications of nanotechnology. This third edition addresses that need by giving particular attention to the widening audience of readers. It also includes a discussion of the social, ethical and political issues that tend to surround any emerging technology.

The editor and his team are to be warmly congratulated for bringing together this exclusive, timely, and useful nanotechnology handbook.

Foreword by James R. Heath

Nanotechnology has become an increasingly popular buzzword over the past five years or so, a trend that has been fueled by a global set of publicly funded nanotechnology initiatives. Even as researchers have been struggling to demonstrate some of the most fundamental and simple aspects of this field, the term nanotechnology has entered into the public consciousness through articles in the popular press and popular fiction. As a consequence, the expectations of the public are high for nanotechnology, even while the actual public definition of nanotechnology remains a bit fuzzy.

Why shouldn't those expectations be high? The late 1990s witnessed a major information technology (IT) revolution and a minor biotechnology revolution. The IT revolution impacted virtually every aspect of life in the western world. I am sitting on an airplane at 30 000 feet at the moment, working on my laptop, as are about half of the other passengers on this plane. The plane itself is riddled with computational and communications equipment. As soon as we land, many of us will pull out cell phones, others will check e-mail via wireless modem, some will do both. This picture would be the same if I was landing in Los Angeles, Beijing, or Capetown. I will probably never actually print this text, but will instead submit it electronically. All of this was unthinkable a dozen years ago. It is therefore no wonder that the public expects marvelous things to happen quickly. However, the science that laid the groundwork for the IT revolution dates back 60 years or more, with its origins in fundamental solid-state physics.

By contrast, the biotech revolution was relatively minor and, at least to date, not particularly effective. The major diseases that plagued mankind a quarter century ago are still here. In some third-world countries, the average lifespan of individuals has actually decreased from where it was a full century ago. While the costs of electronics technologies have plummeted, health care costs have continued to rise. The biotech revolution may have a profound impact, but the task at hand is substantially more difficult than what was required for the IT revolution. In effect, the IT revolution was based on the advanced engineering of two-dimensional digital circuits constructed from relatively simple components – extended solids. The biotech revolution is really dependent upon the ability to reverse engineer three-dimensional analog systems constructed from quite complex components – proteins. Given that the basic science behind biotech is substantially younger than the science that has supported IT, it is perhaps not surprising that the biotech revolution has not really been a proper revolution yet, and it likely needs at least another decade or so to come into fruition.

Prof. James R. Heath

Department of Chemistry
California Institute of Technology
Pasadena, California

Worked in the group of Nobel Laureate Richard E. Smalley at Rice University (1984–88) and co-invented Fullerene molecules which led to a revolution in Chemistry including the realization of nanotubes. The work on Fullerene molecules was cited for the 1996 Nobel Prize in Chemistry. Later he joined the University of California at Los Angeles (1994–2002), and co-founded and served as a Scientific Director of The California Nanosystems Institute.

Where does nanotechnology fit into this picture? In many ways, nanotechnology depends upon the ability to engineer two- and three-dimensional systems constructed from complex components such as macromolecules, biomolecules, nanostructured solids, etc. Furthermore, in terms of patents, publications, and other metrics that can be used to gauge the birth and evolution of a field, nanotech lags some 15–20 years behind biotech. Thus, now is the time that the fundamental science behind nanotechnology is being explored and developed. Nevertheless, progress with that science is moving forward at a dramatic pace. If the scientific community can keep up this pace and if the public sector will continue to support this science, then it is possible, and even perhaps likely, that in 20 years we may be speaking of the nanotech revolution.

The first edition of Springer Handbook of Nanotechnology was timely to assemble chapters in the broad field of nanotechnology. Given the fact that the second edition was in press one year after the publication of the first edition in April 2004, it is clear that the handbook has shown to be a valuable reference for experienced researchers as well as for a novice in the field. The third edition has one Part added and an expanded scope should have a wider appeal.

Preface to the 3rd Edition

On December 29, 1959 at the California Institute of Technology, Nobel Laureate Richard P. Feynman gave at talk at the Annual meeting of the American Physical Society that has become one of the 20th century classic science lectures, titled *There's Plenty of Room at the Bottom*. He presented a technological vision of extreme miniaturization in 1959, several years before the word *chip* became part of the lexicon. He talked about the problem of manipulating and controlling things on a small scale. Extrapolating from known physical laws, Feynman envisioned a technology using the ultimate toolbox of nature, building nanoobjects atom by atom or molecule by molecule. Since the 1980s, many inventions and discoveries in fabrication of nanoobjects have been testament to his vision. In recognition of this reality, National Science and Technology Council (NSTC) of the White House created the Interagency Working Group on Nanoscience, Engineering and Technology (IWGN) in 1998. In a January 2000 speech at the same institute, former President W.J. Clinton talked about the exciting promise of *nanotechnology* and the importance of expanding research in nanoscale science and technology, more broadly. Later that month, he announced in his State of the Union Address an ambitious US$ 497 million federal, multi-agency national nanotechnology initiative (NNI) in the fiscal year 2001 budget, and made the NNI a top science and technology priority. The objective of this initiative was to form a broad-based coalition in which the academe, the private sector, and local, state, and federal governments work together to push the envelop of nanoscience and nanoengineering to reap nanotechnology's potential social and economic benefits.

The funding in the US has continued to increase. In January 2003, the US senate introduced a bill to establish a National Nanotechnology Program. On December 3, 2003, President George W. Bush signed into law the 21st Century Nanotechnology Research and Development Act. The legislation put into law programs and activities supported by the National Nanotechnology Initiative. The bill gave nanotechnology a permanent home in the federal government and authorized US$ 3.7 billion to be spent in the four year period beginning in October 2005, for nanotechnology initiatives at five federal agencies. The funds would provide grants to researchers, coordinate R&D across five federal agencies (National Science Foundation (NSF), Department of Energy (DOE), NASA, National Institute of Standards and Technology (NIST), and Environmental Protection Agency (EPA)), establish interdisciplinary research centers, and accelerate technology transfer into the private sector. In addition, Department of Defense (DOD), Homeland Security, Agriculture and Justice as well as the National Institutes of Health (NIH) also fund large R&D activities. They currently account for more than one-third of the federal budget for nanotechnology.

European Union (EU) made nanosciences and nanotechnologies a priority in Sixth Framework Program (FP6) in 2002 for a period of 2003–2006. They had dedicated small funds in FP4 and FP5 before. FP6 was tailored to help better structure European research and to cope with the strategic objectives set out in Lisbon in 2000. Japan identified nanotechnology as one of its main research priorities in 2001. The funding levels increases sharply from US$ 400 million in 2001 to around US$ 950 million in 2004. In 2003, South Korea embarked upon a ten-year program with around US$ 2 billion of public funding, and Taiwan has committed around US$ 600 million of public funding over six years. Singapore and China are also investing on a large scale. Russia is well funded as well.

Nanotechnology literally means any technology done on a nanoscale that has applications in the real world. Nanotechnology encompasses production and application of physical, chemical and biological systems at scales, ranging from individual atoms or molecules to submicron dimensions, as well as the integration of the resulting nanostructures into larger systems. Nanotechnology is likely to have a profound impact on our economy and society in the early 21st century, comparable to that of semiconductor technology, information technology, or cellular and molecular biology. Science and technology research in nanotechnology promises breakthroughs in areas such as materials and manufacturing, nanoelectronics, medicine and healthcare, energy, biotechnology, information technology and national security. It is widely felt that nanotechnology will be the next industrial revolution.

There is an increasing need for a multidisciplinary, system-oriented approach to design and manufacturing of micro/nanodevices which function reliably. This can only be achieved through the cross-fertilization of ideas from different disciplines and the systematic flow of information and people among research groups. Reliability is a critical technology for many micro- and nanosystems and nanostructured materials. A broad based handbook was needed, and the first edition of Springer Handbook of Nanotechnology was published in April 2004. It presented an overview of nanomaterial synthesis, micro/nanofabrication, micro- and nanocomponents and systems, scanning probe microscopy, reliability issues (including nanotribology and nanomechanics) for nanotechnology, and industrial applications. When the handbook went for sale in Europe, it was sold out in ten days. Reviews on the handbook were very flattering.

Given the explosive growth in nanoscience and nanotechnology, the publisher and the editor decided to develop a second edition after merely six months of publication of the first edition. The second edition (2007) came out in December 2006. The publisher and the editor again decided to develop a third edition after six month of publication of the second edition. This edition of the handbook integrates the knowledge from nanostructures, fabrication, materials science, devices, and reliability point of view. It covers various industrial applications. It also addresses social, ethical, and political issues. Given the significant interest in biomedical applications, and biomimetics a number of additional chapters in this arena have been added. The third edition consists of 53 chapters (new 10, revised 28, and as is 15). The chapters have been written by 139 internationally recognized experts in the field, from academia, national research labs, and industry, and from all over the world.

This handbook is intended for three types of readers: graduate students of nanotechnology, researchers in academia and industry who are active or intend to become active in this field, and practicing engineers and scientists who have encountered a problem and hope to solve it as expeditiously as possible. The handbook should serve as an excellent text for one or two semester graduate courses in nanotechnology in mechanical engineering, materials science, applied physics, or applied chemistry.

We embarked on the development of third edition in June 2007, and we worked very hard to get all the chapters to the publisher in a record time of about 12 months. I wish to sincerely thank the authors for offering to write comprehensive chapters on a tight schedule. This is generally an added responsibility in the hectic work schedules of researchers today. I depended on a large number of reviewers who provided critical reviews. I would like to thank Dr. Phillip J. Bond, Chief of Staff and Under Secretary for Technology, US Department of Commerce, Washington, D.C. for suggestions for chapters as well as authors in the handbook. Last but not the least, I would like to thank my secretary Caterina Runyon-Spears for various administrative duties and her tireless efforts are highly appreciated.

I hope that this handbook will stimulate further interest in this important new field, and the readers of this handbook will find it useful.

February 2010

Bharat Bhushan
Editor

Preface to the 2nd Edition

On 29 December 1959 at the California Institute of Technology, Nobel Laureate Richard P. Feynman gave at talk at the Annual meeting of the American Physical Society that has become one of the 20th century classic science lectures, titled "There's Plenty of Room at the Bottom." He presented a technological vision of extreme miniaturization in 1959, several years before the word "chip" became part of the lexicon. He talked about the problem of manipulating and controlling things on a small scale. Extrapolating from known physical laws, Feynman envisioned a technology using the ultimate toolbox of nature, building nanoobjects atom by atom or molecule by molecule. Since the 1980s, many inventions and discoveries in the fabrication of nanoobjects have been a testament to his vision. In recognition of this reality, the National Science and Technology Council (NSTC) of the White House created the Interagency Working Group on Nanoscience, Engineering and Technology (IWGN) in 1998. In a January 2000 speech at the same institute, former President W. J. Clinton talked about the exciting promise of "nanotechnology" and the importance of expanding research in nanoscale science and, more broadly, technology. Later that month, he announced in his State of the Union Address an ambitious $497 million federal, multiagency national nanotechnology initiative (NNI) in the fiscal year 2001 budget, and made the NNI a top science and technology priority. The objective of this initiative was to form a broad-based coalition in which the academe, the private sector, and local, state, and federal governments work together to push the envelope of nanoscience and nanoengineering to reap nanotechnology's potential social and economic benefits.

The funding in the U.S. has continued to increase. In January 2003, the U. S. senate introduced a bill to establish a National Nanotechnology Program. On 3 December 2003, President George W. Bush signed into law the 21st Century Nanotechnology Research and Development Act. The legislation put into law programs and activities supported by the National Nanotechnology Initiative. The bill gave nanotechnology a permanent home in the federal government and authorized $3.7 billion to be spent in the four year period beginning in October 2005, for nanotechnology initiatives at five federal agencies. The funds would provide grants to researchers, coordinate R&D across five federal agencies (National Science Foundation (NSF), Department of Energy (DOE), NASA, National Institute of Standards and Technology (NIST), and Environmental Protection Agency (EPA)), establish interdisciplinary research centers, and accelerate technology transfer into the private sector. In addition, Department of Defense (DOD), Homeland Security, Agriculture and Justice as well as the National Institutes of Health (NIH) would also fund large R&D activities. They currently account for more than one-third of the federal budget for nanotechnology.

The European Union made nanosciences and nanotechnologies a priority in the Sixth Framework Program (FP6) in 2002 for the period of 2003-2006. They had dedicated small funds in FP4 and FP5 before. FP6 was tailored to help better structure European research and to cope with the strategic objectives set out in Lisbon in 2000. Japan identified nanotechnology as one of its main research priorities in 2001. The funding levels increased sharply from $400 million in 2001 to around $950 million in 2004. In 2003, South Korea embarked upon a ten-year program with around $2 billion of public funding, and Taiwan has committed around $600 million of public funding over six years. Singapore and China are also investing on a large scale. Russia is well funded as well.

Nanotechnology literally means any technology done on a nanoscale that has applications in the real world. Nanotechnology encompasses production and application of physical, chemical and biological systems at scales, ranging from individual atoms or molecules to submicron dimensions, as well as the integration of the resulting nanostructures into larger systems. Nanotechnology is likely to have a profound impact on our economy and society in the early 21st century, comparable to that of semiconductor technology, information technology, or cellular and molecular biology. Science and technology research in nanotechnology promises breakthroughs in areas such as materials and manufacturing, nanoelectronics, medicine and healthcare, energy, biotechnology, information technology and national security. It is widely felt that nanotechnology will be the next industrial revolution.

There is an increasing need for a multidisciplinary, system-oriented approach to design and manufacturing of micro/nanodevices that function reliably. This can only be achieved through the cross-fertilization of ideas from different disciplines and the systematic flow of information and people among research groups. Reliability is a critical technology for many micro- and nanosystems and nanostructured materials. A broad-based handbook was needed, and thus the first edition of Springer Handbook of Nanotechnology was published in April 2004. It presented an overview of nanomaterial synthesis, micro/nanofabrication, micro- and nanocomponents and systems, scanning probe microscopy, reliability issues (including nanotribology and nanomechanics) for nanotechnology, and industrial applications. When the handbook went for sale in Europe, it sold out in ten days. Reviews on the handbook were very flattering.

Given the explosive growth in nanoscience and nanotechnology, the publisher and the editor decided to develop a second edition merely six months after publication of the first edition. This edition of the handbook integrates the knowledge from the nanostructure, fabrication, materials science, devices, and reliability point of view. It covers various industrial applications. It also addresses social, ethical, and political issues. Given the significant interest in biomedical applications, a number of chapters in this arena have been added. The second edition consists of 59 chapters (new: 23; revised: 27; unchanged: 9). The chapters have been written by 154 internationally recognized experts in the field, from academia, national research labs, and industry.

This book is intended for three types of readers: graduate students of nanotechnology, researchers in academia and industry who are active or intend to become active in this field, and practicing engineers and scientists who have encountered a problem and hope to solve it as expeditiously as possible. The handbook should serve as an excellent text for one or two semester graduate courses in nanotechnology in mechanical engineering, materials science, applied physics, or applied chemistry.

We embarked on the development of the second edition in October 2004, and we worked very hard to get all the chapters to the publisher in a record time of about 7 months. I wish to sincerely thank the authors for offering to write comprehensive chapters on a tight schedule. This is generally an added responsibility to the hectic work schedules of researchers today. I depended on a large number of reviewers who provided critical reviews. I would like to thank Dr. Phillip J. Bond, Chief of Staff and Under Secretary for Technology, US Department of Commerce, Washington, D.C. for chapter suggestions as well as authors in the handbook. I would also like to thank my colleague, Dr. Zhenhua Tao, whose efforts during the preparation of this handbook were very useful. Last but not the least, I would like to thank my secretary Caterina Runyon-Spears for various administrative duties; her tireless efforts are highly appreciated.

I hope that this handbook will stimulate further interest in this important new field, and the readers of this handbook will find it useful.

May 2005   Bharat Bhushan
                                                                                            Editor

Preface to the 1st Edition

On December 29, 1959 at the California Institute of Technology, Nobel Laureate Richard P. Feynman gave a talk at the Annual meeting of the American Physical Society that has become one classic science lecture of the 20th century, titled "There's Plenty of Room at the Bottom." He presented a technological vision of extreme miniaturization in 1959, several years before the word "chip" became part of the lexicon. He talked about the problem of manipulating and controlling things on a small scale. Extrapolating from known physical laws, Feynman envisioned a technology using the ultimate toolbox of nature, building nanoobjects atom by atom or molecule by molecule. Since the 1980s, many inventions and discoveries in fabrication of nanoobjects have been a testament to his vision. In recognition of this reality, in a January 2000 speech at the same institute, former President W. J. Clinton talked about the exciting promise of "nanotechnology" and the importance of expanding research in nanoscale science and engineering. Later that month, he announced in his State of the Union Address an ambitious $ 497 million federal, multi-agency national nanotechnology initiative (NNI) in the fiscal year 2001 budget, and made the NNI a top science and technology priority. Nanotechnology literally means any technology done on a nanoscale that has applications in the real world. Nanotechnology encompasses production and application of physical, chemical and biological systems at size scales, ranging from individual atoms or molecules to submicron dimensions as well as the integration of the resulting nanostructures into larger systems. Nanofabrication methods include the manipulation or self-assembly of individual atoms, molecules, or molecular structures to produce nanostructured materials and sub-micron devices. Micro- and nanosystems components are fabricated using top-down lithographic and nonlithographic fabrication techniques. Nanotechnology will have a profound impact on our economy and society in the early 21st century, comparable to that of semiconductor technology, information technology, or advances in cellular and molecular biology. The research and development in nanotechnology will lead to potential breakthroughs in areas such as materials and manufacturing, nanoelectronics, medicine and healthcare, energy, biotechnology, information technology and national security. It is widely felt that nanotechnology will lead to the next industrial revolution.

Reliability is a critical technology for many micro- and nanosystems and nanostructured materials. No book exists on this emerging field. A broad based handbook is needed. The purpose of this handbook is to present an overview of nanomaterial synthesis, micro/nanofabrication, micro- and nanocomponents and systems, reliability issues (including nanotribology and nanomechanics) for nanotechnology, and industrial applications. The chapters have been written by internationally recognized experts in the field, from academia, national research labs and industry from all over the world.

The handbook integrates knowledge from the fabrication, mechanics, materials science and reliability points of view. This book is intended for three types of readers: graduate students of nanotechnology, researchers in academia and industry who are active or intend to become active in this field, and practicing engineers and scientists who have encountered a problem and hope to solve it as expeditiously as possible. The handbook should serve as an excellent text for one or two semester graduate courses in nanotechnology in mechanical engineering, materials science, applied physics, or applied chemistry.

We embarked on this project in February 2002, and we worked very hard to get all the chapters to the publisher in a record time of about 1 year. I wish to sincerely thank the authors for offering to write comprehensive chapters on a tight schedule. This is generally an added responsibility in the hectic work schedules of researchers today. I depended on a large number of reviewers who provided critical reviews. I would like to thank Dr. Phillip J. Bond, Chief of Staff and Under Secretary for Technology, US Department of Commerce, Washington, D.C. for suggestions for chapters as well as authors in the handbook. I would also like to thank my colleague, Dr. Huiwen Liu, whose efforts during the preparation of this handbook were very useful.

I hope that this handbook will stimulate further interest in this important new field, and the readers of this handbook will find it useful.

September 2003   Bharat Bhushan
                                                                                                 Editor

Editors Vita

Dr. Bharat Bhushan received an M.S. in mechanical engineering from the Massachusetts Institute of Technology in 1971, an M.S. in mechanics and a Ph.D. in mechanical engineering from the University of Colorado at Boulder in 1973 and 1976, respectively, an MBA from Rensselaer Polytechnic Institute at Troy, NY in 1980, Doctor Technicae from the University of Trondheim at Trondheim, Norway in 1990, a Doctor of Technical Sciences from the Warsaw University of Technology at Warsaw, Poland in 1996, and Doctor Honouris Causa from the National Academy of Sciences at Gomel, Belarus in 2000. He is a registered professional engineer. He is presently an Ohio Eminent Scholar and The Howard D. Winbigler Professor in the College of Engineering, and the Director of the Nanoprobe Laboratory for Bio- and Nanotechnology and Biomimetics (NLB²) at the Ohio State University, Columbus, Ohio. His research interests include fundamental studies with a focus on scanning probe techniques in the interdisciplinary areas of bio/nanotribology, bio/nanomechanics and bio/nanomaterials characterization, and applications to bio/nanotechnology and biomimetics. He is an internationally recognized expert of bio/nanotribology and bio/nanomechanics using scanning probe microscopy, and is one of the most prolific authors. He is considered by some a pioneer of the tribology and mechanics of magnetic storage devices. He has authored 6 scientific books, more than 90 handbook chapters, more than 700 scientific papers (h factor – 45+; ISI Highly Cited in Materials Science, since 2007), and more than 60 technical reports, edited more than 45 books, and holds 17 US and foreign patents. He is co-editor of Springer NanoScience and Technology Series and co-editor of Microsystem Technologies. He has given more than 400 invited presentations on six continents and more than 140 keynote/plenary addresses at major international conferences.

Dr. Bhushan is an accomplished organizer. He organized the first symposium on Tribology and Mechanics of Magnetic Storage Systems in 1984 and the first international symposium on Advances in Information Storage Systems in 1990, both of which are now held annually. He is the founder of an ASME Information Storage and Processing Systems Division founded in 1993 and served as the founding chair during 1993–1998. His biography has been listed in over two dozen Who's Who books including Who's Who in the World and has received more than two dozen awards for his contributions to science and technology from professional societies, industry, and US government agencies. He is also the recipient of various international fellowships including the Alexander von Humboldt Research Prize for Senior Scientists, Max Planck Foundation Research Award for Outstanding Foreign Scientists, and the Fulbright Senior Scholar Award. He is a foreign member of the International Academy of Engineering (Russia), Byelorussian Academy of Engineering and Technology and the Academy of Triboengineering of Ukraine, an honorary member of the Society of Tribologists of Belarus, a fellow of ASME, IEEE, STLE, and the New York Academy of Sciences, and a member of ASEE, Sigma Xi and Tau Beta Pi.

Dr. Bhushan has previously worked for the R&D Division of Mechanical Technology Inc., Latham, NY; the Technology Services Division of SKF Industries Inc., King of Prussia, PA; the General Products Division Laboratory of IBM Corporation, Tucson, AZ; and the Almaden Research Center of IBM Corporation, San Jose, CA. He has held visiting professor appointments at University of California at Berkeley, University of Cambridge, UK, Technical University Vienna, Austria, University of Paris, Orsay, ETH Zurich and EPFL Lausanne.

List of Authors

Chong H. Ahn
University of Cincinnati
Department of Electrical
and Computer Engineering
Cincinnati, OH 45221, USA
e-mail: *chong.ahn@uc.edu*

Boris Anczykowski
nanoAnalytics GmbH
Münster, Germany
e-mail: *anczykowski@nanoanalytics.com*

W. Robert Ashurst
Auburn University
Department of Chemical Engineering
Auburn, AL 36849, USA
e-mail: *ashurst@auburn.edu*

Massood Z. Atashbar
Western Michigan University
Department of Electrical
and Computer Engineering
Kalamazoo, MI 49008-5329, USA
e-mail: *massood.atashbar@wmich.edu*

Wolfgang Bacsa
University of Toulouse III (Paul Sabatier)
Laboratoire de Physique des Solides (LPST),
UMR 5477 CNRS
Toulouse, France
e-mail: *bacsa@ramansco.ups-tlse.fr;*
bacsa@lpst.ups-tlse.fr

Kelly Bailey
University of Adelaide
CSIRO Human Nutrition
Adelaide SA 5005, Australia
e-mail: *kelly.bailey@csiro.au*

William Sims Bainbridge
National Science Foundation
Division of Information, Science and Engineering
Arlington, VA, USA
e-mail: *wsbainbridge@yahoo.com*

Antonio Baldi
Institut de Microelectronica de Barcelona (IMB)
Centro National Microelectrónica (CNM-CSIC)
Barcelona, Spain
e-mail: *antoni.baldi@cnm.es*

Wilhelm Barthlott
University of Bonn
Nees Institute for Biodiversity of Plants
Meckenheimer Allee 170
53115 Bonn, Germany
e-mail: *barthlott@uni-bonn.de*

Roland Bennewitz
INM – Leibniz Institute for New Materials
66123 Saarbrücken, Germany
e-mail: *roland.bennewitz@inm-gmbh.de*

Bharat Bhushan
Ohio State University
Nanoprobe Laboratory for Bio- and
Nanotechnology and Biomimetics (NLB²)
201 W. 19th Avenue
Columbus, OH 43210-1142, USA
e-mail: *bhushan.2@osu.edu*

Gerd K. Binnig
Definiens AG
Trappentreustr. 1
80339 Munich, Germany
e-mail: *gbinnig@definiens.com*

Marcie R. Black
Bandgap Engineering Inc.
1344 Main St.
Waltham, MA 02451, USA
e-mail: *marcie@alum.mit.edu;*
marcie@bandgap.com

Donald W. Brenner
Department of Materials Science and Engineering
Raleigh, NC, USA
e-mail: *brenner@ncsu.edu*

Jean-Marc Broto
Institut National des Sciences Appliquées
of Toulouse
Laboratoire National
des Champs Magnétiques Pulsés (LNCMP)
Toulouse, France
e-mail: *broto@lncmp.fr*

Guozhong Cao
University of Washington
Dept. of Materials Science and Engineering
302M Roberts Hall
Seattle, WA 98195-2120, USA
e-mail: *gzcao@u.washington.edu*

Edin (I-Chen) Chen
National Central University
Institute of Materials Science and Engineering
Department of Mechanical Engineering
Chung-Li, 320, Taiwan
e-mail: *ichen@ncu.edu.tw*

Yu-Ting Cheng
National Chiao Tung University
Department of Electronics Engineering
& Institute of Electronics
1001, Ta-Hsueh Rd.
Hsinchu, 300, Taiwan, R.O.C.
e-mail: *ytcheng@mail.nctu.edu.tw*

Giovanni Cherubini
IBM Zurich Research Laboratory
Tape Technologies
8803 Rüschlikon, Switzerland
e-mail: *cbi@zurich.ibm.com*

Mu Chiao
Department of Mechanical Engineering
6250 Applied Science Lane
Vancouver, BC V6T 1Z4, Canada
e-mail: *muchiao@mech.ubc.ca*

Jin-Woo Choi
Louisiana State University
Department of Electrical
and Computer Engineering
Baton Rouge, LA 70803, USA
e-mail: *choi@ece.lsu.edu*

Tamara H. Cooper
University of Adelaide
CSIRO Human Nutrition
Adelaide SA 5005, Australia
e-mail: *tamara.cooper@csiro.au*

Alex D. Corwin
GE Global Research
1 Research Circle
Niskayuna, NY 12309, USA
e-mail: *corwin@ge.com*

Maarten P. de Boer
Carnegie Mellon University
Department of Mechanical Engineering
5000 Forbes Avenue
Pittsburgh, PA 15213, USA
e-mail: *mpdebo@andrew.cmu.edu*

Dietrich Dehlinger
Lawrence Livermore National Laboratory
Engineering
Livermore, CA 94551, USA
e-mail: *dehlinger1@llnl.gov*

Frank W. DelRio
National Institute of Standards and Technology
100 Bureau Drive, Stop 8520
Gaithersburg, MD 20899-8520, USA
e-mail: *frank.delrio@nist.gov*

Michel Despont
IBM Zurich Research Laboratory
Micro- and Nanofabrication
8803 Rüschlikon, Switzerland
e-mail: *dpt@zurich.ibm.com*

Lixin Dong
Michigan State University
Electrical and Computer Engineering
2120 Engineering Building
East Lansing, MI 48824-1226, USA
e-mail: *ldong@egr.msu.edu*

Gene Dresselhaus
Massachusetts Institute of Technology
Francis Bitter Magnet Laboratory
Cambridge, MA 02139, USA
e-mail: *gene@mgm.mit.edu*

Mildred S. Dresselhaus
Massachusetts Institute of Technology
Department of Electrical Engineering
and Computer Science
Department of Physics
Cambridge, MA, USA
e-mail: *millie@mgm.mit.edu*

Urs T. Dürig
IBM Zurich Research Laboratory
Micro-/Nanofabrication
8803 Rüschlikon, Switzerland
e-mail: *drg@zurich.ibm.com*

Andreas Ebner
Johannes Kepler University Linz
Institute for Biophysics
Altenberger Str. 69
4040 Linz, Austria
e-mail: *andreas.ebner@jku.at*

Evangelos Eleftheriou
IBM Zurich Research Laboratory
8803 Rüschlikon, Switzerland
e-mail: *ele@zurich.ibm.com*

Emmanuel Flahaut
Université Paul Sabatier
CIRIMAT, Centre Interuniversitaire de Recherche
et d'Ingénierie des Matériaux, UMR 5085 CNRS
118 Route de Narbonne
31062 Toulouse, France
e-mail: *flahaut@chimie.ups-tlse.fr*

Anatol Fritsch
University of Leipzig
Institute of Experimental Physics I
Division of Soft Matter Physics
Linnéstr. 5
04103 Leipzig, Germany
e-mail: *anatol.fritsch@uni-leipzig.de*

Harald Fuchs
Universität Münster
Physikalisches Institut
Münster, Germany
e-mail: *fuchsh@uni-muenster.de*

Christoph Gerber
University of Basel
Institute of Physics
National Competence Center for Research
in Nanoscale Science (NCCR) Basel
Klingelbergstr. 82
4056 Basel, Switzerland
e-mail: *christoph.gerber@unibas.ch*

Franz J. Giessibl
Universität Regensburg
Institute of Experimental and Applied Physics
Universitätsstr. 31
93053 Regensburg, Germany
e-mail: *franz.giessibl@physik.uni-regensburg.de*

Enrico Gnecco
University of Basel
National Center of Competence in Research
Department of Physics
Klingelbergstr. 82
4056 Basel, Switzerland
e-mail: *enrico.gnecco@unibas.ch*

Stanislav N. Gorb
Max Planck Institut für Metallforschung
Evolutionary Biomaterials Group
Heisenbergstr. 3
70569 Stuttgart, Germany
e-mail: *s.gorb@mf.mpg.de*

Hermann Gruber
University of Linz
Institute of Biophysics
Altenberger Str. 69
4040 Linz, Austria
e-mail: *hermann.gruber@jku.at*

Jason Hafner
Rice University
Department of Physics and Astronomy
Houston, TX 77251, USA
e-mail: *hafner@rice.edu*

Judith A. Harrison
U.S. Naval Academy
Chemistry Department
572 Holloway Road
Annapolis, MD 21402-5026, USA
e-mail: *jah@usna.edu*

Martin Hegner
CRANN – The Naughton Institute
Trinity College, University of Dublin
School of Physics
Dublin, 2, Ireland
e-mail: *martin.hegner@tcd.ie*

Thomas Helbling
ETH Zurich
Micro and Nanosystems
Department of Mechanical
and Process Engineering
8092 Zurich, Switzerland
e-mail: *thomas.helbling@micro.mavt.ethz.ch*

Michael J. Heller
University of California San Diego
Department of Bioengineering
Dept. of Electrical and Computer Engineering
La Jolla, CA, USA
e-mail: *mjheller@ucsd.edu*

Seong-Jun Heo
Lam Research Corp.
4650 Cushing Parkway
Fremont, CA 94538, USA
e-mail: *seongjun.heo@lamrc.com*

Christofer Hierold
ETH Zurich
Micro and Nanosystems
Department of Mechanical
and Process Engineering
8092 Zurich, Switzerland
e-mail: *christofer.hierold@micro.mavt.ethz.ch*

Peter Hinterdorfer
University of Linz
Institute for Biophysics
Altenberger Str. 69
4040 Linz, Austria
e-mail: *peter.hinterdorfer@jku.at*

Dalibor Hodko
Nanogen, Inc.
10498 Pacific Center Court
San Diego, CA 92121, USA
e-mail: *dhodko@nanogen.com*

Hendrik Hölscher
Forschungszentrum Karlsruhe
Institute of Microstructure Technology
Linnéstr. 5
76021 Karlsruhe, Germany
e-mail: *hendrik.hoelscher@imt.fzk.de*

Hirotaka Hosoi
Hokkaido University
Creative Research Initiative Sousei
Kita 21, Nishi 10, Kita-ku
Sapporo, Japan
e-mail: *hosoi@cris.hokudai.ac.jp*

Katrin Hübner
Staatliche Fachoberschule Neu-Ulm
89231 Neu-Ulm, Germany
e-mail: *katrin.huebner1@web.de*

Douglas L. Irving
North Carolina State University
Materials Science and Engineering
Raleigh, NC 27695-7907, USA
e-mail: *doug_irving@ncsu.edu*

Jacob N. Israelachvili
University of California
Department of Chemical Engineering
and Materials Department
Santa Barbara, CA 93106-5080, USA
e-mail: *jacob@engineering.ucsb.edu*

Guangyao Jia
University of California, Irvine
Department of Mechanical
and Aerospace Engineering
Irvine, CA, USA
e-mail: *gjia@uci.edu*

Sungho Jin
University of California, San Diego
Department of Mechanical
and Aerospace Engineering
9500 Gilman Drive
La Jolla, CA 92093-0411, USA
e-mail: *jin@ucsd.edu*

Anne Jourdain
Interuniversity Microelectronics Center (IMEC)
Leuven, Belgium
e-mail: *jourdain@imec.be*

Yong Chae Jung
Samsung Electronics C., Ltd.
Senior Engineer Process Development Team
San #16 Banwol-Dong, Hwasung-City
Gyeonggi-Do 445-701, Korea
e-mail: yc423.jung@samsung.com

Harold Kahn
Case Western Reserve University
Department of Materials Science and Engineering
Cleveland, OH, USA
e-mail: kahn@cwru.edu

Roger Kamm
Massachusetts Institute of Technology
Department of Biological Engineering
77 Massachusetts Avenue
Cambridge, MA 02139, USA
e-mail: rdkamm@mit.edu

Ruti Kapon
Weizmann Institute of Science
Department of Biological Chemistry
Rehovot 76100, Israel
e-mail: ruti.kapon@weizmann.ac.il

Josef Käs
University of Leipzig
Institute of Experimental Physics I
Division of Soft Matter Physics
Linnéstr. 5
04103 Leipzig, Germany
e-mail: jkaes@physik.uni-leipzig.de

Horacio Kido
University of California at Irvine
Mechanical and Aerospace Engineering
Irvine, CA, USA
e-mail: hkido@uci.edu

Tobias Kießling
University of Leipzig
Institute of Experimental Physics I
Division of Soft Matter Physics
Linnéstr. 5
04103 Leipzig, Germany
e-mail: Tobias.Kiessling@uni-leipzig.de

Jitae Kim
University of California at Irvine
Department of Mechanical
and Aerospace Engineering
Irvine, CA, USA
e-mail: jitaekim@uci.edu

Jongbaeg Kim
Yonsei University
School of Mechanical Engineering
1st Engineering Bldg.
Seoul, 120-749, South Korea
e-mail: kimjb@yonsei.ac.kr

Nahui Kim
Samsung Advanced Institute of Technology
Research and Development
Seoul, South Korea
e-mail: nahui.kim@samsung.com

Kerstin Koch
Rhine-Waal University of Applied Science
Department of Life Science, Biology
and Nanobiotechnology
Landwehr 4
47533 Kleve, Germany
e-mail: kerstin.koch@hochschule.rhein-waal.de

Jing Kong
Massachusetts Institute of Technology
Department of Electrical Engineering
and Computer Science
Cambridge, MA, USA
e-mail: jingkong@mit.edu

Tobias Kraus
Leibniz-Institut für Neue Materialien gGmbH
Campus D2 2
66123 Saarbrücken, Germany
e-mail: tobias.kraus@inm-gmbh.de

Anders Kristensen
Technical University of Denmark
DTU Nanotech
2800 Kongens Lyngby, Denmark
e-mail: anders.kristensen@nanotech.dtu.dk

Ratnesh Lal
University of Chicago
Center for Nanomedicine
5841 S Maryland Av
Chicago, IL 60637, USA
e-mail: rlal@uchicago.edu

Jan Lammerding
Harvard Medical School
Brigham and Women's Hospital
65 Landsdowne St
Cambridge, MA 02139, USA
e-mail: jlammerding@rics.bwh.harvard.edu

Hans Peter Lang
University of Basel
Institute of Physics, National Competence Center
for Research in Nanoscale Science (NCCR) Basel
Klingelbergstr. 82
4056 Basel, Switzerland
e-mail: hans-peter.lang@unibas.ch

Carmen LaTorre
Owens Corning Science and Technology
Roofing and Asphalt
2790 Columbus Road
Granville, OH 43023, USA
e-mail: carmen.latorre@owenscorning.com

Christophe Laurent
Université Paul Sabatier
CIRIMAT UMR 5085 CNRS
118 Route de Narbonne
31062 Toulouse, France
e-mail: laurent@chimie.ups-tlse.fr

Abraham P. Lee
University of California Irvine
Department of Biomedical Engineering
Department of Mechanical
and Aerospace Engineering
Irvine, CA 92697, USA
e-mail: aplee@uci.edu

Stephen C. Lee
Ohio State University
Biomedical Engineering Center
Columbus, OH 43210, USA
e-mail: lee@bme.ohio-state.edu

Wayne R. Leifert
Adelaide Business Centre
CSIRO Human Nutrition
Adelaide SA 5000, Australia
e-mail: wayne.leifert@csiro.au

Liwei Lin
UC Berkeley
Mechanical Engineering Department
5126 Etcheverry
Berkeley, CA 94720-1740, USA
e-mail: lwlin@me.berkeley.edu

Yu-Ming Lin
IBM T.J. Watson Research Center
Nanometer Scale Science & Technology
1101 Kitchawan Road
Yorktown Heigths, NY 10598, USA
e-mail: yming@us.ibm.com

Marc J. Madou
University of California Irvine
Department of Mechanical and Aerospace
and Biomedical Engineering
Irvine, CA, USA
e-mail: mmadou@uci.edu

Othmar Marti
Ulm University
Institute of Experimental Physics
Albert-Einstein-Allee 11
89069 Ulm, Germany
e-mail: othmar.marti@uni-ulm.de

Jack Martin
66 Summer Street
Foxborough, MA 02035, USA
e-mail: jack.martin@alumni.tufts.edu

Shinji Matsui
University of Hyogo
Laboratory of Advanced Science
and Technology for Industry
Hyogo, Japan
e-mail: matsui@lasti.u-hyogo.ac.jp

Mehran Mehregany
Case Western Reserve University
Department of Electrical Engineering
and Computer Science
Cleveland, OH 44106, USA
e-mail: *mxm31@cwru.edu*

Etienne Menard
Semprius, Inc.
4915 Prospectus Dr.
Durham, NC 27713, USA
e-mail: *etienne.menard@semprius.com*

Ernst Meyer
University of Basel
Institute of Physics
Basel, Switzerland
e-mail: *ernst.meyer@unibas.ch*

Robert Modliñski
Baolab Microsystems
Terrassa 08220, Spain
e-mail: *rmodlinski@gmx.com*

Mohammad Mofrad
University of California, Berkeley
Department of Bioengineering
Berkeley, CA 94720, USA
e-mail: *mofrad@berkeley.edu*

Marc Monthioux
CEMES – UPR A-8011 CNRS
Carbones et Matériaux Carbonés,
Carbons and Carbon-Containing Materials
29 Rue Jeanne Marvig
31055 Toulouse 4, France
e-mail: *monthiou@cemes.fr*

Markus Morgenstern
RWTH Aachen University
II. Institute of Physics B and JARA-FIT
52056 Aachen, Germany
e-mail: *mmorgens@physik.rwth-aachen.de*

Seizo Morita
Osaka University
Department of Electronic Engineering
Suita-City
Osaka, Japan
e-mail: *smorita@ele.eng.osaka-u.ac.jp*

Koichi Mukasa
Hokkaido University
Nanoelectronics Laboratory
Sapporo, Japan
e-mail: *mukasa@nano.eng.hokudai.ac.jp*

Bradley J. Nelson
Swiss Federal Institute of Technology (ETH)
Institute of Robotics and Intelligent Systems
8092 Zurich, Switzerland
e-mail: *bnelson@ethz.ch*

Michael Nosonovsky
University of Wisconsin-Milwaukee
Department of Mechanical Engineering
3200 N. Cramer St.
Milwaukee, WI 53211, USA
e-mail: *nosonovs@uwm.edu*

Hiroshi Onishi
Kanagawa Academy of Science and Technology
Surface Chemistry Laboratory
Kanagawa, Japan
e-mail: *oni@net.ksp.or.jp*

Alain Peigney
Centre Inter-universitaire de Recherche
sur l'Industrialisation des Matériaux (CIRIMAT)
Toulouse 4, France
e-mail: *peigney@chimie.ups-tlse.fr*

Oliver Pfeiffer
Individual Computing GmbH
Ingelsteinweg 2d
4143 Dornach, Switzerland
e-mail: *oliver.pfeiffer@gmail.com*

Haralampos Pozidis
IBM Zurich Research Laboratory
Storage Technologies
Rüschlikon, Switzerland
e-mail: *hap@zurich.ibm.com*

Robert Puers
Katholieke Universiteit Leuven
ESAT/MICAS
Leuven, Belgium
e-mail: *bob.puers@esat.kuleuven.ac.be*

Calvin F. Quate
Stanford University
Edward L. Ginzton Laboratory
450 Via Palou
Stanford, CA 94305-4088, USA
e-mail: *quate@stanford.edu*

Oded Rabin
University of Maryland
Department of Materials Science and Engineering
College Park, MD, USA
e-mail: *oded@umd.edu*

Françisco M. Raymo
University of Miami
Department of Chemistry
1301 Memorial Drive
Coral Gables, FL 33146-0431, USA
e-mail: *fraymo@miami.edu*

Manitra Razafinimanana
University of Toulouse III (Paul Sabatier)
Centre de Physique des Plasmas
et leurs Applications (CPPAT)
Toulouse, France
e-mail: *razafinimanana@cpat.ups-tlse.fr*

Ziv Reich
Weizmann Institute of Science Ha'Nesi Ha'Rishon
Department of Biological Chemistry
Rehovot 76100, Israel
e-mail: *ziv.reich@weizmann.ac.il*

John A. Rogers
University of Illinois
Department of Materials Science and Engineering
Urbana, IL, USA
e-mail: *jrogers@uiuc.edu*

Cosmin Roman
ETH Zurich
Micro and Nanosystems Department of Mechanical
and Process Engineering
8092 Zurich, Switzerland
e-mail: *cosmin.roman@micro.mavt.ethz.ch*

Marina Ruths
University of Massachusetts Lowell
Department of Chemistry
1 University Avenue
Lowell, MA 01854, USA
e-mail: *marina_ruths@uml.edu*

Ozgur Sahin
The Rowland Institute at Harvard
100 Edwin H. Land Blvd
Cambridge, MA 02142, USA
e-mail: *sahin@rowland.harvard.edu*

Akira Sasahara
Japan Advanced Institute
of Science and Technology
School of Materials Science
1-1 Asahidai
923-1292 Nomi, Japan
e-mail: *sasahara@jaist.ac.jp*

Helmut Schift
Paul Scherrer Institute
Laboratory for Micro- and Nanotechnology
5232 Villigen PSI, Switzerland
e-mail: *helmut.schift@psi.ch*

André Schirmeisen
University of Münster
Institute of Physics
Wilhelm-Klemm-Str. 10
48149 Münster, Germany
e-mail: *schirmeisen@uni-muenster.de*

Christian Schulze
Beiersdorf AG
Research & Development
Unnastr. 48
20245 Hamburg, Germany
e-mail: *christian.schulze@beiersdorf.com;*
christian.schulze@uni-leipzig.de

Alexander Schwarz
University of Hamburg
Institute of Applied Physics
Jungiusstr. 11
20355 Hamburg, Germany
e-mail: *aschwarz@physnet.uni-hamburg.de*

Udo D. Schwarz
Yale University
Department of Mechanical Engineering
15 Prospect Street
New Haven, CT 06520-8284, USA
e-mail: udo.schwarz@yale.edu

Philippe Serp
Ecole Nationale Supérieure d'Ingénieurs
en Arts Chimiques et Technologiques
Laboratoire de Chimie de Coordination (LCC)
118 Route de Narbonne
31077 Toulouse, France
e-mail: philippe.serp@ensiacet.fr

Huamei (Mary) Shang
GE Healthcare
4855 W. Electric Ave.
Milwaukee, WI 53219, USA
e-mail: huamei.shang@ge.com

Susan B. Sinnott
University of Florida
Department of Materials Science and Engineering
154 Rhines Hall
Gainesville, FL 32611-6400, USA
e-mail: ssinn@mse.ufl.edu

Anisoara Socoliuc
SPECS Zurich GmbH
Technoparkstr. 1
8005 Zurich, Switzerland
e-mail: socoliuc@nanonis.com

Olav Solgaard
Stanford University
E.L. Ginzton Laboratory
450 Via Palou
Stanford, CA 94305-4088, USA
e-mail: solgaard@stanford.edu

Dan Strehle
University of Leipzig
Institute of Experimental Physics I
Division of Soft Matter Physics
Linnéstr. 5
04103 Leipzig, Germany
e-mail: dan.strehle@uni-leipzig.de

Carsten Stüber
University of Leipzig
Institute of Experimental Physics I
Division of Soft Matter Physics
Linnéstr. 5
04103 Leipzig, Germany
e-mail: stueber@rz.uni-leipzig.de

Yu-Chuan Su
ESS 210
Department of Engineering and System Science 101
Kuang-Fu Road
Hsinchu, 30013, Taiwan
e-mail: ycsu@ess.nthu.edu.tw

Kazuhisa Sueoka
Graduate School of Information Science
and Technology
Hokkaido University
Nanoelectronics Laboratory
Kita-14, Nishi-9, Kita-ku
060-0814 Sapporo, Japan
e-mail: sueoka@nano.isthokudai.ac.jp

Yasuhiro Sugawara
Osaka University
Department of Applied Physics
Yamada-Oka 2-1, Suita
565-0871 Osaka, Japan
e-mail: sugawara@ap.eng.osaka-u.ac.jp

Benjamin Sullivan
TearLab Corp.
11025 Roselle Street
San Diego, CA 92121, USA
e-mail: bdsulliv@TearLab.com

Paul Swanson
Nexogen, Inc.
Engineering
8360 C Camino Santa Fe
San Diego, CA 92121, USA
e-mail: pswanson@nexogentech.com

Yung-Chieh Tan
Washington University School of Medicine
Department of Medicine
Division of Dermatology
660 S. Euclid Ave.
St. Louis, MO 63110, USA
e-mail: *ytanster@gmail.com*

Shia-Yen Teh
University of California at Irvine
Biomedical Engineering Department
3120 Natural Sciences II
Irvine, CA 92697-2715, USA
e-mail: *steh@uci.edu*

W. Merlijn van Spengen
Leiden University
Kamerlingh Onnes Laboratory
Niels Bohrweg 2
Leiden, CA 2333, The Netherlands
e-mail: *spengen@physics.leidenuniv.nl*

Peter Vettiger
University of Neuchâtel
SAMLAB
Jaquet-Droz 1
2002 Neuchâtel, Switzerland
e-mail: *peter.vettiger@unine.ch*

Franziska Wetzel
University of Leipzig
Institute of Experimental Physics I
Division of Soft Matter Physics
Linnéstr. 5
04103 Leipzig, Germany
e-mail: *franziska.wetzel@uni-leipzig.de*

Heiko Wolf
IBM Research GmbH
Zurich Research Laboratory
Säumerstr. 4
8803 Rüschlikon, Switzerland
e-mail: *hwo@zurich.ibm.com*

Darrin J. Young
Case Western Reserve University
Department of EECS, Glennan 510
10900 Euclid Avenue
Cleveland, OH 44106, USA
e-mail: *djy@po.cwru.edu*

Babak Ziaie
Purdue University
Birck Nanotechnology Center
1205 W. State St.
West Lafayette, IN 47907-2035, USA
e-mail: *bziaie@purdue.edu*

Christian A. Zorman
Case Western Reserve University
Department of Electrical Engineering
and Computer Science
10900 Euclid Avenue
Cleveland, OH 44106, USA
e-mail: *caz@case.edu*

Jim V. Zoval
Saddleback College
Department of Math and Science
28000 Marguerite Parkway
Mission Viejo, CA 92692, USA
e-mail: *jzoval@saddleback.edu*

Acknowledgements

B.12 MEMS/NEMS Devices and Applications
by Darrin J. Young, Christian A. Zorman, Mehran Mehregany

The authors wish to thank Wen H. Ko for the helpful discussions and suggestions, Peng Cong for updating the references, Michael Suster and Joseph Seeger for preparing the figures.

B.16 Biological Molecules in Therapeutic Nanodevices
by Stephen C. Lee, Bharat Bhushan

The authors acknowledge the contribution to the chapter by Philip D. Barnes. This work was supported in part by grants from the Program for Homeland Security of the Ohio State University (program no. 14 525) and award GRT00011123 (project no. 60 015 533) to S.C.L. from the National Science Foundation.

B.19 Centrifuge-Based Fluidic Platforms
by Jim V. Zoval, Guangyao Jia, Horacio Kido, Jitae Kim, Nahui Kim, Marc J. Madou

The authors thank Sue Cresswell, Gyros AB; Gregory J. Kellogg, Tecan Boston; Raj Barathur, Burstein Technologies; Jim Lee, Ohio State University; John Hines, Tony Ricco, and Michael Flynn, NASA Ames; Sylvia Danuert; University of Kentucky, Regis Peytavi, Dominic Gagné, Francious J. Picard and Michel G. Bergeron, Laval University.

目 录

缩略语

Part B 微纳机电系统和生物微纳机电系统

12. 微纳机电系统器件及应用 ⋯ 3
    12.1 微机电系统器件及应用 ⋯ 5
    12.2 纳机电系统（NEMS） ⋯ 24
    12.3 当前挑战与未来趋势 ⋯ 27
    参考文献 ⋯ 28

13. 新一代DNA混合和自组装纳米制造器件 ⋯ 33
    13.1 电子微阵列技术 ⋯ 35
    13.2 电场辅助纳米制造工艺 ⋯ 41
    13.3 结 论 ⋯ 43
    参考文献 ⋯ 44

14. 单壁碳纳米管传感器的概念 ⋯ 47
    14.1 单壁碳纳米管传感器的设计 ⋯ 48
    14.2 单壁碳纳米管传感器的制造 ⋯ 56
    14.3 最新应用举例 ⋯ 60
    14.4 结束语 ⋯ 65
    参考文献 ⋯ 65

15. 纳米机械悬臂阵列传感器 ⋯ 71
    15.1 技 术 ⋯ 71
    15.2 悬臂阵列传感器 ⋯ 73
    15.3 工作模型 ⋯ 74
    15.4 微制造 ⋯ 78
    15.5 测量设备 ⋯ 78
    15.6 功能技术 ⋯ 82
    15.7 应 用 ⋯ 83
    15.8 结论与展望 ⋯ 89
    参考文献 ⋯ 90

16. 医疗纳米器件中的生物分子 ⋯ 97
    16.1 定义和领域 ⋯ 98

16.2 组装方法 105
　　16.3 传感器件 115
　　16.4 结束语：应用障碍 122
　　参考文献 124

17. G-蛋白耦合受体：表面显示和生物传感器技术进展 129
　　17.1 GPCR:G-蛋白激活周期 132
　　17.2 GPCRs和G-蛋白的制备 133
　　17.3 GPCR信号的蛋白工程 134
　　17.4 GPCR生物传感 135
　　17.5 GPCRs的未来 143
　　参考文献 143

18. 微流体器件及其在实验室芯片上的应用 147
　　18.1 微流体器件的材料和微/纳米制造技术 148
　　18.2 有源微流体器件 151
　　18.3 无源微流体智能器件 157
　　18.4 生物化学分析的实验室芯片 164
　　参考文献 171

19. 基于流体平台的离心分离机 175
　　19.1 为什么流体推动源于向心力？ 176
　　19.2 致密的光盘或微离心机流体 178
　　19.3 光盘的应用 182
　　19.4 结论 193
　　参考文献 194

20. 微流体器件中的微/纳米液滴 197
　　20.1 有源的或者可编程的液滴系统 198
　　20.2 无源的液滴控制技术 201
　　20.3 应用 208
　　20.4 结论 210
　　参考文献 210

Contents

List of Abbreviations

Part B MEMS/NEMS and BioMEMS/NEMS

12 MEMS/NEMS Devices and Applications
Darrin J. Young, Christian A. Zorman, Mehran Mehregany 359
- 12.1 MEMS Devices and Applications 361
- 12.2 Nanoelectromechanical Systems (NEMS) 380
- 12.3 Current Challenges and Future Trends 383
- **References** 384

13 Next-Generation DNA Hybridization and Self-Assembly Nanofabrication Devices
Michael J. Heller, Benjamin Sullivan, Dietrich Dehlinger, Paul Swanson, Dalibor Hodko 389
- 13.1 Electronic Microarray Technology 391
- 13.2 Electric Field-Assisted Nanofabrication Processes 397
- 13.3 Conclusions 399
- **References** 400

14 Single-Walled Carbon Nanotube Sensor Concepts
Cosmin Roman, Thomas Helbling, Christofer Hierold 403
- 14.1 Design Considerations for SWNT Sensors 404
- 14.2 Fabrication of SWNT Sensors 412
- 14.3 Example State-of-the-Art Applications 416
- 14.4 Concluding Remarks 421
- **References** 421

15 Nanomechanical Cantilever Array Sensors
Hans Peter Lang, Martin Hegner, Christoph Gerber 427
- 15.1 Technique 427
- 15.2 Cantilever Array Sensors 429
- 15.3 Modes of Operation 430
- 15.4 Microfabrication 434
- 15.5 Measurement Setup 434
- 15.6 Functionalization Techniques 438
- 15.7 Applications 439
- 15.8 Conclusions and Outlook 445
- **References** 446

16 Biological Molecules in Therapeutic Nanodevices
Stephen C. Lee, Bharat Bhushan 453
- 16.1 Definitions and Scope 454
- 16.2 Assembly Approaches 461
- 16.3 Sensing Devices 471
- 16.4 Concluding Remarks: Barriers to Practice 478
- **References** 480

17 G-Protein Coupled Receptors: Progress in Surface Display and Biosensor Technology
Wayne R. Leifert, Tamara H. Cooper, Kelly Bailey 485
- 17.1 The GPCR:G-Protein Activation Cycle 488
- 17.2 Preparation of GPCRs and G-Proteins 489
- 17.3 Protein Engineering in GPCR Signaling 490
- 17.4 GPCR Biosensing 491
- 17.5 The Future of GPCRs 499
- **References** 499

18 Microfluidic Devices and Their Applications to Lab-on-a-Chip
Chong H. Ahn, Jin-Woo Choi 503
- 18.1 Materials for Microfluidic Devices and Micro/Nanofabrication Techniques 504
- 18.2 Active Microfluidic Devices 507
- 18.3 Smart Passive Microfluidic Devices 513
- 18.4 Lab-on-a-Chip for Biochemical Analysis 520
- **References** 527

19 Centrifuge-Based Fluidic Platforms
Jim V. Zoval, Guangyao Jia, Horacio Kido, Jitae Kim, Nahui Kim, Marc J. Madou 531
- 19.1 Why Centripetal Force for Fluid Propulsion? 532
- 19.2 Compact Disc or Microcentrifuge Fluidics 534
- 19.3 CD Applications 538
- 19.4 Conclusion 549
- **References** 550

20 Micro-/Nanodroplets in Microfluidic Devices
Yung-Chieh Tan, Shia-Yen Teh, Abraham P. Lee 553
- 20.1 Active or Programmable Droplet Systems 554
- 20.2 Passive Droplet Control Techniques 557
- 20.3 Applications 564
- 20.4 Conclusions 566
- **References** 566

List of Abbreviations

μCP	microcontact printing
1-D	one-dimensional
18-MEA	18-methyl eicosanoic acid
2-D	two-dimensional
2-DEG	two-dimensional electron gas
3-APTES	3-aminopropyltriethoxysilane
3-D	three-dimensional

A

a-BSA	anti-bovine serum albumin
a-C	amorphous carbon
A/D	analog-to-digital
AA	amino acid
AAM	anodized alumina membrane
ABP	actin binding protein
AC	alternating-current
AC	amorphous carbon
ACF	autocorrelation function
ADC	analog-to-digital converter
ADXL	analog devices accelerometer
AFAM	atomic force acoustic microscopy
AFM	atomic force microscope
AFM	atomic force microscopy
AKD	alkylketene dimer
ALD	atomic layer deposition
AM	amplitude modulation
AMU	atomic mass unit
AOD	acoustooptical deflector
AOM	acoustooptical modulator
AP	alkaline phosphatase
APB	actin binding protein
APCVD	atmospheric-pressure chemical vapor deposition
APDMES	aminopropyldimethylethoxysilane
APTES	aminopropyltriethoxysilane
ASIC	application-specific integrated circuit
ASR	analyte-specific reagent
ATP	adenosine triphosphate

B

BAP	barometric absolute pressure
BAPDMA	behenyl amidopropyl dimethylamine glutamate
bcc	body-centered cubic
BCH	brucite-type cobalt hydroxide
BCS	Bardeen–Cooper–Schrieffer
BD	blu-ray disc
BDCS	biphenyldimethylchlorosilane
BE	boundary element
BFP	biomembrane force probe
BGA	ball grid array
BHF	buffered HF
BHPET	1,1'-(3,6,9,12,15-pentaoxapentadecane-1,15-diyl)bis(3-hydroxyethyl-1H-imidazolium-1-yl) di[bis(trifluoromethanesulfonyl)imide]
BHPT	1,1'-(pentane-1,5-diyl)bis(3-hydroxyethyl-1H-imidazolium-1-yl) di[bis(trifluoromethanesulfonyl)imide]
BiCMOS	bipolar CMOS
bioMEMS	biomedical microelectromechanical system
bioNEMS	biomedical nanoelectromechanical system
BMIM	1-butyl-3-methylimidazolium
BP	bit pitch
BPAG1	bullous pemphigoid antigen 1
BPT	biphenyl-4-thiol
BPTC	cross-linked BPT
BSA	bovine serum albumin
BST	barium strontium titanate
BTMAC	behentrimonium chloride

C

CA	constant amplitude
CA	contact angle
CAD	computer-aided design
CAH	contact angle hysteresis
cAMP	cyclic adenosine monophosphate
CAS	Crk-associated substrate
CBA	cantilever beam array
CBD	chemical bath deposition
CCD	charge-coupled device
CCVD	catalytic chemical vapor deposition
CD	compact disc
CD	critical dimension
CDR	complementarity determining region
CDW	charge density wave
CE	capillary electrophoresis
CE	constant excitation
CEW	continuous electrowetting
CG	controlled geometry
CHO	Chinese hamster ovary
CIC	cantilever in cantilever
CMC	cell membrane complex
CMC	critical micelle concentration
CMOS	complementary metal–oxide–semiconductor
CMP	chemical mechanical polishing

CNF	carbon nanofiber		DOS	density of states
CNFET	carbon nanotube field-effect transistor		DP	decylphosphonate
CNT	carbon nanotube		DPN	dip-pen nanolithography
COC	cyclic olefin copolymer		DRAM	dynamic random-access memory
COF	chip-on-flex		DRIE	deep reactive ion etching
COF	coefficient of friction		ds	double-stranded
COG	cost of goods		DSC	differential scanning calorimetry
CoO	cost of ownership		DSP	digital signal processor
COS	CV-1 in origin with SV40		DTR	discrete track recording
CP	circularly permuted		DTSSP	3,3'-dithio-bis(sulfosuccinimidylproprionate)
CPU	central processing unit		DUV	deep-ultraviolet
CRP	C-reactive protein		DVD	digital versatile disc
CSK	cytoskeleton		DWNT	double-walled CNT
CSM	continuous stiffness measurement			
CTE	coefficient of thermal expansion			
Cu-TBBP	Cu-tetra-3,5 di-tertiary-butyl-phenyl porphyrin			
CVD	chemical vapor deposition			

E

EAM	embedded atom method
EB	electron beam
EBD	electron beam deposition
EBID	electron-beam-induced deposition
EBL	electron-beam lithography
ECM	extracellular matrix
ECR-CVD	electron cyclotron resonance chemical vapor deposition
ED	electron diffraction
EDC	1-ethyl-3-(3-diamethylaminopropyl) carbodiimide
EDL	electrostatic double layer
EDP	ethylene diamine pyrochatechol
EDTA	ethylenediamine tetraacetic acid
EDX	energy-dispersive x-ray
EELS	electron energy loss spectra
EFM	electric field gradient microscopy
EFM	electrostatic force microscopy
EHD	elastohydrodynamic
EO	electroosmosis
EOF	electroosmotic flow
EOS	electrical overstress
EPA	Environmental Protection Agency
EPB	electrical parking brake
ESD	electrostatic discharge
ESEM	environmental scanning electron microscope
EU	European Union
EUV	extreme ultraviolet
EW	electrowetting
EWOD	electrowetting on dielectric

D

DBR	distributed Bragg reflector
DC-PECVD	direct-current plasma-enhanced CVD
DC	direct-current
DDT	dichlorodiphenyltrichloroethane
DEP	dielectrophoresis
DFB	distributed feedback
DFM	dynamic force microscopy
DFS	dynamic force spectroscopy
DGU	density gradient ultracentrifugation
DI	FESPdigital instrument force modulation etched Si probe
DI	TESPdigital instrument tapping mode etched Si probe
DI	digital instrument
DI	deionized
DIMP	diisopropylmethylphosphonate
DIP	dual inline packaging
DIPS	industrial postpackaging
DLC	diamondlike carbon
DLP	digital light processing
DLVO	Derjaguin–Landau–Verwey–Overbeek
DMD	deformable mirror display
DMD	digital mirror device
DMDM	1,3-dimethylol-5,5-dimethyl
DMMP	dimethylmethylphosphonate
DMSO	dimethyl sulfoxide
DMT	Derjaguin–Muller–Toporov
DNA	deoxyribonucleic acid
DNT	2,4-dinitrotoluene
DOD	Department of Defense
DOE	Department of Energy
DOE	diffractive optical element
DOF	degree of freedom
DOPC	1,2-dioleoyl-sn-glycero-3-phosphocholine

F

F-actin	filamentous actin
FA	focal adhesion
FAA	formaldehyde–acetic acid–ethanol
FACS	fluorescence-activated cell sorting

FAK	focal adhesion kinase		HDT	hexadecanethiol
FBS	fetal bovine serum		HDTV	high-definition television
FC	flip-chip		HEK	human embryonic kidney 293
FCA	filtered cathodic arc		HEL	hot embossing lithography
fcc	face-centered cubic		HEXSIL	hexagonal honeycomb polysilicon
FCP	force calibration plot		HF	hydrofluoric
FCS	fluorescence correlation spectroscopy		HMDS	hexamethyldisilazane
FD	finite difference		HNA	hydrofluoric-nitric-acetic
FDA	Food and Drug Administration		HOMO	highest occupied molecular orbital
FE	finite element		HOP	highly oriented pyrolytic
FEM	finite element method		HOPG	highly oriented pyrolytic graphite
FEM	finite element modeling		HOT	holographic optical tweezer
FESEM	field emission SEM		HP	hot-pressing
FESP	force modulation etched Si probe		HPI	hexagonally packed intermediate
FET	field-effect transistor		HRTEM	high-resolution transmission electron microscope
FFM	friction force microscope			
FFM	friction force microscopy		HSA	human serum albumin
FIB-CVD	focused ion beam chemical vapor deposition		HtBDC	hexa-*tert*-butyl-decacyclene
			HTCS	high-temperature superconductivity
FIB	focused ion beam		HTS	high throughput screening
FIM	field ion microscope		HUVEC	human umbilical venous endothelial cell
FIP	feline coronavirus			
FKT	Frenkel–Kontorova–Tomlinson			
FM	frequency modulation			
FMEA	failure-mode effect analysis			
FP6	Sixth Framework Program			
FP	fluorescence polarization			
FPR	*N*-formyl peptide receptor			
FS	force spectroscopy			
FTIR	Fourier-transform infrared			
FV	force–volume			

I

			IBD	ion beam deposition
			IC	integrated circuit
			ICA	independent component analysis
			ICAM-1	intercellular adhesion molecules 1
			ICAM-2	intercellular adhesion molecules 2
			ICT	information and communication technology
			IDA	interdigitated array
			IF	intermediate filament
			IF	intermediate-frequency
			IFN	interferon
			IgG	immunoglobulin G
			IKVAV	isoleucine–lysine–valine–alanine–valine
			IL	ionic liquid
			IMAC	immobilized metal ion affinity chromatography
			IMEC	Interuniversity MicroElectronics Center
			IR	infrared
			ISE	indentation size effect
			ITO	indium tin oxide
			ITRS	International Technology Roadmap for Semiconductors
			IWGN	Interagency Working Group on Nanoscience, Engineering, and Technology

G

GABA	γ-aminobutyric acid
GDP	guanosine diphosphate
GF	gauge factor
GFP	green fluorescent protein
GMR	giant magnetoresistive
GOD	glucose oxidase
GPCR	G-protein coupled receptor
GPS	global positioning system
GSED	gaseous secondary-electron detector
GTP	guanosine triphosphate
GW	Greenwood and Williamson

H

HAR	high aspect ratio
HARMEMS	high-aspect-ratio MEMS
HARPSS	high-aspect-ratio combined poly- and single-crystal silicon
HBM	human body model
hcp	hexagonal close-packed
HDD	hard-disk drive

J

JC	jump-to-contact
JFIL	jet-and-flash imprint lithography
JKR	Johnson–Kendall–Roberts

K

KASH	Klarsicht, ANC-1, Syne Homology
KPFM	Kelvin probe force microscopy

L

LA	lauric acid
LAR	low aspect ratio
LB	Langmuir–Blodgett
LBL	layer-by-layer
LCC	leadless chip carrier
LCD	liquid-crystal display
LCoS	liquid crystal on silicon
LCP	liquid-crystal polymer
LDL	low-density lipoprotein
LDOS	local density of states
LED	light-emitting diode
LFA-1	leukocyte function-associated antigen-1
LFM	lateral force microscope
LFM	lateral force microscopy
LIGA	Lithographie Galvanoformung Abformung
LJ	Lennard-Jones
LMD	laser microdissection
LMPC	laser microdissection and pressure catapulting
LN	liquid-nitrogen
LoD	limit-of-detection
LOR	lift-off resist
LPC	laser pressure catapulting
LPCVD	low-pressure chemical vapor deposition
LSC	laser scanning cytometry
LSN	low-stress silicon nitride
LT-SFM	low-temperature scanning force microscope
LT-SPM	low-temperature scanning probe microscopy
LT-STM	low-temperature scanning tunneling microscope
LT	low-temperature
LTM	laser tracking microrheology
LTO	low-temperature oxide
LTRS	laser tweezers Raman spectroscopy
LUMO	lowest unoccupied molecular orbital
LVDT	linear variable differential transformer

M

MALDI	matrix assisted laser desorption ionization
MAP	manifold absolute pressure
MAPK	mitogen-activated protein kinase
MAPL	molecular assembly patterning by lift-off
MBE	molecular-beam epitaxy
MC	microcantilever
MC	microcapillary
MCM	multi-chip module
MD	molecular dynamics
ME	metal-evaporated
MEMS	microelectromechanical system
MExFM	magnetic exchange force microscopy
MFM	magnetic field microscopy
MFM	magnetic force microscope
MFM	magnetic force microscopy
MHD	magnetohydrodynamic
MIM	metal–insulator–metal
MIMIC	micromolding in capillaries
MLE	maximum likelihood estimator
MOCVD	metalorganic chemical vapor deposition
MOEMS	microoptoelectromechanical system
MOS	metal–oxide–semiconductor
MOSFET	metal–oxide–semiconductor field-effect transistor
MP	metal particle
MPTMS	mercaptopropyltrimethoxysilane
MRFM	magnetic resonance force microscopy
MRFM	molecular recognition force microscopy
MRI	magnetic resonance imaging
MRP	molecular recognition phase
MscL	mechanosensitive channel of large conductance
MST	microsystem technology
MT	microtubule
mTAS	micro total analysis system
MTTF	mean time to failure
MUMP	multiuser MEMS process
MVD	molecular vapor deposition
MWCNT	multiwall carbon nanotube
MWNT	multiwall nanotube
MYD/BHW	Muller–Yushchenko–Derjaguin/Burgess–Hughes–White

N

NA	numerical aperture
NADIS	nanoscale dispensing
NASA	National Aeronautics and Space Administration
NC-AFM	noncontact atomic force microscopy
NEMS	nanoelectromechanical system
NGL	next-generation lithography
NHS	N-hydroxysuccinimidyl
NIH	National Institute of Health
NIL	nanoimprint lithography
NIST	National Institute of Standards and Technology
NMP	no-moving-part
NMR	nuclear magnetic resonance
NMR	nuclear mass resonance
NNI	National Nanotechnology Initiative

NOEMS	nanooptoelectromechanical system		PET	poly(ethyleneterephthalate)
NP	nanoparticle		PETN	pentaerythritol tetranitrate
NP	nanoprobe		PFDA	perfluorodecanoic acid
NSF	National Science Foundation		PFDP	perfluorodecylphosphonate
NSOM	near-field scanning optical microscopy		PFDTES	perfluorodecyltriethoxysilane
NSTC	National Science and Technology Council		PFM	photonic force microscope
			PFOS	perfluorooctanesulfonate
NTA	nitrilotriacetate		PFPE	perfluoropolyether
nTP	nanotransfer printing		PFTS	perfluorodecyltricholorosilane
			PhC	photonic crystal
			PI3K	phosphatidylinositol-3-kinase

O

			PI	polyisoprene
			PID	proportional–integral–differential
ODA	octadecylamine		PKA	protein kinase
ODDMS	n-octadecyldimethyl(dimethylamino)silane		PKC	protein kinase C
			PKI	protein kinase inhibitor
ODMS	n-octyldimethyl(dimethylamino)silane		PL	photolithography
ODP	octadecylphosphonate		PLC	phospholipase C
ODTS	octadecyltrichlorosilane		PLD	pulsed laser deposition
OLED	organic light-emitting device		PMAA	poly(methacrylic acid)
OM	optical microscope		PML	promyelocytic leukemia
OMVPE	organometallic vapor-phase epitaxy		PMMA	poly(methyl methacrylate)
OS	optical stretcher		POCT	point-of-care testing
OT	optical tweezers		POM	polyoxy-methylene
OTRS	optical tweezers Raman spectroscopy		PP	polypropylene
OTS	octadecyltrichlorosilane		PPD	p-phenylenediamine
oxLDL	oxidized low-density lipoprotein		PPMA	poly(propyl methacrylate)
			PPy	polypyrrole
			PS-PDMS	poly(styrene-b-dimethylsiloxane)

P

			PS/clay	polystyrene/nanoclay composite
			PS	polystyrene
P–V	peak-to-valley		PSA	prostate-specific antigen
PAA	poly(acrylic acid)		PSD	position-sensitive detector
PAA	porous anodic alumina		PSD	position-sensitive diode
PAH	poly(allylamine hydrochloride)		PSD	power-spectral density
PAPP	p-aminophenyl phosphate		PSG	phosphosilicate glass
Pax	paxillin		PSGL-1	P-selectin glycoprotein ligand-1
PBC	periodic boundary condition		PTFE	polytetrafluoroethylene
PBS	phosphate-buffered saline		PUA	polyurethane acrylate
PC	polycarbonate		PUR	polyurethane
PCB	printed circuit board		PVA	polyvinyl alcohol
PCL	polycaprolactone		PVD	physical vapor deposition
PCR	polymerase chain reaction		PVDC	polyvinylidene chloride
PDA	personal digital assistant		PVDF	polyvinyledene fluoride
PDMS	polydimethylsiloxane		PVS	polyvinylsiloxane
PDP	2-pyridyldithiopropionyl		PWR	plasmon-waveguide resonance
PDP	pyridyldithiopropionate		PZT	lead zirconate titanate
PE	polyethylene			
PECVD	plasma-enhanced chemical vapor deposition			

Q

PEEK	polyetheretherketone			
PEG	polyethylene glycol		QB	quantum box
PEI	polyethyleneimine		QCM	quartz crystal microbalance
PEN	polyethylene naphthalate		QFN	quad flat no-lead
PES	photoemission spectroscopy		QPD	quadrant photodiode
PES	position error signal		QWR	quantum wire

R

RBC	red blood cell
RCA	Radio Corporation of America
RF	radiofrequency
RFID	radiofrequency identification
RGD	arginine–glycine–aspartic
RH	relative humidity
RHEED	reflection high-energy electron diffraction
RICM	reflection interference contrast microscopy
RIE	reactive-ion etching
RKKY	Ruderman–Kittel–Kasuya–Yoshida
RMS	root mean square
RNA	ribonucleic acid
ROS	reactive oxygen species
RPC	reverse phase column
RPM	revolutions per minute
RSA	random sequential adsorption
RT	room temperature
RTP	rapid thermal processing

S

SAE	specific adhesion energy
SAM	scanning acoustic microscopy
SAM	self-assembled monolayer
SARS-CoV	syndrome associated coronavirus
SATI	self-assembly, transfer, and integration
SATP	(S-acetylthio)propionate
SAW	surface acoustic wave
SB	Schottky barrier
SCFv	single-chain fragment variable
SCM	scanning capacitance microscopy
SCPM	scanning chemical potential microscopy
SCREAM	single-crystal reactive etching and metallization
SDA	scratch drive actuator
SEcM	scanning electrochemical microscopy
SEFM	scanning electrostatic force microscopy
SEM	scanning electron microscope
SEM	scanning electron microscopy
SFA	surface forces apparatus
SFAM	scanning force acoustic microscopy
SFD	shear flow detachment
SFIL	step and flash imprint lithography
SFM	scanning force microscope
SFM	scanning force microscopy
SGS	small-gap semiconducting
SICM	scanning ion conductance microscopy
SIM	scanning ion microscope
SIP	single inline package
SKPM	scanning Kelvin probe microscopy
SL	soft lithography
SLIGA	sacrificial LIGA
SLL	sacrificial layer lithography
SLM	spatial light modulator
SMA	shape memory alloy
SMM	scanning magnetic microscopy
SNOM	scanning near field optical microscopy
SNP	single nucleotide polymorphisms
SNR	signal-to-noise ratio
SOG	spin-on-glass
SOI	silicon-on-insulator
SOIC	small outline integrated circuit
SoS	silicon-on-sapphire
SP-STM	spin-polarized STM
SPM	scanning probe microscope
SPM	scanning probe microscopy
SPR	surface plasmon resonance
sPROM	structurally programmable microfluidic system
SPS	spark plasma sintering
SRAM	static random access memory
SRC	sampling rate converter
SSIL	step-and-stamp imprint lithography
SSRM	scanning spreading resistance microscopy
STED	stimulated emission depletion
SThM	scanning thermal microscope
STM	scanning tunneling microscope
STM	scanning tunneling microscopy
STORM	statistical optical reconstruction microscopy
STP	standard temperature and pressure
STS	scanning tunneling spectroscopy
SUN	Sad1p/UNC-84
SWCNT	single-wall carbon nanotube
SWCNT	single-walled carbon nanotube
SWNT	single wall nanotube
SWNT	single-wall nanotube

T

TA	tilt angle
TASA	template-assisted self-assembly
TCM	tetracysteine motif
TCNQ	tetracyanoquinodimethane
TCP	tricresyl phosphate
TEM	transmission electron microscope
TEM	transmission electron microscopy
TESP	tapping mode etched silicon probe
TGA	thermogravimetric analysis
TI	Texas Instruments
TIRF	total internal reflection fluorescence
TIRM	total internal reflection microscopy
TLP	transmission-line pulse
TM	tapping mode
TMAH	tetramethyl ammonium hydroxide
TMR	tetramethylrhodamine
TMS	tetramethylsilane

TMS	trimethylsilyl
TNT	trinitrotoluene
TP	track pitch
TPE-FCCS	two-photon excitation fluorescence cross-correlation spectroscopy
TPI	threads per inch
TPMS	tire pressure monitoring system
TR	torsional resonance
TREC	topography and recognition
TRIM	transport of ions in matter
TSDC	thermally stimulated depolarization current
TTF	tetrathiafulvalene
TV	television

U

UAA	unnatural AA
UHV	ultrahigh vacuum
ULSI	ultralarge-scale integration
UML	unified modeling language
UNCD	ultrananocrystalline diamond
UV	ultraviolet
UVA	ultraviolet A

V

VBS	vinculin binding site
VCO	voltage-controlled oscillator
VCSEL	vertical-cavity surface-emitting laser
vdW	van der Waals
VHH	variable heavy–heavy
VLSI	very large-scale integration
VOC	volatile organic compound
VPE	vapor-phase epitaxy
VSC	vehicle stability control

X

XPS	x-ray photon spectroscopy
XRD	x-ray powder diffraction

Y

YFP	yellow fluorescent protein

Z

Z-DOL	perfluoropolyether

Part B MEMS/NEMS and BioMEMS/NEMS

12 MEMS/NEMS Devices and Applications
Darrin J. Young, Cleveland, USA
Christian A. Zorman, Cleveland, USA
Mehran Mehregany, Cleveland, USA

13 Next-Generation DNA Hybridization and Self-Assembly Nanofabrication Devices
Michael J. Heller, La Jolla, USA
Benjamin Sullivan, San Diego, USA
Dietrich Dehlinger, Livermore, USA
Paul Swanson, San Diego, USA
Dalibor Hodko, San Diego, USA

14 Single-Walled Carbon Nanotube Sensor Concepts
Cosmin Roman, Zurich, Switzerland
Thomas Helbling, Zurich, Switzerland
Christofer Hierold, Zurich, Switzerland

15 Nanomechanical Cantilever Array Sensors
Hans Peter Lang, Basel, Switzerland
Martin Hegner, Dublin, Ireland
Christoph Gerber, Basel, Switzerland

16 Biological Molecules in Therapeutic Nanodevices
Stephen C. Lee, Columbus, USA
Bharat Bhushan, Columbus, USA

17 G-Protein Coupled Receptors: Progress in Surface Display and Biosensor Technology
Wayne R. Leifert, Adelaide, Australia
Tamara H. Cooper, Adelaide, Australia
Kelly Bailey, Adelaide, Australia

18 Microfluidic Devices and Their Applications to Lab-on-a-Chip
Chong H. Ahn, Cincinnati, USA
Jin-Woo Choi, Baton Rouge, USA

19 Centrifuge-Based Fluidic Platforms
Jim V. Zoval, Mission Viejo, USA
Guangyao Jia, Irvine, USA
Horacio Kido, Irvine, USA
Jitae Kim, Irvine, USA
Nahui Kim, Seoul, South Korea
Marc J. Madou, Irvine, USA

20 Micro-/Nanodroplets in Microfluidic Devices
Yung-Chieh Tan, St. Louis, USA
Shia-Yen Teh, Irvine, USA
Abraham P. Lee, Irvine, USA

12. MEMS/NEMS Devices and Applications

Darrin J. Young, Christian A. Zorman, Mehran Mehregany

Microelectromechanical systems (MEMS) have played key roles in many important areas, for example transportation, communication, automated manufacturing, environmental monitoring, health care, defense systems, and a wide range of consumer products. MEMS are inherently small, thus offering attractive characteristics such as reduced size, weight, and power dissipation and improved speed and precision compared to their macroscopic counterparts. Integrated circuit (IC) fabrication technology has been the primary enabling technology for MEMS besides a few special etching, bonding and assembly techniques. Microfabrication provides a powerful tool for batch processing and miniaturizing electromechanical devices and systems to a dimensional scale that is not accessible by conventional machining techniques. As IC fabrication technology continues to scale toward deep submicrometer and nanometer feature sizes, a variety of nanoelectromechanical systems (NEMS) can be envisioned in the foreseeable future. Nanoscale mechanical devices and systems integrated with nanoelectronics will open a vast number of new exploratory research areas in science and engineering. NEMS will most likely serve as an enabling technology, merging engineering with the life sciences in ways that are not currently feasible with microscale tools and technologies.

MEMS has been applied to a wide range of fields. Hundreds of microdevices have been developed for specific applications. It is thus difficult to provide an overview covering every aspect of the

12.1	MEMS Devices and Applications	361
	12.1.1 Pressure Sensor	361
	12.1.2 Inertial Sensor	364
	12.1.3 Optical MEMS	369
	12.1.4 RF MEMS	373
12.2	Nanoelectromechanical Systems (NEMS)	380
	12.2.1 Materials and Fabrication Techniques	381
	12.2.2 Transduction Techniques	382
	12.2.3 Application Areas	383
12.3	Current Challenges and Future Trends	383
References		384

topic. In this chapter, key aspects of MEMS technology and applications are illustrated by selecting a few demonstrative device examples, such as pressure sensors, inertial sensors, optical and wireless communication devices. Microstructure examples with dimensions on the order of submicrometer are presented with fabrication technologies for future NEMS applications.

Although MEMS has experienced significant growth over the past decade, many challenges still remain. In broad terms, these challenges can be grouped into three general categories: (1) fabrication challenges; (2) packaging challenges; and (3) application challenges. Challenges in these areas will, in large measure, determine the commercial success of a particular MEMS device in both technical and economic terms. This chapter presents a brief discussion of some of these challenges as well as possible approaches to addressing them.

Microelectromechanical Systems, generally referred to as MEMS, has had a history of research and development over a few decades. Besides the traditional microfabricated sensors and actuators, the field covers micromechanical components and systems integrated or microassembled with electronics on the same substrate or package, achieving high-performance functional systems. These devices and systems have played key roles in many important areas such as transportation, communication, automated manufacturing, environmental

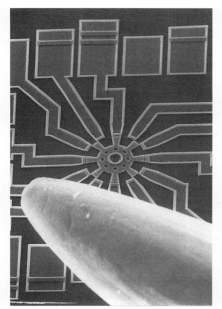

Fig. 12.1 SEM micrograph of a polysilicon microelectromechanical motor (after [12.1])

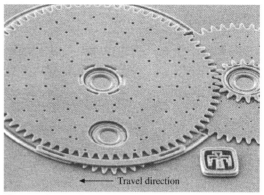

Fig. 12.2 SEM micrograph of polysilicon microgears (after [12.3])

monitoring, health care, defense systems, and a wide range of consumer products. MEMS are inherently small, thus offering attractive characteristics such as reduced size, weight, and power dissipation and improved speed and precision compared to their macroscopic counterparts. The development of MEMS requires appropriate fabrication technologies that enable the definition of small geometries, precise dimension control, design flexibility, interfacing with microelectronics, repeatability, reliability, high yield, and low cost. Integrated circuits (IC) fabrication technology meets all of the above criteria and has been the primary enabling fabrication technology for MEMS besides a few special etching, bonding and assembly techniques. Microfabrication provides a powerful tool for batch processing and miniaturization of electromechanical devices and systems into a dimensional scale, which is not accessible by conventional machining techniques. Most MEMS devices exhibit a length or width ranging from micrometers to several hundreds of micrometers with a thickness from submicrometer up to tens of micrometers depending upon fabrication technique employed. A physical displacement of a sensor or an actuator is typically on the same order of magnitude. Figure 12.1 shows an SEM micrograph of a microelectromechanical motor developed in late 1980s [12.1]. Polycrystalline silicon (polysilicon) surface micromachining technology was used to fabricate the micromotor achieving a diameter of 150 μm and a minimum vertical feature size on the order of a micrometer. A probe tip is also shown in the micrograph for a size comparison. This device example and similar others [12.2] demonstrated at that time what MEMS technology could accomplish in microscale machining and served as a strong technology indicator for continued MEMS development. The field has expanded greatly in recent years along with rapid technology advances. Figure 12.2, for example, shows a photo of microgears fabricated in mid-1990s using a five-level polysilicon surface micromachining technology [12.3]. This device represents one of the most advanced surface micromachining fabrication process developed to date. One can imagine that a wide range of sophisticated microelectromechanical devices and systems can be realized through applying such technology in the future. As IC fabrication technology continues to scale toward deep submicrometer and nanometer feature sizes, a variety of nanoelectromechanical systems (NEMS) can be envisioned in the foreseeable future. Nanoscale mechanical devices and systems integrated with nanoelectronics will open a vast number of new exploratory research areas in science and engineering. NEMS will most likely serve as an enabling technology merging engineering with the life sciences in ways that are not currently feasible with the microscale tools and technologies.

This chapter will provide a general overview on MEMS and NEMS devices along with their applications. MEMS technology has been applied to a wide range of fields. Over hundreds of microdevices have been developed for specific applications. Thus, it is difficult to provide an overview covering every aspect of the topic. It is the authors' intent to illustrate key aspects of MEMS technology and its impact to specific applications by selecting a few demonstrative device examples in this chapter. For a wide-ranging discussion of nearly all types of micromachined sensors and actuators, books by *Kovacs* [12.4] and *Senturia* [12.5] are recommended.

12.1 MEMS Devices and Applications

MEMS devices have played key roles in many areas of development. Microfabricated sensors, actuators, and electronics are the most critical components required to implement a complete system for a specific function. Microsensors and actuators can be fabricated by various micromachining processing technologies. In this section, a number of selected MEMS devices are presented to illustrate the basic device operating principles as well as to demonstrate key aspects of the microfabrication technology and application impact.

12.1.1 Pressure Sensor

Pressure sensors are one of the early devices realized by silicon micromachining technologies and have become successful commercial products. The devices have been widely used in various industrial and biomedical applications. The sensors can be based on piezoelectric, piezoresistive, capacitive, and resonant sensing mechanisms. Silicon bulk and surface micromachining techniques have been used for sensor batch fabrication, thus achieving size miniaturization and low cost. Two types of pressure sensors, piezoresistive and capacitive, are described here for an illustration purpose.

Piezoresistive Sensor
The piezoresisitve effect in silicon has been widely used for implementing pressure sensors. A pressure-induced strain deforms the silicon band structure, thus changing the resistivity of the material. The piezoresistive effect is typically crystal orientation dependent and is also affected by doping and temperature. A practical piezoresistive pressure sensor can be implemented by fabricating four sensing resistors along the edges of a thin silicon diaphragm, which acts as a mechanical amplifier to increase the stress and strain at the sensor site. The four sensing elements are connected in a bridge configuration with push–pull signals to increase the sensitivity. The measurable pressure range for such a sensor can be from 10^{-3} to 10^6 Torr depending upon the design. An example of a piezoresistive pressure sensor is shown in Fig. 12.3. The device consists of a silicon diaphragm suspended over a reference vacuum cavity to form an absolute pressure sensor. An external pressure applied over the diaphragm introduces a stress on the sensing resistors, thus resulting in a resistance value change corresponding to the pressure. The fabrication sequence is outlined as follows. The piezoresistors are typically first formed through a boron diffusion process followed by a high-temperature annealing step in order to achieve a resistance value on the order of a few kiloohms. The wafer is then passivated with a silicon dioxide layer and contact windows are opened for metallization. At this point, the wafer is patterned on the backside, followed by a timed silicon wet etch to form the diaphragm, typically having a thickness around a few tens of micrometers. The diaphragm can have a length of several hundreds of micrometers. A second silicon wafer is then bonded to the device wafer in vacuum to form a reference vacuum cavity, thus completing the fabrication process. The second wafer can also be further etched through to form an inlet port, implementing a gauge pressure sensor [12.6]. The piezoresistive sensors are simple to fabricate and can be readily interfaced with electronic systems. However, the resistors are temperature dependent and consume DC power. Long-term characteristic drift and resistor thermal noise ultimately limit the sensor resolution.

Capacitive Sensor
Capacitive pressure sensors are attractive because they are virtually temperature independent and consume zero

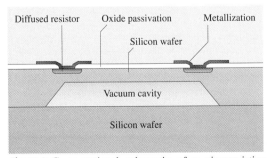

Fig. 12.3 Cross-sectional schematic of a piezoresistive pressure sensor

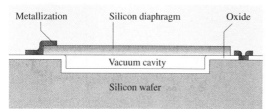

Fig. 12.4 Cross-sectional schematic of a capacitive pressure sensor

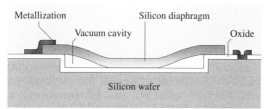

Fig. 12.5 Cross-sectional schematic of a touch-mode capacitive pressure sensor

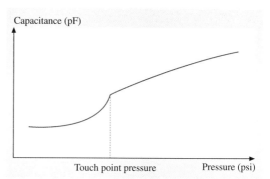

Fig. 12.6 Touch-mode capacitive pressure sensor characteristic response

DC power. The devices do not exhibit initial turn-on drift and are stable over time. Furthermore, CMOS microelectronic circuits can be readily interfaced with the sensors to provide advanced signal conditioning and processing, thus improving overall system performance. An example of a capacitive pressure sensor is shown in Fig. 12.4. The device consists of an edge clamped silicon diaphragm suspended over a vacuum cavity. The diaphragm can be square or circular with a typical thickness of a few micrometers and a length or radius of a few hundreds micrometers, respectively. The vacuum cavity typically has a depth of a few micrometers. The diaphragm and substrate form a pressure dependent air-gap variable capacitor. An increased external pressure causes the diaphragm to deflect towards the substrate, thus resulting in an increase in the capacitance value. A simplified fabrication process can be outlined as follows. A silicon wafer is first patterned and etched to form the cavity. The wafer is then oxidized followed by bonding to a second silicon wafer with a heavily-doped boron layer, which defines the diaphragm thickness, at the surface. The bonding process can be performed in vacuum to realize the vacuum cavity. If the vacuum bonding is not performed at this stage, a low pressure sealing process can be used to form the vacuum cavity after patterning the sensor diaphragm, provided that sealing channels are available. The silicon substrate above the boron layer is then removed through a wet etching process, followed by patterning to form the sensor diaphragm, which serves as the device top electrode. Contact pads are formed by metallization and patterning. This type of pressure sensor exhibits a nonlinear characteristic and a limited dynamic range. These phenomena, however, can be alleviated through applying an electrostatic force-balanced feedback architecture. A common practice is to introduce another electrode above the sensing diaphragm through wafer bonding [12.7], thus forming two capacitors in series with the diaphragm being the middle electrode. The capacitors are interfaced with electronic circuits, which convert the sensor capacitance value to an output voltage corresponding to the diaphragm position. This voltage is further processed to generate a feedback signal to the top electrode, thus introducing an electrostatic pull up force to maintain the deflectable diaphragm at its nominal position. This negative feedback loop would substantially minimize the device nonlinearity and also extend the sensor dynamic range.

A capacitive pressure sensor achieving an inherent linear characteristic response and a wide dynamic range can be implemented by employing a touch-mode architecture [12.8]. Figure 12.5 shows the cross-sectional view of a touch-mode pressure sensor. The device consists of an edge-clamped silicon diaphragm suspended over a vacuum cavity. The diaphragm deflects under an increasing external pressure and touches the substrate, causing a linear increase in the sensor capacitance value beyond the touch point pressure. Figure 12.6 shows a typical device characteristic curve. The touch point pressure can be designed through engineering the sensor geometric parameters such as the diaphragm size, thickness, cavity depth, etc., for various application requirements. The device can be fabricated using a process flow similar to the flow outlined for the basic

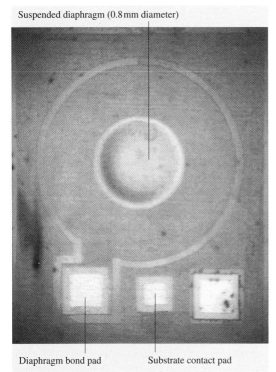

Fig. 12.7 Photo of a touch-mode capacitive pressure sensor (after [12.8])

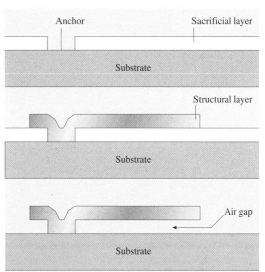

Fig. 12.8 Simplified fabrication sequence of surface micromachining technology

capacitive pressure sensor. Figure 12.7 presents a photo of a fabricated touch-mode sensor employing a circular diaphragm with a diameter of 800 μm and a thickness of 5 μm suspended over a 2.5 μm vacuum cavity. The device achieves a touch point pressure of 8 psi and exhibits a linear capacitance range from 33 pF at 10 psi to 40 pF at 32 psi (absolute pressures). Similar sensor structures have been demonstrated by using single-crystal 3C-SiC diaphragm achieving a high-temperature pressure sensing capability up to 400 °C [12.9].

The above processes use bulk silicon materials for machining and are usually referred to as bulk micromachining. The same devices can also be fabricated using so called surface micromachining. Surface micromachining technology is attractive for integrating MEMS sensors with on-chip electronic circuits. As a result, advanced signal processing capabilities such as data conversion, offset and noise cancellation, digital calibration, temperature compensation, etc. can be implemented adjacent to microsensors on a same substrate, providing a complete high-performance microsystem solution. The single chip approach also eliminates external wiring, which is critical for minimizing noise pick up and enhancing system performance. Surface micromachining, simply stated, is a method of fabricating MEMS through depositing, patterning, and etching a sequence of thin films with thickness on the order of a micrometer. Figure 12.8 illustrates a typical surface micromachining process flow [12.11]. The process

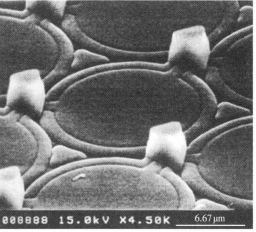

Fig. 12.9 SEM micrograph of polysilicon surface-micromachined capacitive pressure sensors (after [12.10])

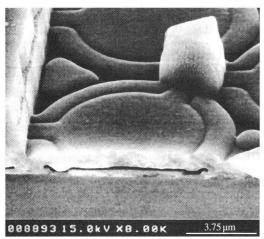

Fig. 12.10 SEM micrograph of a close-up view of a polysilicon surface-micromachined capacitive pressure sensor (after [12.10])

tion. A structural layer, typically a polysilicon film, is deposited and patterned. The underlying sacrificial layer is then removed to freely release the suspended microstructure and to complete the fabrication sequence. The processing materials and steps are compatible with standard integrated circuit process, thus can be readily incorporated as an add-on module to an IC process [12.11–13]. A similar surface micromachining technology has been developed to produce monolithic pressure sensor systems [12.10]. Figure 12.9 shows an SEM micrograph of an array of MEMS capacitive pressure sensors fabricated with BiCMOS electronics on the same substrate. Each sensor consists of a 0.8 μm thick circular polysilicon membrane with a diameter on the order of 20 μm suspended over a 0.3 μm deep vacuum cavity. The devices operate using the same principle as the sensor shown in Fig. 12.4. A close view of the sensor cross-section is shown in Fig. 12.10, which shows the suspended membrane and underneath air gap. These sensors have demonstrated operations in pressure ranges up to 400 bar with an accuracy of 1.5%.

12.1.2 Inertial Sensor

Micromachined inertial sensors consist of accelerometers and gyroscopes. These devices are one of the important types of silicon-based MEMS sensors that have been successfully commercialized. MEMS accelerometers alone have the second largest sales volume after pressure sensors. Gyroscopes are expected to reach a comparable sales volume in a foreseeable future. Accelerometers have been used in a wide range of applications including automotive application for safety systems, active suspension and stability control, biomedical application for activity monitoring, and numerous consumer products such as head-mount displays, camcorders, three-dimensional mouse, etc. High-sensitivity accelerometers are crucial for implementing self-contained navigation and guidance systems. A gyroscope is another type of inertial sensor that measures rate or angle of rotation. The devices can be used along with accelerometers to provide heading information in an inertial navigation system. Gyroscopes also are useful in applications such as automotive ride stabilization and rollover detection, camcorder stabilization, virtual reality, etc. Inertial sensors fabricated by micromachining technology can achieve reduced size, weight, and cost, all which are critical for consumer applications. More importantly, these sensors can be integrated with microelectronic circuits to achieve a functional microsystem with high performance.

starts by depositing a layer of sacrificial material such as silicon dioxide over a wafer followed by anchor forma-

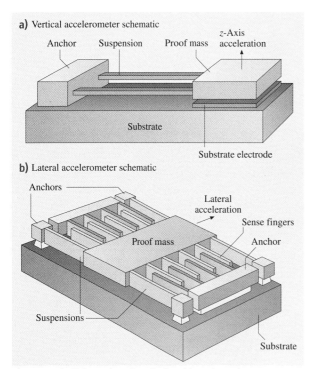

Fig. 12.11a,b Schematics of vertical and lateral accelerometers

Accelerometer

An accelerometer generally consists of a proof mass suspended by compliant mechanical suspensions anchored to a fixed frame. An external acceleration displaces the support frame relative to the proof mass. The displacement can result in an internal stress change in the suspension, which can be detected by piezoresistive sensors as a measure of the external acceleration. The displacement can also be detected as a capacitance change in capacitive accelerometers. Capacitive sensors are attractive for various applications because they exhibit high sensitivity and low temperature dependence, turn-on drift, power dissipation, and noise. The sensors can also be readily integrated with CMOS electronics to perform advanced signal processing for high system performance. Capacitive accelerometers may be divided into two categories as vertical and lateral type sensors. Figure 12.11 shows sensor structures for the two versions. In a vertical device, the proof mass is suspended above the substrate electrode by a small gap typically on the order of a micrometer, forming a parallel-plate sense capacitance. The proof mass moves in the direction perpendicular to the substrate (z-axis) upon a vertical input acceleration, thus changing the gap and hence the capacitance value. The lateral accelerometer consists of a number of movable fingers attached to the proof mass, forming a sense capacitance with an array of fixed parallel fingers. The sensor proof mass moves

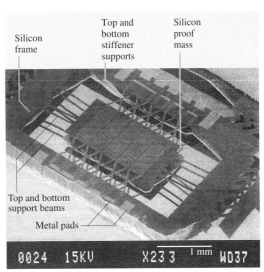

Fig. 12.13 SEM micrograph of a MEMS z-axis accelerometer fabricated by using a combined surface and bulk micromachining technology (after [12.15])

in a plane parallel to the substrate when subjected to a lateral input acceleration, thus changing the overlap area of these fingers; hence the capacitance value. Figure 12.12 shows an SEM top view of a surface-micromachined polysilion z-axis accelerometer [12.14]. The device consists of a $400 \times 400\,\mu m^2$ proof mass with a thickness of $2\,\mu m$ suspended above the substrate electrode by four folded beam suspensions with an air gap around $2\,\mu m$, thus achieving a sense capacitance of $\approx 500\,fF$. The visible holes are used to ensure complete removal of the sacrificial oxide underneath the proof mass at the end of the fabrication process. The sensor can be interfaced with a microelectronic charge amplifier converting the capacitance value to an output voltage for further signal processing and analysis. Force feedback architecture can be applied to stabilize the proof mass position. The combs around the periphery of the proof mass can exert an electrostatic levitation force on the proof mass to achieve the position control, thus improving the system frequency response and linearity performance [12.14].

Surface micromachined accelerometers typically suffer from severe mechanical thermal vibration, commonly referred to as Brownian motion [12.16], due to the small proof mass, thus resulting in a high mechanical noise floor which ultimately limits the sensor resolution. Vacuum packaging can be employed to minimize this adverse effect but with a penalty of increasing

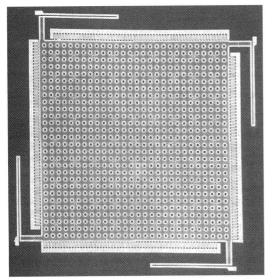

Fig. 12.12 SEM micrograph of a polysilicon surface-micromachined z-axis accelerometer (after [12.14])

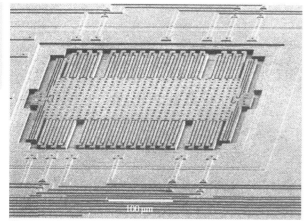

Fig. 12.14 SEM micrograph of a polysilicon surface-micromachined lateral accelerometer (© Analog Devices Inc.)

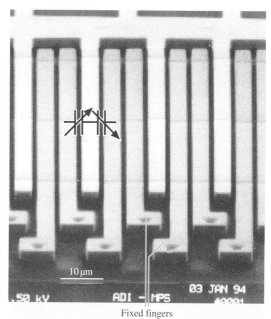

Fig. 12.15 SEM micrograph of a capacitive sensing finger structure

system complexity and cost. Accelerometers using large proof masses fabricated by bulk micromachining or a combination of surface and bulk micromachining techniques are attractive for circumventing this problem. Figure 12.13 shows an SEM micrograph of an all-silicon z-axis accelerometer fabricated through a single silicon wafer by using combined surface and bulk micromachining process to obtain a large proof mass with dimensions of $\approx 2 \times 1 \times 450\,\mu m^3$ [12.15]. The large mass suppresses the Brownian motion effect, achieving a high performance with a resolution on the order of several μg. Similar fabrication techniques have been used to demonstrate a three-axis capacitive accelerometer achieving a noise floor of $\approx 1\,\mu g/\sqrt{Hz}$ [12.17].

A surface-micromachined lateral accelerometer developed by Analog Devices Inc. is shown in Fig. 12.14. The sensor consists of a center proof mass supported by folded beam suspensions with arrays of attached movable fingers, forming a sense capacitance with the fixed parallel fingers. The device is fabricated using a $6\,\mu m$ thick polysilicon structural layer with a small air gap on the order of a micrometer to increase the sensor capacitance value, thus improving the device resolution. Figure 12.15 shows a close-up view of the finger structure for a typical lateral accelerometer. Each movable finger forms differen-

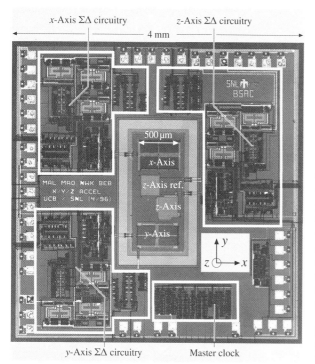

Fig. 12.16 Photo of a monolithic three-axis polysilicon surface-micromachined accelerometer with integrated interface and control electronics (after [12.18]) ◄

tial capacitances with two adjacent fixed fingers. This sensing capacitance configuration is attractive for interfacing with differential electronic detection circuits to suppress common-mode noise and other undesirable signal coupling. Monolithic accelerometers with a three-axis sensing capability integrated with on-chip electronic detection circuits have been realized using surfacing micromachining and CMOS microelectronics fabrication technologies [12.18]. Figure 12.16 shows a photo of one of these microsystem chips, which has an area of 4×4 mm^2. One vertical accelerometer and two lateral accelerometers are placed at the chip center with corresponding detection electronics along the periphery. A z-axis reference device, which is not movable, is used with the vertical sensor for electronic interfacing. The prototype system achieves a sensing resolution on the order of 1 mG with a 100 Hz bandwidth along each axis. The level of performance is adequate for automobile safety activation systems, vehicle stability and active suspension control, and various consumer products. Recently, monolithic MEMS accelerometers fabricated by using post-CMOS surface micromachining fabrication technology have been developed to achieve an acceleration noise floor of $50\,\mu g/\sqrt{Hz}$ [12.19]. This technology can enable MEMS capacitive inertial sensors to be integrated with interface electronics in a commercial CMOS process, thus minimizing prototyping cost.

Gyroscope

Most of micromachined gyroscopes employ vibrating mechanical elements to sense rotations. The sensors rely on energy transfer between two vibration modes of a structure caused by Coriolis acceleration. Figure 12.17 presents a schematic of a z-axis vibratory rate gyroscope. The device consists of an oscillating mass electrostatically driven into resonance along the drive-mode axis using comb fingers. An angular rotation along the vertical axis (z-axis) introduces a Coriolis acceleration, which results in a structure deflection along the sense-mode axis, shown in the figure. The deflection changes the differential sense capacitance value, which can be detected as a measure of input angular rotation. A z-axis vibratory rate gyroscope operating upon this principle is fabricated using surface micromachining technology and integrated together with electronic detection circuits, as illustrated in Fig. 12.18 [12.20]. The micromachined sensor is fabricated using polysilicon structural material with a thickness around $2\,\mu$m and occupies an area of 1×1 mm^2. The sensor achieves a resolution

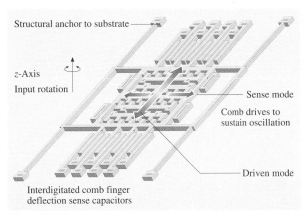

Fig. 12.17 Schematic of a vibratory rate gyroscope

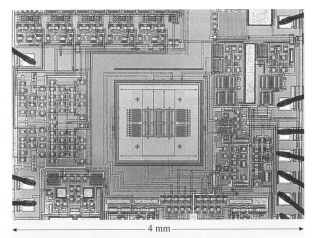

Fig. 12.18 Photo of a monolithic polysilicon surface-micromachined z-axis vibratory gyroscope with integrated interface and control electronics (after [12.20])

of $\approx 1°/(s\sqrt{Hz})$ under a vacuum pressure around 50 mTorr. Other MEMS single-axis gyroscopes integrated in commercial IC processes were demonstrated recently achieving an enhance performance [12.21, 22].

A dual-axis gyroscope based on a rotational disk at its resonance can be used to sense angular rotation along two lateral axes (x- and y-axis). Figure 12.19 shows a device schematic demonstrating the operating principle. A rotor disk supported by four mechanical suspensions can be driven into angular resonance along the z-axis. An input angular rotation along the x-axis will generate a Coriolis acceleration causing the disk to rotate along the y-axis, and vice versa. This

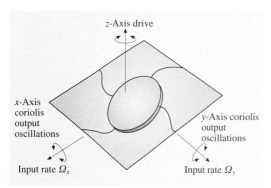

Fig. 12.19 Schematic of a dual-axis gyroscope

Coriolis-acceleration-induced rotation will change the sensor capacitance values between the disk and different sensing electrodes underneath. The capacitance change can be detected and processed by electronic interface circuits. Angular rotations along the two lateral axes can be measured simultaneously using this device architecture. Figure 12.20 shows a photo of a dual-axis gyroscope fabricated using a 2 μm thick polysilicon surface micromachining technology [12.23]. As shown in the figure, curved electrostatic drive combs are positioned along the circumference of the rotor dick to drive it into resonance along the vertical axis. The gyroscope exhibits a low random walk of $1°/\sqrt{h}$ under a vacuum pressure around 60 mTorr. With accelerometers and gyroscopes each capable of three-axis sensing, a micormachine-based inertial measurement system providing a six-degree-of-freedom sensing capability

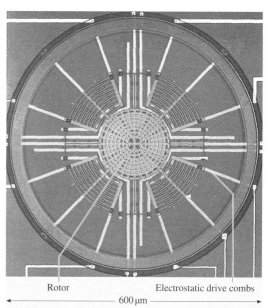

Fig. 12.20 Photo of a polysilicon surface-micromachined dual-axis gyroscope (after [12.23])

can be realized. Figure 12.21 presents a photo of such a system containing a dual-axis gyroscope, a z-axis gyroscope, and a three-axis accelerometer chip integrated with microelectronic circuitry. Due to the precision in device layout and fabrication, the system can measure angular rotation and acceleration without the need to align individual sensors.

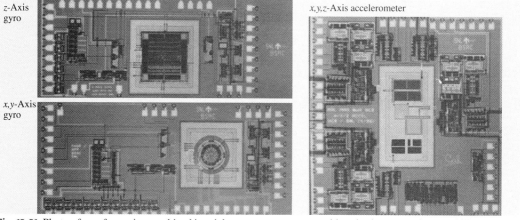

Fig. 12.21 Photo of a surface-micromachined inertial measurement system with a six-degree sensing capability

12.1.3 Optical MEMS

Surface micromachining has served as a key enabling technology to realize microeletromechancal optical devices for various applications ranging from sophisticated visual information displays and fiber-optic telecommunication to bar-code reading. Most of the existing optical systems are implemented using conventional optical components, which suffer from bulky size, high cost, large power consumption, poor efficiency and reliability issues. MEMS technology is promising for producing miniaturized, reliable, inexpensive optical components to revolutionize conventional optical systems [12.25]. In this section of the chapter, a few selected MEMS optical devices will be presented to illustrate their impact in the fields of visual display, precision optical platform, and data switching for optical communication.

Visual Display

An early MEMS device successfully used for various display applications is the Texas Instruments Digital Micromirror Device (DMD). The DMD technology can achieve higher performance in terms of resolution, brightness, contrast ratio, and convergence than the conventional cathode ray tube and is critical for digital high-definition television applications. A DMD consists of a large array of small mirrors with a typical area of $16 \times 16 \,\mu m^2$ as illustrated in Fig. 12.22. A probe tip is shown in the figure for a size comparison. Fig-

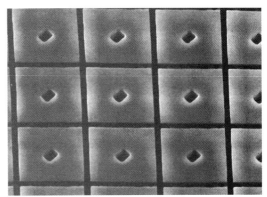

Fig. 12.23 SEM micrograph of a close-up view of a DMD pixel array (after [12.24])

ure 12.23 shows an SEM micrograph of a close-up view of a DMD pixel array [12.24]. Each mirror is capable of rotating by $\pm 10°$ corresponding to either the *on* or *off* position due to an electrostatic actuation force. Light reflected from any on-mirrors passes through a projection lens and creates images on a large screen. Light from the remaining off-mirrors is reflected away from the projection lens to an absorber. The proportion of time during each video frame that a mirror remains in the on-state determines shades of gray, from black for zero on-time to white for a hundred percent on-time. Color can be added by a color wheel or a three-DMD chip setup. The three DMD chips are used for projecting red, green and blue colors. Each DMD pixel consists of a mirror connected by a mirror support post to an underlying yoke. The yoke in turn is connected by torsion hinges to hinge support posts, as shown in Fig. 12.24 [12.26]. The support post and hinges are hidden under the mirror to avoid light diffraction and thus improve contrast ratio and optical efficiency. There are two gaps on the order of a micrometer, one between the mirror and the underlying hinges and address electrodes, and a second between the coplanar address electrodes and hinges and an underlying metal layer from the CMOS static random access memory (SRAM) structure. The yoke is tilted over the second gap by an electrostatic actuation force, thus rotating the mirror plate. The SRAM determines which angle the mirror needs to be tilted by applying proper actuation voltages to the mirror and address electrodes. The DMD is fabricated using an aluminum-based surface micromachining technology. Three layers of aluminum thin film are deposited and patterned to form the mirror and its suspension system. Polymer material is used as sacrifi-

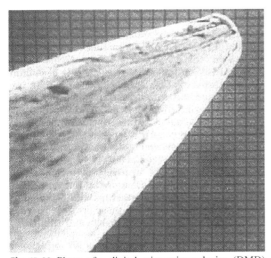

Fig. 12.22 Photo of a digital micromirror device (DMD) array (© Texas Instruments)

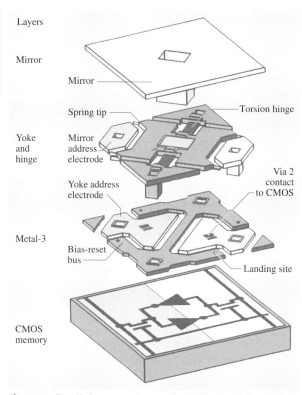

Fig. 12.24 Detailed structure layout of a DMD pixel (after [12.26])

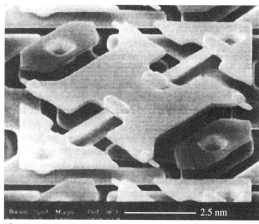

Fig. 12.26 SEM micrograph of a close-up of a DMD yoke and hinges (after [12.26])

cial layer and is removed by a plasma etch at the end of the process to freely release the micromirror structure. The micromachining process is compatible with standard CMOS fabrication, allowing the DMD to be monolithically integrated with a mature CMOS address circuit technology, thus achieving high yield and low cost. Figure 12.25 shows an SEM micrograph of a fabricated DMD pixel revealing its cross section after an ion

Fig. 12.25 SEM micrograph of a DMD pixel after removing half of the mirror plate using ion milling (© Texas Instruments)

milling. A close-up view on the yoke and hinge support under the mirror is shown in Fig. 12.26.

Precision Optical Platform

The growing optical communication and measurement industry require low-cost, high-performance optoelectronic modules such as laser-to-fiber couplers, scanners, interferometers, etc. A precision alignment and the ability to actuate optical components such as mirrors, gratings, and lenses with sufficient accuracy are critical for high-performance optical applications. Conventional hybrid optical integration approaches, such as the silicon-optical-bench, suffers from a limited alignment tolerance of $\pm 1\,\mu m$ and also lacks of component actuation capability [12.27, 28]. As a result, only simple optical systems can be constructed with no more than a few components, thus severely limiting the performance. Micromachining, however, provides a critical enabling technology, allowing movable optical components to be fabricated on a silicon substrate. Component movement with high precision can be achieved through electrostatic actuation. By combining micromachined movable optical components with lasers, lenses, and fibers on the same substrate, an on-chip complex self-aligning optical system can be realized. Figure 12.27a shows an SEM micrograph of a surface-micromachined, electrostatically actuated microreflector for laser-to-fiber coupling and external-cavity-laser applications [12.29]. The device consists of a polysilicon mirror plate hinged to a support beam. The mirror and the support, in turn, are hinged to a vibromotor-actuator slider. The microhinge technol-

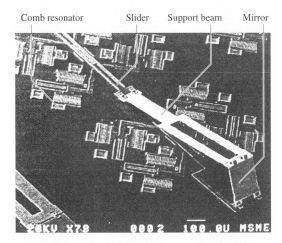

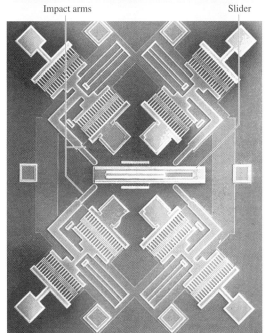

Fig. 12.27 (a) SEM micrograph of a surface-micromachined, electrostatically actuated microreflector; (b) SEM micrograph of a surface-micromachined vibromotor (after [12.29])

ogy [12.30] allows the joints to rotate out of the substrate plane to achieve large aspect ratios. Common-mode actuation of the sliders results in a translational motion, while differential slider motion produces an out-of-plane mirror rotation. These motions permit the microreflector to redirect an optical beam in a desirable location. Each of the two slides is actuated with an integrated microvibromotor shown in Fig. 12.27b. The vibromotor consists of four electrostatic comb resonators with attached impact arms driving a slider through oblique impact. The two opposing impacters are used for each travel direction to balance the forces. The resonator is a capacitively driven mass anchored to the substrate through a folded beam flexure. The flexure compliance determines the resonant frequency and travel range of the resonator. When the comb structures are driven at their resonant frequency (around 8 kHz), the slider exhibits a maximum velocity of over 1 mm/s. Characterization of the vibromotor also shows that a slider step resolution of less than 0.3 μm can be achieved [12.31], making it attractive for precision alignment of various optical components. The prototype microreflector can obtain an angular travel range over 90° and a translational travel range of 60 μm. By using this device, beam steering, fiber coupling, and optical scanning have been demonstrated.

Optical Data Switching

High-speed communication infrastructures are highly desirable for transferring and processing real-time voice and video information. Optical fiber communication technology has been identified as the critical backbone to support such systems. A high-performance optical data switching network, which routes various optical signals from sources to destinations, is one of the key building blocks for system implementation. At present, optical signal switching is performed by using hybrid optical-electronic-optical (O-E-O) switches. These devices first convert incoming light from input fibers to electrical signals first and then route the electrical signals to the proper output ports after signal analyses. At the output ports, the electrical signals are converted back to streams of photons or optical signals for further transmission over the fibers. The O-E-O switches are expensive to build, integrate, and maintain. Furthermore, they consume substantial amount of power and introduce additional latency. It is therefore highly desirable to develop an all-optical switching network in which optical signals can be routed without intermediate conversion into electrical form, thus minimizing power dissipation and system delay. While a number of approaches are being considered for building all-optical switches, MEMS technology is attractive because it can provide arrays of tiny movable mirrors which can redirect incoming beams from input fibers

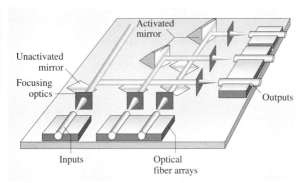

Fig. 12.28 Schematic of a two-dimensional micromirror-based fiber optic switching matrix

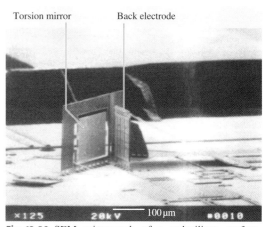

Fig. 12.30 SEM micrograph of a polysilicon surface-micromachined vertical torsion mirror (after [12.32])

to corresponding output fibers. As described in the previous sections, these micromirrors can be batch fabricated using silicon micromachining technologies, thus achieving an integrated solution with the potential for low cost. A significant reduction in power dissipation is also expected.

Figure 12.28 shows an architecture of a two-dimensional micromirror array forming a switching matrix with rows of input fibers and columns of output fibers (or vice versa). An optical beam from an input fiber can be directed to an output fiber through activating the corresponding reflecting micromirror. Switches with eight inputs and eight outputs can be readily implemented using this technique, which can be further

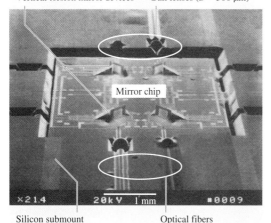

Fig. 12.29 SEM micrograph of a 2×2 MEMS fiber optic switching network (after [12.32])

extended to a 64×64 matrix. The micromirrors are moved between two fixed stops by digital control, thus eliminating the need for precision motion control. Figure 12.29 presents an SEM micrograph of a simple 2×2 MEMS fiber optic switching network prototype for an illustration purpose [12.32]. The network includes a mirror chip passively integrated with a silicon submount, which contains optical fibers and ball lenses. The mirror chip consists of four surface-micromachined vertical torsion mirrors. The four mirrors are arranged such that in the *reflection* mode, the input beams are reflected by two 45° vertical torsion mirrors and coupled into the output fibers located on the same side of the chip. In the *transmission* mode, the vertical torsion mirrors are rotated out of the optical paths, thus allowing the input beams to be coupled into the opposing output fibers. Figure 12.30 shows an SEM micrograph of a polysilicon vertical torsion mirror. The device consists of a mirror plate attached to a vertical supporting frame by torsion beams and a vertical back electrode plate. The mirror plate is $\approx 200\,\mu m$ wide, $160\,\mu m$ long, and $1.5\,\mu m$ thick. The mirror surface is coated with a thin layer of gold to improve the optical reflectivity. The back plate is used to electrostatically actuate the mirror plate so that the mirror can be rotated out of the optical path in the *transmission* mode. Surface micromachining with microhinge technology is used to realize the overall structure. The back electrode plate is integrated with a scratch drive actuator array [12.33] for self-assembly. The self-assembly approach is critical when multiple vertical torsion mirrors are used to implement more advanced functions.

A more sophisticated optical switching network with a large scaling potential can be implemented by using a three-dimensional (3-D) switching architecture as shown in Fig. 12.31. The network consists of arrays of two-axis mirrors to steer optical beams from input fibers to output fibers. A precision analog closed-loop mirror position control is required to accurately direct a beam along two angles so that one input fiber can be optically connected to any output fiber. The optical length depends little on which set of fibers are connected, thus achieving a more uniform switching characteristic, which is critical for implementing large scale network. Two-axis mirrors are the crucial components for implementing the 3-D architecture. Figure 12.32 shows an SEM micrograph of a surface-micromachined two-axis beam-steering mirror positioned by using self-assembly technique [12.34]. The self-assembly is accomplished during the final release step of the mirror processing sequence. Mechanical energy is stored in a special high-stress layer during the deposition, which is put on top of the four assembly arms. Immediately after the assembly arms are released, the tensile stress in this layer causes the arms to bend up, pushing the mirror frame and lifting it above the silicon substrate. All mirrors used in the switching network can be fabricated simultaneously without any human intervention or external power supply.

12.1.4 RF MEMS

The increasing demand for wireless communication applications, such as cellular and cordless telephony, wireless data networks, two-way paging, global positioning system, etc., motivates a growing interest in building miniaturized wireless transceivers with multistandard capabilities. Such transceivers will greatly enhance the convenience and accessibility of various wireless services independent of geographic location. Miniaturizing current single-standard transceivers, through a high-level of integration, is a critical step towards building transceivers that are compatible with multiple standards. Highly integrated transceivers will also result in reduced package complexity, power consumption, and cost. At present, most radio transceivers rely on a large number of discrete frequency-selection components, such as radio-frequency (RF) and intermediate-frequency (IF) band-pass filters, RF voltage-controlled oscillators (VCOs), quartz crystal oscillators, solid-state switches, etc. to perform the necessary analog signal processing. Figure 12.33 shows a schematic of

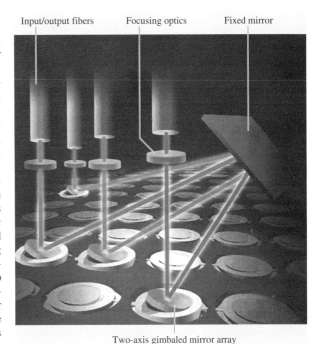

Fig. 12.31 Schematic of a three-dimensional micromirror-based fiber optic switching matrix

Fig. 12.32 SEM micrograph of a surface-micromachined two-axis beam-steering micromirror positioned using a self-assembly technique (after [12.34])

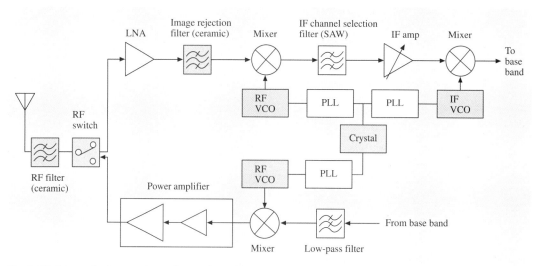

Fig. 12.33 Schematic of a superheterodyne radio architecture

a superheterodyne radio architecture, in which discrete components are shaded in dark color. Theses off-chip devices occupy the majority of the system area, thus severely hindering transceiver miniaturization. MEMS technology, however, offers a potential solution to integrate these discrete components onto silicon substrates with microelectronics, achieving a size reduction of a few orders of magnitude. It is therefore expected to become an enabling technology to ultimately miniaturize radio transceivers for future wireless communications.

MEMS Variable Capacitors

Integrated high-performance variable capacitors are critical for low noise VCOs, antenna tuning, tunable matching networks, etc. Capacitors with high quality factors (Q), large tuning range and linear characteristics are crucial for achieving system performance requirements. On-chip silicon PN junction and MOS based variable capacitors suffer from low quality factors (below 10 at 1 GHz), limited tuning range and poor linearity, thus are inadequate for building high-performance transceivers. MEMS technology has demonstrated monolithic variable capacitors achieving stringent performance requirements. These devices typically reply on an electrostatic actuation method to vary the air gap between a set of parallel plates [12.35–38] or vary the capacitance area between a set of conductors [12.39] or mechanically displace a dielectric layer

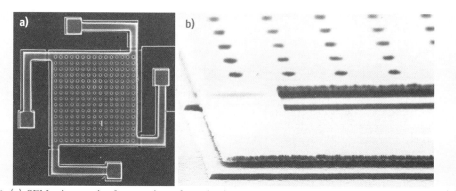

Fig. 12.34 (a) SEM micrograph of a top view of an aluminum surface-micromachined variable capacitor; (b) SEM micrograph of a cross-sectional view of the variable capacitor (after [12.35])

in an air-gap capacitor [12.40]. Improved tuning ranges have been achieved with various device configurations. Figure 12.34 shows SEM micrographs of an aluminum micromachined variable capacitor fabricated on a silicon substrate [12.35]. The device consists of a $200 \times 200\,\mu m^2$ aluminum plate with a thickness of $1\,\mu m$ suspended above the bottom electrode by an air gap of $1.5\,\mu m$. Aluminum is selected as the structural material due to its low resistivity, critical for achieving a high quality factor at high frequencies. A DC voltage applied across the top and bottom electrodes introduces an electrostatic pull-down force, which pulls the top plate towards the bottom electrode, thus changing the device capacitance value. The capacitors are fabricated using aluminum-based surface micromachining technology. Sputtered aluminum is used for building the capacitor top and bottom electrodes. Photoresist is served as the sacrificial layer, which is then removed through an oxygen-based plasma dry etch to release the microstructure. The processing technology requires a low thermal budget, thus allowing the variable capacitors to be fabricated on top of wafers with completed electronic circuits without degrading the performance of active devices. Figure 12.35 presents an SEM micrograph of four MEMS tunable capacitors connected in parallel. This device achieves a nominal capacitance value of $2\,pF$ and a tuning range of 15% with 3 V. A quality factor

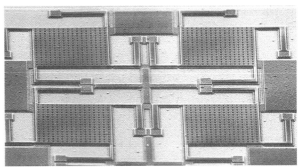

Fig. 12.35 SEM micrograph of four MEMS aluminum variable capacitors connected in parallel (after [12.35])

of 62 has been demonstrated at 1 GHz, which matches or exceeds that of discrete varactor diodes and is at least an order of magnitude larger than that of a typical junction capacitor implemented in a standard IC process.

MEMS tunable capacitors based upon varying capacitance area between a set of conductors have been demonstrated. Figure 12.36 shows an SEM micrograph of a such device [12.39]. The capacitor comprises arrays of interdigitated electrodes, which can be electrostatically actuated to vary the electrode overlap area. A close-up view of the electrodes is shown in Fig. 12.37. The capacitor is fabricated using a silicon-

Fig. 12.36 SEM micrograph of a silicon tunable capacitor using a comb drive actuator (after [12.39])

Fig. 12.37 SEM micrograph of a close view of a tunable capacitor comb fingers (after [12.39])

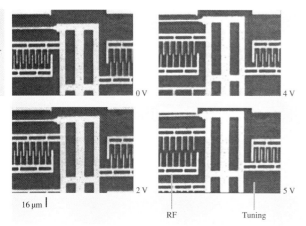

Fig. 12.38 Photos of comb fingers at different actuation voltages (after [12.39])

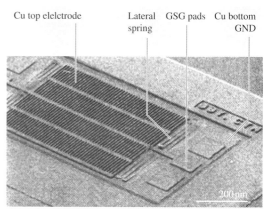

Fig. 12.39 SEM micrograph of a copper surface-micromachined tunable capacitor with a movable dielectric layer (after [12.40])

on-insulator (SOI) substrate with a top silicon layer thickness around 20 μm to obtain a high aspect ratio for the electrodes, critical for achieving a large capacitance density and reduced tuning voltage. The silicon layer is etched to form the device structure followed by removing the underneath oxide to release the capacitor. A thin aluminum layer is then sputtered over the capacitor to reduce the series resistive loss. The device exhibits a quality factor of 34 at 500 MHz and can be tuned between 2.48 and 5.19 pF with an actuation voltage under 5 V, corresponding to a tuning range over 100%. Figure 12.38 shows the variation of electrode overlap area under different tuning voltages.

Tunable capacitors relying on a movable dielectric layer have been fabricated using MEMS technology. Figure 12.39 presents an SEM micrograph of a copper-based micromachined tunable capacitor [12.40]. The device consists of an array of copper top electrodes suspended above a bottom copper plate with an air gap of $\approx$ 1 μm. A thin nitride layer is deposited, patterned, and suspended between the two copper layers by lateral mechanical spring suspensions after sacrificial release. A DC voltage applied across the copper layers introduces a lateral electrostatic pull-in force on the nitride, thus resulting in a movement which changes the overlapping area between each copper electrode and the bottom plate, and hence the device capacitance. The tunable capacitor achieves a quality factor over 200 at 1 GHz with 1 pF capacitance due to the highly conductive copper layers and a tuning range around 8% with 10 V.

Micromachined Inductors

Integrated inductors with high quality factors are as critical as the tunable capacitors for high performance RF system implementation. They are the key components for building low-noise oscillators, low-loss matching networks, etc. Conventional on-chip spiral inductors suffer from limited quality factors of around 5 at 1 GHz, an order of magnitude lower than the required val-

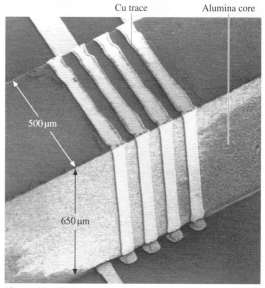

Fig. 12.40 SEM micrograph of a 3-D coil inductor fabricated on a silicon substrate (after [12.41])

ues from discrete counterparts. The poor performance is mainly caused by substrate loss and metal resistive loss at high frequencies. Micromachining technology provides an attractive solution to minimize these loss contributions; hence enhancing the device quality factors. Figure 12.40 shows an SEM micrograph of a 3-D coil inductor fabricated on a silicon substrate [12.41]. The device consists of 4-turn 5 μm thick copper traces electroplated around an insulting core with a 650 μm by 500 μm cross section. Compared to spiral inductors, this geometry minimizes the coil area which is in close proximity to the substrate and hence the eddy-current loss, resulting in a maximized Q-factor and device self-resonant frequency. Copper is selected as the interconnect metal because of its low sheet resistance, critical for achieving a high Q-factor. The inductor achieves a 14 nH inductance value with a quality factor of 16 at 1 GHz. A single-turn 3-D device exhibits a Q-factor of 30 at 1 GHz, which matches the performance of discrete counterparts. The high-Q 3-D inductor and MEMS tunable capacitors, shown in Fig. 12.35, have been employed to implement a RF CMOS VCO achieving a low phase noise performance suitable for typical wireless communication applications such as GMS cellular telephony [12.42].

Other 3-D inductor structures such as the levitated spiral inductors have been demonstrated using micromachining fabrication technology. Figure 12.41 shows an SEM micrograph of a levitated copper inductor, which is suspended above the substrate through supporting posts [12.44]. The levitated geometry can minimize the substrate loss, thus achieving an improved quality factor. The inductor shown in the figure

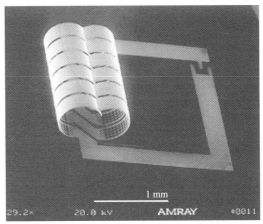

Fig. 12.42 SEM micrograph of a self-assembled out-of-plane coil inductor (after [12.43])

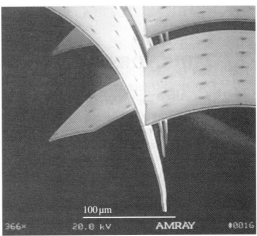

Fig. 12.43 SEM micrograph of an interlocking trace from a self-assembled out-of-plane oil inductor (after [12.43])

achieves a 1.4 nH inductance value with a Q-factor of 38 at 1.8 GHz using a glass substrate. Similar inductor structures have been demonstrated on standard silicon substrates achieving a nominal inductance value of ≈ 1.4 nH with a Q-factor of 70 measured at 6 GHz [12.45].

A self-assembled out-of-plane coil has been fabricated using micromachining technology. The inductor winding traces are made of refractory metals with controlled built-in stress such that the traces can curl out of the substrate surface upon release and interlock into each other to form coil windings. Figure 12.42 shows

Fig. 12.41 SEM micrograph of a levitated spiral inductor fabricated on a glass substrate (after [12.40])

an SEM micrograph of a self-assembled out-of-plane coil inductor [12.43]. A close-up view of an interlocking trace is shown in Fig. 12.43. Copper is plated on the interlocked traces to form highly conductive windings at the end of processing sequence. The inductor shown in Fig. 12.42 achieves a quality factor around 40 at 1 GHz.

MEMS Switches

The microelectromechanical switch is another potentially attractive miniaturized component enabled by micromachining technologies. These switches offer superior electrical performance in terms of insertion loss, isolation, linearity, etc., and are intended to replace off-chip solid-state counterparts, which provide switching between the receiver and transmitter signal paths. They are also critical for building phase shifters, tunable antennas, and filters. The MEMS switches can be characterized into two categories: capacitive and metal-to-metal contact types. Figure 12.44 presents a cross-sectional schematic of an RF MEMS capacitive switch. The device consists of a conductive membrane, typically made of aluminum or gold alloy suspended above a coplanar electrode by an air gap of a few micrometers. For RF or microwave applications, actual metal-to-metal contact is not necessary; rather, a step change in the plate-to-plate capacitance realizes the switching function. A thin silicon nitride layer with a thickness on the order of 1000 Å is typically deposited above the bottom electrode. When the switch is in the on-state, the membrane is high resulting in a small plate-to-plate capacitance; hence, a minimum high-frequency signal coupling (high isolation) between the two electrodes. In the off-state (with a large enough applied DC voltage), the switch provides a large capacitance due to the thin dielectric layer, thus causing

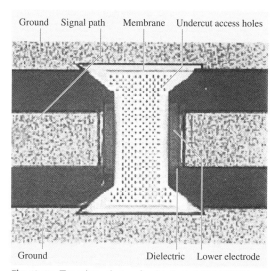

Fig. 12.45 Top view photo of a fabricated RF MEMS capacitive switch (after [12.46])

a strong signal coupling (low insertion loss). The capacitive switch consumes near-zero power, which is attractive for low power portable applications. Switching cycles over millions for this type of device have been demonstrated. Figure 12.45 shows a top view photo of a fabricated MEMS capacitive switch [12.46]. Surface micromachining technology, using metal for the electrodes and polymer as the sacrificial layer, is used to fabricate the device. The switch can be actuated with a DC voltage on the order of 50 V and exhibits a low insertion loss of ≈ -0.28 dB at 35 GHz and a high isolation of -35 dB at the same frequency.

Metal-to-metal contact switches are important for interfacing large bandwidth signals including DC. This type of device typically consists of a cantilever beam or clamped-clamped bridge with a metallic contact pad positioned at the beam tip or underneath bridge center. Through an electrostatic actuation, a contact can be formed between the suspended contact pad and an underlying electrode on the substrate [12.47–49]. Figure 12.46 shows a cross-sectional schematic of a metal-to-metal contact switch [12.49]. The top view of the fabricated device is presented in Fig. 12.47. The switch exhibits an actuation voltage of 30 V, a response time of 20 µs, and mechanical strength to withstand 10^9 actuations. An isolation greater than 50 dB below 2 GHz and insertion loss less than 0.2 dB from DC through 40 GHz has been demonstrated. Metal-to-metal contact switches relying on electrothermal actuation method have also

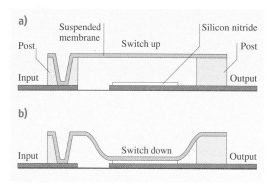

Fig. 12.44 Cross-sectional schematics of an RF MEMS capacitive switch

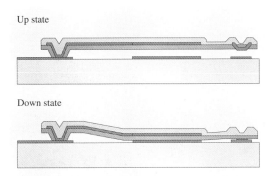

Fig. 12.46 Cross-sectional schematic of a metal-to-metal contact switch (after [12.49])

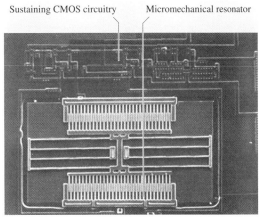

Fig. 12.48 SEM micrograph of a surface-micromachined comb drive resonator integrated with CMOS sustaining electronics (after [12.52])

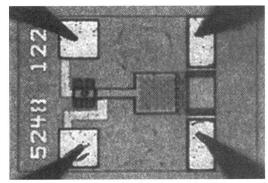

Fig. 12.47 Top view photo of a fabricated metal-to-metal contact switch (after [12.49])

been developed to demonstrate a low actuation voltage around 3 V, however, at an expense of reduced switching speed of 300 μs and increased power dissipation in the range of 60–100 mW [12.50, 51]. The fabricated switches achieve an off-state isolation of −20 dB at 40 GHz and an insertion loss of −0.1 dB up to 50 GHz.

MEMS Resonators

Microelectromechanical resonators based upon polysilicon comb-drive structures, suspended beams, and center-pivoted disk configurations have been demonstrated for performing analog signal processing [12.50, 52–56]. These microresonators can be excited into mechanical resonance through an electrostatic drive. The mechanical motion causes a change of device capacitance resulting in an output electrical current when a proper DC bias voltage is used. The output current exhibits the same frequency as the mechanical resonance, thus achieving an electrical filtering function through the electromechanical coupling. Micromachined polysilicon flexural-mode mechanical resonators have demonstrated a quality factor greater than 80 000 in a 50 μTorr vacuum [12.57]. This level of performance is comparable to a typical quartz crystal and is thus attractive for implementing monolithic low-noise and low-drift reference signal sources. Figure 12.48 shows an SEM micrograph of a surface-micromachined comb drive resonator integrated with CMOS sustaining electronics on a same substrate to form a monolithic high-Q MEMS resonator-based oscillator [12.52]. The oscillator achieves an operating frequency of 16.5 KHz with a clean spectral purity. A chip area of $\approx 420 \times 230\,\mu m^2$ is consumed for fabricating the overall system, representing a size reduction by orders of magnitude compared to conventional quartz crystal oscillators.

Micromachined high-Q resonators can be coupled to implement low-loss frequency selection filters. Figure 12.49 shows an SEM micrograph of a surface-micromachined polysilicon two-resonator, spring-coupled bandpass micromechanical filter [12.54]. The filter consists of two silicon micromechanical clamped-clamped beam resonators, coupled mechanically by a soft spring, all suspended 0.1 μm above the substrate. Polysilicon strip lines underlie the central regions of each resonator and serve as capacitive transducer electrodes positioned to induce resonator vibration in a direction perpendicular to the substrate. Under a normal operation, the device is excited capacitively by a signal voltage applied to the input electrode.

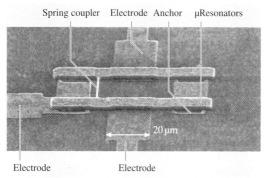

Fig. 12.49 SEM micrograph of a polysilicon surface-micromachined two-resonator spring-coupled bandpass micromechanical filter (after [12.54])

The output is taken at the other end of the structure, also by capacitive transduction. The filter achieves a center frequency of 7.81 MHz, a bandwidth of 0.23%, and an insertion loss less than 2 dB. The achieved performance is attractive for implementing filters in the low MHz range.

To obtain a higher mechanical resonant frequency with low losses, a surface-micromachined contour-mode disk resonator has been proposed, as shown in Fig. 12.50 [12.56]. The resonator consists of a polysilicon disk suspended 5000 Å above the substrate with a single anchor at its center. Plated metal input electrodes surround the perimeter of the disk with a narrow separation of around 1000 Å, which defines the capacitive, electromechanical transducer of the device. To operate the device, a DC bias voltage is applied to the structure with an AC input signal applied to the electrodes, resulting in a time varying electrostatic force acting radially on the disk. When the input signal matches the device resonant frequency, the resulting electrostatic force is amplified by the Q-factor of the resonator, pro-

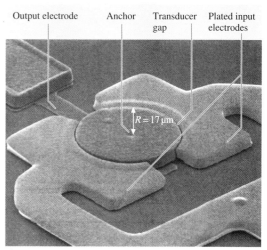

Fig. 12.50 SEM micrograph of a polysilicon surface-micromachined contour-mode disk resonator (after [12.56])

ducing expansion and contraction of the disk along its radius. This motion, in turn, produces a time-varying output current at the same frequency, thus achieving the desirable filtering. The prototype resonator demonstrates an operating frequency of 156 MHz with a Q-factor of 9400 in vacuum. The increased resonant frequency is comparable to the first intermediate frequency used in a typical wireless transceiver design and is thus suitable for implementing IF bandpass filters. Recently, self-aligned MEMS fabrication technique was developed to demonstrate vibrating radial-contour mode polysilicon micromechanical disk resonators with resonant frequencies up to 1.156 GHz and measured Q's close to 3000 in both vacuum and air [12.51]. The achieved performance is attractive for potentially replacing RF frequency selection filters in current wireless transceivers with MEMS versions.

12.2 Nanoelectromechanical Systems (NEMS)

Unlike their microscale counterparts, nanoelectromechanical systems (NEMS) are made of electromechanical devices that have critical structural dimensions at or below 100 nm. These devices are attractive for applications where structures of very small mass and/or very large surface area-to-volume ratios provide essential functionality, such as force sensors, chemical sensors, biological sensors, and ultrahigh frequency resonators to name a few. NEMS fabrication processes can be classified into two general categories based on the approach used to create the structures. *Top-down* approaches utilize submicrometer lithographic techniques to fabricate device structures from *bulk* material, either thin films or thick substrates. *Bottom-up* approaches involve the fabrication of nanoscale devices in much the same way that nature constructs objects, by sequential assembly using atomic and/or molecular building blocks. While advancements in bottom-up approaches are developing at a very rapid pace, most advanced NEMS devices are currently created utilizing top-down techniques that combine existing process technologies, such as electron-beam lithography, conventional film growth and chemical etching. Top-down approaches make integration with microscale packaging relatively straightforward since the only significant difference between the nanoscale and microscale processing steps is the method used to pattern the various features.

In large measure, NEMS has followed a developmental path similar to the route taken in the development of MEMS in that both have leveraged existing processing techniques from the IC industry. For instance, the electron-beam lithographic techniques used in top-down NEMS fabrication are the same techniques that have become standard in the fabrication of submicrometer transistors. Furthermore, the materials used in many of the first generation, top-down NEMS devices, (Si, GaAs, Si_3N_4, SiC) were first used in ICs and then in MEMS. Like the first MEMS devices, the first generation NEMS structures consisted of free-standing nanomechanical beams, paddle oscillators, and tethered plates made using simple bulk and single layer surface *nano*machining processes. Recent advancements have focused on incorporating *nano*materials such as nanotubes and nanowires synthesized using bottom-up approaches into NEMS devices by integrating these materials into top-down nano- and micromachining processes. The following text serves only at a brief introduction to the technology, highlighting the key materials, fabrication approaches, and emerging application areas. For additional details and perspectives, readers are encouraged to consult an excellent review on the subject [12.58].

12.2.1 Materials and Fabrication Techniques

Like Si MEMS, Si NEMS capitalizes on well-developed processing techniques for Si and the availability of high-quality substrates. *Cleland* and *Roukes* [12.59] reported a relatively simple process to fabricate nanomechanical clamped-clamped beams directly from single-crystal (100) Si substrates. As illustrated in Fig. 12.1, the process begins with the thermal oxidation of a Si substrate (Fig. 12.1a). Large Ni contact pads were then fabricated using optical lithography and lift-off. A polymethyl methacrylate (PMMA) lift-off mold was then deposited and patterned using electron-beam lithography into the shape of nanomechanical beams (Fig. 12.1b). Ni was then deposited and patterned by lifting off the PMMA (Fig. 12.1c). Next, the underlying oxide film was patterned by RIE using the Ni film as an etch mask. After oxide etching, nanomechanical beams were patterned by etching the Si substrate using RIE, as shown in Fig. 12.1d. Following Si RIE, the Ni etch mask was removed and the sidewalls of the Si nanomechanical beams were lightly oxidized in order to protect them during the release step (Fig. 12.1e). After performing an anisotropic SiO_2 etch to clear any oxide from the field areas, the Si beams were released using an isotropic Si RIE step, as shown in Fig. 12.1f. After release, the protective SiO_2 film was removed by wet etching in HF (Fig. 12.1g). Using this process, the authors reported the successful fabrication of nanomechanical Si beams with micrometer-scale lengths ($\approx 8\,\mu m$) and submicrometer widths (330 nm) and heights (800 nm).

The advent of silicon-on-insulator (SOI) substrates with high quality, submicrometer-thick silicon top layers enables the fabrication of nanomechanical Si beams with fewer processing steps than the aforementioned technique, since the buried oxide layer makes these device structures relatively easy to pattern and release. Additionally, the buried SiO_2 layer electrically isolates the beams from the substrate. *Carr* and *Craighead* [12.60] detail a process that uses SOI substrates to fabricate submicrometer clamped-clamped mechanical beams and suspended plates with submicrometer tethers. The process, presented in Fig. 12.2, begins with the deposition of PMMA on an SOI substrate. The SOI substrate has a top Si layer that was either 50 or 200 nm in thickness. The PMMA is patterned into a metal lift-off mask by electron beam lithography (Fig. 12.2b). An Al film is then deposited and patterned by lift-off into a Si etch mask, as shown in Fig. 12.2c. The nanomechanical beams are then patterned by Si RIE and released by etching the underlying SiO_2 in a buffered hydrofluoric acid solution as shown in Fig. 12.2d,e, respectively. Using this process, nanomechanical beams that were 7–16 μm in length, 120–200 nm in width and 50 or 200 nm in thickness were successfully fabricated.

Fabrication of NEMS structures is not limited to Si. In fact, III–V compounds, such as gallium arsenide (GaAs), make particularly good NEMS materials from a fabrication perspective because thin epitaxial GaAs films can be grown on lattice-matched materials that can be used as sacrificial release layers. A collection of clamped-clamped nanomechanical GaAs beams fabricated on lattice-matched sacrificial layers having micrometer-scale lengths and submicrometer widths and thicknesses is shown in Fig. 12.3. *Tighe* et al. [12.61] reported on the fabrication of GaAs plates suspended with nanomechanical tethers. The structures were made from single-crystal GaAs films that were epitaxially grown on aluminum arsenide (AlAs) sacrificial layers. Ni etch masks were fabricated using electron beam lithography and lift-off as described previously. The GaAs films were patterned into beams using a chemically assisted ion beam etching process and released using a highly selective AlAs etchant. In a second example, *Tang* et al. [12.62] has capitalized on the ability to grow high-quality GaAs layers on ternary compounds such as $Al_xGa_{1-x}As$ to fabricate complex GaAs-based structures, such as submicrometer clamped-clamped beams from GaAs/AlGaAs quantum well heterostructures. As with the process described by *Tighe* et al. [12.61], this process exploits a lattice matched sacrificial layer, in this case $Al_{0.8}Ga_{0.2}As$, which can be selectively etched to release the heterostructure layers.

Silicon carbide and diamond NEMS structures have been developed for applications requiring a material with a higher acoustic velocity and/or a higher degree of chemical inertness than Si. Silicon carbide nanomechanical resonators have been successfully fabricated from both epitaxial 3C-SiC films grown on Si substrates [12.63] and bulk 6H-SiC substrates [12.64]. In the case of the 3C-SiC devices, the ultrathin epitaxial films were grown by atmospheric pressure chemical vapor deposition (APCVD) on (100)Si substrates. Nanomechanical beams were patterned using a metal RIE mask that was itself patterned by e-beam lithography. Reactive ion etching was performed using two NH_3-based plasma chemistries, with the first recipe performing an anisotropic SiC etch down to the Si substrate and the second performing an isotropic Si etch used to release the SiC beams. The two etches were performed sequentially, thereby eliminating a separate wet or dry release step. For the 6H-SiC structures, a suitable sacrificial layer was not available since the structures were fabricated directly on commercially available bulk wafers. To fabricate the structures, a metal etch mask was lithographically patterned by e-beam techniques on the 6H-SiC surface. The anisotropic SiC etch mentioned above was then performed, but with the substrate tilted roughly 45° with respect to the direction of the plasma using a special fixture to hold the wafer. A second such etch was performed on the substrate tilted back 90° with respect to the first etch, resulting in released beams with triangular cross sections.

Nanomechanical resonators have also been fashioned out of thin nanocrystalline diamond thin films [12.65]. In this case, the diamond films were deposited on SiO_2-coated Si substrates by microwave plasma chemical vapor deposition using CH_4 and H_2 as feedstock. The diamond films were patterned by RIE using metal masks patterned by e-beam lithography. The plasma chemistry in this case was based on CF_4 and O_2. The devices were then released in a buffered HF solution. It is noteworthy that the structures did not require a critical-point drying step after the wet chemical release, owing to the chemical inertness of the diamond surface.

NEMS structures are not restricted to those that can be made from patterned thin films using top-down techniques. In fact, carbon nanotubes (CNT) have been incorporated into NEMS devices using an approach that combines both bottom-up and top-down processing techniques. An example illustrating the promise and challenges of merging bottom-up with top-down techniques is the CNT-based electrostatic rotational actuator developed by *Fennimore* et al. [12.66]. In this example, multiwalled CNTs (MWCNT) are grown using a conventional arc discharge process, which typically produces an assortment of CNTs. The CNTs are then transferred to a suitable SiO_2 coated Si in a 1,2-dichlorobenzene suspension. An AFM or SEM is then used to select a properly positioned CNT as determined by prefabricated alignment marks on the substrate. Conventional electron-beam lithography and lift-off techniques are then used to pattern an Au film into contact/anchor pads on the two ends of the CNT, a rotor pad at its center and two counter electrodes at 90° to the anchor pads. Anchoring is accomplished by sandwiching the CNT between the Au contact and the underlying SiO_2 film. The rotor is released by simply etching the sacrificial SiO_2 layer, taking care not to completely undercut the anchors yet allowing for adequate clearance for the rotor. Under proper conditions, the outer wall of the MWCNTs could be detached from the inner walls in order to allow for free rotation of the rotor plate.

12.2.2 Transduction Techniques

Several unique approaches have been developed to actuate and sense the motion of NEMS devices. Electrostatic actuation can be used to actuate beams [12.67], tethered meshes [12.68], and paddle oscillators [12.69]. *Sekaric* et al. [12.70] has shown that low power lasers can be used to drive paddle oscillators into self-oscillation by induced thermal effects on the structures. In these examples, an optical detection scheme based on the modulation of incident laser light by a vibrating beam is used to detect the motion of the beams. *Cleland* and *Roukes* [12.59] describe a magnetomotive transduction technique that capitalizes on a time-varying Lorentz force created by an alternating current in the presence of a strong magnetic field. In this case, the nanomechanical beam is positioned in the magnetic field so that an AC current passing through the beam is transverse to the field lines. The resulting Lorentz force causes the beam to oscillate, which creates an electromotive force along the beam that can be detected as a voltage. Thus, in this method, the excitation and detection are performed electrically. In all of the above-mentioned cases, the measurements were performed in vacuum, presumably to minimize the effects of squeeze-film damping as well as mass loading due to adsorbates from the environment.

12.2.3 Application Areas

For the most part, NEMS technology is still in the initial stage of development. Technological challenges related to fabrication and packaging will require innovative solutions before such devices make a significant commercial impact. Nevertheless, NEMS devices have already been used for precision measurements [12.71] enabling researchers to probe the properties of matter on a nanoscopic level [12.72, 73]. Sensor technologies based on NEMS structures, most notably for attogram scale mass detection [12.74, 75], attonewton force detection [12.76], virus detection [12.77], and gaseous chemical detection [12.78] have emerged and will continue to mature. Without question, NEMS structures will prove to be useful platforms for a host of experiments and scientific discoveries in fields ranging from physics to biology, and with advancements in process integration and packaging, there is little doubt that NEMS technology will find its way into commercial micro/nanosystems as well.

12.3 Current Challenges and Future Trends

Although the field of MEMS has experienced significant growth over the past decade, many challenges still remain. In broad terms, these challenges can be grouped into three general categories:

1. Fabrication challenges
2. Packaging challenges
3. Application challenges.

Challenges in these areas will, in large measure, determine the commercial success of a particular MEMS device both in technical and economic terms. The following presents a brief discussion of some of these challenges as well as possible approaches to address them.

In terms of fabrication, MEMS is currently dominated by planar processing techniques which find their roots in silicon IC fabrication. The planar approach and the strong dependence on silicon worked well in the early years, since many of the processing tools and methodologies commonplace in IC fabrication could be directly utilized in the fabrication of MEMS devices. This approach lends itself to the integration of MEMS with silicon ICs. Therefore, it still is popular for various applications. However, modular process integration of micromachining with standard IC fabrication is not straightforward and represents a great challenge in terms of processing material compatibility, thermal budget requirements, etc. Furthermore, planar processing places significant geometric restrictions on device designs, especially for complex mechanical components requiring high aspect ratio three-dimensional geometries, which are certain to increase as the application areas for MEMS continue to grow. Along the same lines, new applications will likely demand materials other than silicon, which may not be compatible with the conventional microfabrication approach, posing a significant challenge if integration with silicon microelectronics is required. Microassembly technique can become an attractive solution to alleviate these issues. Multifunctional microsystems can be implemented by assembling various MEMS devices and electronic building blocks fabricated through disparate processing technologies. Microsystems on a common substrate will likely become the ultimate solution. Development of sophisticated modeling programs for device design and performance will become increasingly important as fabrication processes and device designs become more complex. In terms of NEMS, the most significant challenge is likely the integration of nano- and microfabrication techniques into a unified process, since NEMS devices are likely to consist of both nanoscale and microscale structures. Integration will be particularly challenging for nanoscale devices fabricated using a *bottom-up* approach, since no analog is found in microfabrication. Nevertheless, hybrid systems consisting of nanoscale and microscale components will become increasingly common as the field continues to expand.

Fabrication issues notwithstanding, packaging is and will continue to be a significant challenge to the implementation of MEMS. MEMS is unlike IC packaging which benefits from a high degree of standardization. MEMS devices inherently require interaction with the environment, and since each application has in some way a unique environment, standardization of packaging becomes extremely difficult. This lack of standardization tends to drive up the costs associated with packaging, making MEMS less competitive with alternative approaches. In addition, packaging tends to negate the effects of miniaturization based upon microfabrication, especially for MEMS devices requiring protection from certain environmental conditions. Moreover, packaging can cause performance degradation of MEMS devices, especially in situations where the environment exerts mechanical stresses on the package, which in turn results in a long-term device performance drift. To address many of these issues, wafer level packaging schemes that are customized to the device of interest will likely become more common. In essence, packaging of MEMS will move away from the conventional IC methods that utilize independently manufactured packages toward custom packages, which are created specifically for the device as a part of the batch fabrication process.

Without question, the increasing advancement of MEMS will open many new potential application areas to the technology. In most cases, MEMS will be one of several alternatives available for implementation. For cost sensitive applications, the trade off between technical capabilities and cost will challenge those who desire to commercialize the technology. The biggest challenge to the field will be to identify application areas that are well suited for MEMS/NEMS technology and have no serious challengers. As MEMS technology moves away from component level and more towards microsystems solutions, it is likely that such application areas will come to the fore.

References

12.1 M. Mehregany, S.F. Bart, L.S. Tavrow, J.H. Lang, S.D. Senturia: Principles in design and microfabrication of variable-capacitance side-drive motors, J. Vac. Sci. Technol. A **8**, 3614–3624 (1990)

12.2 Y.-C. Tai, R.S. Muller: IC-processed electrostatic synchronous micromotors, Sens. Actuators **20**, 49–55 (1989)

12.3 J.J. Sniegowski, S.L. Miller, G.F. LaVigne, M.S. Roders, P.J. McWhorter: Monolithic geared-mechanisms driven by a polysilicon surface-micromachined on-chip electrostatic microengine, IEEE Solid-State Sens. Actuators Workshop (1996) pp. 178–182

12.4 G.T.A. Kovacs: *Micromachined Transducer Sourcebook* (McGraw-Hill, Boston 1998)

12.5 S.D. Senturia: *Microsystem Design* (Kluwer, Dordrecht 1998)

12.6 J.E. Gragg, W.E. McCulley, W.B. Newton, C.E. Derrington: Compensation and calibration of a monolithic four terminal silicon pressure transducer, IEEE Solid-State Sens. Actuators Workshop (1984) pp. 21–27

12.7 Y. Wang, M. Esashi: A novel electrostatic servo capacitive vacuum sensor, IEEE Int. Conf. Solid-State Sens. Actuators (1997) pp. 1457–1460

12.8 W.H. Ko, Q. Wang: Touch mode capacitive pressure densors, Sens. Actuators **75**, 242–251 (1999)

12.9 D.J. Young, J. Du, C.A. Zorman, W.H. Ko: High-temperature single-crystal 3C-SiC capacitive pressure sensor, IEEE Sens. J. **4**, 464–470 (2004)

12.10 H. Kapels, R. Aigner, C. Kolle: Monolithic surface-micromachined sensor system for high pressure applications, Int. Conf. Solid-State Sens. Actuators (2001) pp. 56–59

12.11 J.M. Bustillo, R.T. Howe, R.S. Muller: Surface micromachining for microelectromechanical systems, Proc. IEEE **86**(8), 1552–1574 (1998)

12.12 J.H. Smith, S. Montague, J.J. Sniegowski, J.R. Murray, P.J. McWhorter: Embedded micromechanical devices for the monolithic integration of MEMS with CMOS, IEEE Int. Electron Dev. Meet. (1993) pp. 609–612

12.13 T.A. Core, W.K. Tsang, S.J. Sherman: Fabrication technology for an integrated surface-micromachined sensor, Solid State Technol. **36**(10), 39–40, 42, 46–47 (1993)

12.14 C. Lu, M. Lemkin, B.E. Boser: A monolithic surface micromachined accelerometer with digital output, IEEE Int. Solid-State Circuits Conf. (1995) pp. 160–161

12.15 N. Yazdi, K. Najafi: An all-silicon single-wafer fabrication technology for precision microaccelerometers, IEEE Int. Conf. Solid-State Sens. Actuators (1997) pp. 1181–1184

12.16 T.B. Gabrielson: Mechanical-thermal noise in micromachined acoustic and vibration sensors, IEEE Trans. Electron Dev. **40**(5), 903–909 (1993)

12.17 J. Chae, H. Kulah, K. Najafi: A monolithic three-axis micro-g micromachined silicon capacitive accelerometer, IEEE J. Solid-State Circuits **14**, 235–242 (2005)

12.18 M. Lemkin, M.A. Ortiz, N. Wongkomet, B.E. Boser, J.H. Smith: A 3-axis surface micromachined $\sum\Delta$ accelerometer, IEEE Int. Solid-State Circuits Conf. (1997) pp. 202–203

12.19 J. Wu, G.K. Fedder, L.R. Carley: A low-noise low-offset capacitive sensing amplifier for a 50 μg/$\sqrt{Hz}$ monolithic CMOS MEMS accelerometer, IEEE J. Solid-State Circuits **39**, 722–730 (2004)

12.20 W.A. Clark, R.T. Howe: Surface micromachined Z-axis vibratory rate gyroscope, IEEE Solid-State Sens. Actuators Workshop (1996) pp. 283–287

12.21 J.A. Geen, S.J. Sherman, J.F. Chang, S.R. Lewis: Single-chip surface micromachined integrated gyroscope with 50°/h Allan deviation, IEEE J. Solid-State Circuits **37**, 1860–1866 (2002)

12.22 H. Xie, G.K. Fedder: Fabrication, characterization, and analysis of a DRIE CMOS-MEMS gyroscope, IEEE Sens. J. **3**, 622–631 (2003)

12.23 T. Juneau, A.P. Pisano: Micromachined dual input axis angular rate sensor, IEEE Solid-State Sens. Actuators Workshop (1996) pp. 299–302

12.24 L.J. Hornbeck: Current status of the digital micromirror device (DMD) for projection television applications, IEEE Int. Electron Dev. Meet. (1993) pp. 381–384

12.25 R.S. Muller, K.Y. Lau: Surface-micromachined microoptical elements and systems, Proc. IEEE **86**(8), 1705–1720 (1998)

12.26 P.F. Van Kessel, L.J. Hornbeck, R.E. Meier, M.R. Douglass: A MEMS-based projection display, Proc. IEEE **86**(8), 1687–1704 (1998)

12.27 M.S. Cohen, M.F. Cina, E. Bassous, M.M. Opyrsko, J.L. Speidell, F.J. Canora, M.J. DeFranza: Packaging of high density fiber/laser modules using passive alignment techniques, IEEE Trans. Compon. Hybrids Manuf. Technol. **15**, 944–954 (1992)

12.28 M.J. Wale, C. Edge: Self-aligned flip-chip assembly of photonic devices with electrical and optical connections, IEEE Trans. Compon. Hybrids Manuf. Technol. **13**, 780–786 (1990)

12.29 M.J. Daneman, N.C. Tien, O. Solgaard, K.Y. Lau, R.S. Muller: Linear vibromotor-actuated micromachined microreflector for integrated optical systems, IEEE Solid-State Sens. Actuators Workshop (1996) pp. 109–112

12.30 K.S.J. Pister, M.W. Judy, S.R. Burgett, R.S. Fearing: Microfabricated hinges, Sens. Actuators **33**(3), 249–256 (1992)

12.31 O. Solgaard, M. Daneman, N.C. Tien, A. Friedberger, R.S. Muller, K.Y. Lau: Optoelectronic packaging using silicon surface-micromachined alignment mirrors, IEEE Photon. Technol. Lett. **7**(1), 41–43 (1995)

12.32 S.S. Lee, L.S. Huang, C.J. Kim, M.C. Wu: 2×2 MEMS fiber optic switches with silicon sub-mount for low-cost packaging, IEEE Solid-State Sens. Actuators Workshop (1998) pp. 281–284

12.33 T. Akiyama, H. Fujita: A quantitative analysis of scratch drive actuator using Buckling motion, Tech. Dig., 8th IEEE Int. MEMS Workshop (1995) pp. 310–315

12.34 V.A. Aksyuk, F. Pardo, D.J. Bishop: Stress-induced curvature engineering in surface-micromachined devices, Proc. SPIE **3680**, 984 (1999)

12.35 D.J. Young, B.E. Boser: A micromachined variable capacitor for monolithic low-noise VCOs, IEEE Solid-State Sens. Actuators Workshop (1996) pp. 86–89

12.36 A. Dec, K. Suyama: Micromachined electromechanically tunable capacitors and their applications to RF IC's, IEEE Trans. Microw. Theory Tech. **46**, 2587–2596 (1998)

12.37 Z. Li, N.C. Tien: A high tuning-ratio silicon-micromachined variable capacitor with low driving voltage, IEEE Solid-State Sens. Actuators Workshop (2002) pp. 239–242

12.38 Z. Xiao, W. Peng, R.F. Wolffenbuttel, K.R. Farmer: Micromachined variable capacitor with wide tuning range, IEEE Solid-State Sens. Actuators Workshop (2002) pp. 346–349

12.39 J.J. Yao, S.T. Park, J. DeNatale: High tuning-ratio MEMS-based tunable capacitors for RF communications applications, IEEE Solid-State Sens. Actuators Workshop (1998) pp. 124–127

12.40 J.B. Yoon, C.T.-C. Nguyen: A high-Q tunable micromechanical capacitor with movable dielectric for RF applications, IEEE Int. Electron Dev. Meet. (2000) pp. 489–492

12.41 D.J. Young, V. Malba, J.J. Ou, A.F. Bernhardt, B.E. Boser: Monolithic high-performance three-dimensional coil inductors for wireless communication applications, IEEE Int. Electron Dev. Meet. (1997) pp. 67–70

12.42 D.J. Young, B.E. Boser, V. Malba, A.F. Bernhardt: A micromachined RF low phase noise voltage-controlled oscillator for wireless communication, Int. J. RF Microw. Comput.-Aided Eng. **11**(5), 285–300 (2001)

12.43 C.L. Chua, D.K. Fork, K.V. Schuylenbergh, J.P. Lu: Self-assembled out-of-plane high Q inductors, IEEE Solid-State Sens. Actuators Workshop (2002) pp. 372–373

12.44 J.B. Yoon, C.H. Han, E. Yoon, K. Lee, C.K. Kim: Monolithic high-Q overhang inductors fabricated

12.45 J.B. Yoon, Y. Choi, B. Kim, Y. Eo, E. Yoon: CMOS-compatible surface-micromachined suspended-spiral inductors for multi-GHz silicon RF ICs, IEEE Electron Dev. Lett. **23**, 591–593 (2002)

12.46 C.L. Goldsmith, Z. Yao, S. Eshelman, D. Denniston: Performance of low-loss RF MEMS capacitive switches, IEEE Microw. Guided Wave Lett. **8**(8), 269–271 (1998)

12.47 J.J. Yao, M.F. Chang: A surface micromachined miniature switch for telecommunication applications with signal frequencies from DC up to 40 GHz, 8th Int. Conf. Solid-State Sens. Actuators (1995) pp. 384–387

12.48 P.M. Zavracky, N.E. McGruer, R.H. Morriosn, D. Potter: Microswitches and microrelays with a view toward microwave applications, Int. J. RF Microw. Comput.-Aided Eng. **9**(4), 338–347 (1999)

12.49 D. Hyman, J. Lam, B. Warneke, A. Schmitz, T.Y. Hsu, J. Brown, J. Schaffner, A. Walston, R.Y. Loo, M. Mehregany, J. Lee: Surface-micromachined RF MEMs switches on GaAs substrates, Int. J. RF Microw. Comput.-Aided Eng. **9**(4), 348–361 (1999)

12.50 J. Wang, Z. Ren, C.T.C. Nguyen: 1.156-GHz self-aligned vibrating micromechanical disk resonator, IEEE Trans. Ultrason. Ferr. Freq. Control **51**, 1607–1628 (2004)

12.51 Y. Wang, Z. Li, D.T. McCormick, N.C. Tien: A low-voltage lateral MEMS switch with high RF performance, J. Microelectromech. Syst. **13**, 902–911 (2004)

12.52 C.T.C. Nguyen, R.T. Howe: CMOS microelectromechanical resonator oscillator, IEEE Int. Electron Dev. Meet. (1993) pp. 199–202

12.53 L. Lin, R.T. Howe, A.P. Pisano: Microelectromechanical filters for signal processing, IEEE J. Microelectromech. Syst. **7**(3), 286–294 (1998)

12.54 F.D. Bannon III, J.R. Clark, C.T.C. Nguyen: High frequency micromechanical filter, IEEE J. Solid-State Circuits **35**(4), 512–526 (2000)

12.55 K. Wang, Y. Yu, A.C. Wong, C.T.C. Nguyen: VHF free-free beam high-Q micromechanical resonators, 12th IEEE Int. Conf. Micro Electro Mech. Syst. (1999) pp. 453–458

12.56 J.R. Clark, W.T. Hsu, C.T.C. Nguyen: High-Q VHF micromechanical contour-mode disk resonators, IEEE Int. Electron Dev. Meet. (2000) pp. 493–496

12.57 C.T.C. Nguyen, R.T. Howe: Quality factor control for micromechanical resonator, IEEE Int. Electron Dev. Meet. (1992) pp. 505–508

12.58 M.L. Roukes: Plenty of room, indeed, Sci. Am. **285**, 48–57 (2001)

12.59 A.N. Cleland, M.L. Roukes: Fabrication of high frequency nanometer scale mechanical resonators from bulk Si crystals, Appl. Phys. Lett. **69**, 2653–2655 (1996)

12.60 D.W. Carr, H.G. Craighead: Fabrication of nanoelectromechanical systems in single crystal silicon using silicon on insulator substrates and electron beam lithography, J. Vac. Sci. Technol. B **15**, 2760–2763 (1997)

12.61 T.S. Tighe, J.M. Worlock, M.L. Roukes: Direct thermal conductance measurements on suspended monocrystalline nanostructures, Appl. Phys. Lett. **70**, 2687–2689 (1997)

12.62 H.X. Tang, X.M.H. Huang, M.L. Roukes, M. Bichler, W. Wegscheider: Two-dimensional electron-gas actuation and transduction for GaAs nanoelectromechanical systems, Appl. Phys. Lett. **81**, 3879–3881 (2002)

12.63 Y.T. Yang, K.L. Ekinci, X.M.H. Huang, L.M. Schiavone, M.L. Roukes, C.A. Zorman, M. Mehregany: Monocrystalline silicon carbide nanoelectromechanical systems, Appl. Phys. Lett. **78**, 162–164 (2001)

12.64 X.M.H. Huang, X.L. Feng, M.K. Prakash, S. Kumar, C.A. Zorman, M. Mehregany, M.L. Roukes: Fabrication of suspended nanomechanical structures from bulk 6H-SiC substrates, Mater. Sci. Forum **457–460**, 1531–1534 (2004)

12.65 L. Sekaric, M. Zalalutdinov, S.W. Turner, A.T. Zehnder, J.M. Parpia, H.G. Craighead: Nanomechanical resonant structures as tunable passive modulators, Appl. Phys. Lett. **80**, 3617–3619 (2002)

12.66 A.M. Fennimore, T.D. Yuzvinsky, W.Q. Han, M.S. Fuhrer, J. Cummings, A. Zettl: Rotational actuators based on carbon nanotubes, Nature **424**, 408–410 (2003)

12.67 D.W. Carr, S. Evoy, L. Sekaric, H.G. Craighead, J.M. Parpia: Measurement of mechanical resonance and losses in nanometer scale silicon wires, Appl. Phys. Lett. **75**, 920–922 (1999)

12.68 D.W. Carr, L. Sekaric, H.G. Craighead: Measurement of nanomechanical resonant structures in single-crystal silicon, J. Vac. Sci. Technol. B **16**, 3821–3824 (1998)

12.69 S. Evoy, D.W. Carr, L. Sekaric, A. Olkhovets, J.M. Parpia, H.G. Craighead: Nanofabrication and electrostatic operation of single-crystal silicon paddle oscillators, J. Appl. Phys. **86**, 6072–6077 (1999)

12.70 L. Sekaric, J.M. Parpia, H.G. Craighead, T. Feygelson, B.H. Houston, J.E. Butler: Nanomechanical resonant structures in nanocrystalline diamond, Appl. Phys. Lett. **81**, 4455–4457 (2002)

12.71 A.N. Cleland, M.L. Roukes: A nanometre-scale mechanical electrometer, Nature **392**, 160–162 (1998)

12.72 K. Schwab, E.A. Henriksen, J.M. Worlock, M.L. Roukes: Measurement of the quantum of thermal conductance, Nature **404**, 974–977 (2000)

12.73 S. Evoy, A. Olkhovets, L. Sekaric, J.M. Parpia, H.G. Craighead, D.W. Carr: Temperature-dependent internal friction in silicon nanoelectromechanical systems, Appl. Phys. Lett. **77**, 2397–2399 (2000)

12.74 K.L. Ekinci, X.M.H. Huang, M.L. Roukes: Ultrasensitive nanoelectromechanical mass detection, Appl. Phys. Lett. **84**, 4469–4471 (2004)

12.75 B. Illic, H.G. Craighead, S. Krylov, W. Senaratne, C. Ober, P. Neuzil: Attogram detection using nanoelectromechanical oscillators, J. Appl. Phys. **95**, 3694–3703 (2004)

12.76 T.D. Stowe, K. Yasumura, T.W. Kenny, D. Botkin, K. Wago, D. Rugar: Attonewton force detection using ultrathin silicon cantilevers, Appl. Phys. Lett. **71**, 288–290 (1997)

12.77 B. Illic, Y. Yang, H.G. Craighead: Virus detection using nanoelectromechanical devices, Appl. Phys. Lett. **85**, 2604–2606 (2004)

12.78 H. Liu, J. Kameoka, D.A. Czaplewski, H.G. Craighead: Polymeric nanowire chemical sensor, Nano Lett. **4**, 617–675 (2004)

13. Next-Generation DNA Hybridization and Self-Assembly Nanofabrication Devices

Michael J. Heller, Benjamin Sullivan, Dietrich Dehlinger, Paul Swanson, Dalibor Hodko

The new era of nanotechnology presents many challenges and opportunities. One area of considerable challenge is nanofabrication, in particular the development of fabrication technologies that can evolve into viable manufacturing processes. Considerable efforts are being expended to refine classical top-down approaches, such as photolithography, to produce silicon-based electronics with nanometer-scale features. So-called bottom-up or self-assembly processes are also being researched and developed as new ways of producing heterogeneous nanostructures, nanomaterials and nanodevices. It is also hoped that there are novel ways to combine the best aspects of both top-down and bottom-up processes to create a totally unique paradigm change for the integration of heterogeneous molecules and nanocomponents into higher order structures. Over the past decade, sophisticated microelectrode array devices produced by the top-down process (photolithography) have been developed and commercialized for DNA diagnostic genotyping applications. These devices have the ability to produce electric field geometries on their surfaces that allow DNA molecules to be transported to or from any site on the surface of the array. Such devices are also able to assist in the self-assembly (via hybridization) of DNA molecules at specific locations on the array surface. Now a new generation of these microarray devices are available that contain integrated CMOS components within their underlying silicon structure. The integrated CMOS allows more precise control over the voltages and currents sourced to the individual microelectrode sites. While such microelectronic array devices have been used primarily for DNA diagnostic applications, they do have the intrinsic ability to transport almost any type of charged molecule or other entity to or from any site on the surface of the array. These include other molecules with self-assembling properties such as peptides and proteins, as well as nanoparticles, cells and even micron-scale semiconductor components. Microelectronic arrays thus have the potential to be used in a highly parallel electric field *pick and place* fabrication process allowing a variety of molecules and nanostructures to be organized into higher order two- and three-dimensional structures. This truly represents a synergy of combining the best aspects of *top-down* and *bottom-up* technologies into a novel nanomanufacturing process.

13.1	Electronic Microarray Technology	391
	13.1.1 400 Test Site CMOS Microarray	392
	13.1.2 Electric Field Technology Description	394
	13.1.3 Electronic DNA Hybridization and Assay Design	395
	13.1.4 DNA Genotyping Applications	396
	13.1.5 On-Chip Strand Displacement Amplification	396
	13.1.6 Cell Separation on Microelectronic Arrays	397
13.2	Electric Field-Assisted Nanofabrication Processes	397
	13.2.1 Electric Field-Assisted Self-Assembly Nanofabrication	397
13.3	Conclusions	399
References		400

Nanotechnology and nanoscience are producing a wide range of new ideas and concepts, and are likely to enable novel nanoelectronics, nanophotonics, nanomaterials, energy conversion processes and a new generation of biomaterials, biosensors and other biomedical devices. Many of the challenges and opportunities in

Table 13.1 Some challenges for nanotechnology and nanofabrication (National Nanotechnology Initiative & NSF)

(1)	Better understanding of scaling problems and phenomena
(2)	Better synthetic methods for nano building blocks
(3)	Control of nanoscale building blocks (such as size and shape)
(4)	Enormous complexity, heterogeneous materials and sizes
(5)	Surfaces for nanostructure assembly
(6)	Directed hierarchical self-assembly (mimic biological)
(7)	Need for highly parallel processes
(8)	Equipment for parallel directed self-assembly
(9)	Integrate bottom-up and top-down approaches
(10)	Need better analytical capabilities
(11)	Tools for modeling and simulations
(12)	Scale-up issues for manufacturing

nanotechnology have been identified through the efforts of the National Nanotechnology Initiative [13.1]. While many opportunities exist, there are also considerable challenges that must be met and overcome in order to obtain the benefits (Table 13.1). Most challenging will be those areas that relate to nanofabrication, in particular the development of viable fabrication technologies which will lead to cost-effective nanomanufacturing processes. Enormous efforts are now being carried out to refine classical top-down or photolithography processes to produce silicon (CMOS) integrated electronic devices with nanometer-scale features. While this goal is being achieved, this type of process requires billion-dollar fabrication facilities and it appears to be reaching some fundamental limits. So-called *bottom-up* self-assembly processes are also being studied and developed as possible new ways of producing nanoelectronics as well as new nanomaterials and nanodevices. Generally, self-assembly-based nanoelectronics are envisioned as one of the more revolutionary outcomes of nanotechnology. There are now numerous examples of promising nanocomponents such as organic electron transfer molecules, quantum dots, carbon nanotubes and nanowires, and also some limited success in first-level assembly of such nanocomponents into simple structures with higher order electronic properties [13.2–4]. Nevertheless, the issue of developing a viable cost-effective self-assembly nanofabrication process that allows billions of nanocomponents to be assembled into useful logic and memory devices still remains a considerable challenge. In addition to the nanoelectronic applications, other new nanomaterials and nanodevices with higher order photonic, mechanical, mechanistic, sensory, chemical, catalytic and therapeutic properties are also envisioned as an outcome of nanotechnology efforts [13.1–4]. Again, a key problem in enabling such new materials and devices will most likely be in developing effective nanomanufacturing technologies for organizing and integrating heterogeneous components of different sizes and compositions into these higher level structures and devices.

Living systems provide some of the best examples of self-assembly or self-organization processes that should be considered very closely when developing strategies for *bottom-up* nanofabrication. The molecular biology of living systems includes many molecules which have high fidelity recognition properties such as DNA, RNA, and many types of protein macromolecules. Proteins can serve as structural elements, as binding recognition moieties (antibodies), and as highly efficient chemomechanical catalytic macromolecules (enzymes). Such biomolecules are able to interact and organize into second-order macromolecules and nanostructures which store and translate genetic information (involving DNA and RNA structural proteins as well as enzymes), and perform biomolecular syntheses and energy conversion metabolic processes (involving enzymes and structural proteins). All of these biomolecules, macromolecules, nanostructures and nanoscale processes are integrated and contained within higher order membrane-encased structures called cells. Cells in turn can then replicate and differentiate (via these nanoscale processes) to form and maintain living organisms. Thus, biology has developed the ultimate *bottom-up* nanofabrication processes that allow component biomolecules and nanostructures with intrinsic self-assembly and catalytic properties to be organized into highly intricate living systems.

Of all the different biomolecules that could be useful for nanofabrication, nucleic acids, with their high fidelity recognition and intrinsic self-assembly properties, represent a most promising material that can be used to create nanoelectronic, nanophotonic and many other types of organized nanostructures [13.5–8]. The nucleic acids, which include deoxyribonucleic acid (DNA), ribonucleic acid (RNA) and other synthetic

DNA analogs (peptide nucleic acids and so on) are *programmable* molecules, which have intrinsic molecular recognition and self-assembly properties via their nucleotide base (A, T, G, C) sequence. Short DNA sequences called oligonucleotides are readily synthesized by automated techniques. They can additionally be modified with a variety of functional groups such as amines, biotin moieties, fluorescent or chromophore groups, and charge transfer molecules. Additionally, synthetic DNA molecules can be attached to quantum dots, metallic nanoparticles and carbon nanotubes, as well as surfaces like glass, silicon, gold and semiconductor materials. Synthetic DNA molecules (oligonucleotides) represent an ideal type of *molecular Lego* for the self-assembly of nanocomponents into more complex two- and three-dimensional higher order structures. Initially, DNA sequences can be used as a kind of template for assembly on solid surfaces. The technique involves taking complementary DNA sequences and using them as a kind of selective glue to bind other DNA-modified macromolecules or nanostructures together. The base pairing property of DNA allows one single strand of DNA with a unique base sequence to recognize and bind together with its complementary DNA strand to form a stable double-stranded DNA structures. While high-fidelity recognition molecules like DNA allow one to *self-assemble* higher order structures, the process has some significant limitations. First, for in vitro applications (in a test tube), DNA and other high-fidelity recognition molecules like antibodies, streptavidins, and lectins work most efficiency when the complexity of the system is relatively low. In other words, as the complexity of the system increases (more unrelated DNA sequences, proteins, and other biomolecules), the high-fidelity recognition properties of DNA molecules are overcome by nonspecific binding and other entropy-related factors, and the specificity and efficiency of DNA hybridization is considerably reduced. Under in vivo conditions (inside living cells), the binding interactions of high-fidelity recognition molecules like DNA are much more controlled and compartmentalized, and the DNA hybridization process is assisted by structural protein elements and active dynamic enzyme molecules. Thus, new nanofabrication processes based on self-assembly or self-organization using high-fidelity recognition molecules like DNA should also incorporate strategies for assisting and controlling the overall process.

Active microelectronic arrays have been developed for a number of applications in bioresearch and DNA clinical diagnostics [13.8–17]. These active microarrays are able to produce electric fields on the array surface that allow charged reagent molecules (DNA, RNA, proteins, enzymes), nanostructures, cells and microscale structures to be transported to any of the microscopic sites on the device surface. When DNA hybridization is carried out on the microarray, the device allows electric fields to direct the self-assembly of the DNA hybrid at the test site. In principle, these active microarray devices can serve as *motherboards or hostboards* to assist in the self-assembly of DNA molecules, as well as other moieties such as nanostructures or even microscale components [13.18–25]. Active microarray electric field assembly is thus a type of *pick and place* process that has the potential to be used for heterogeneous integration and nanofabrication of molecular and nanoscale components into higher order materials, structures and devices [13.24].

13.1 Electronic Microarray Technology

In the last decade, the development of microarray technologies has greatly expanded our analytical capabilities of carrying out both DNA and protein analysis [13.26]. Many of these novel microarray technologies now allow us to analyze thousands of DNA sequences with very high specificity and sensitivity. Examples include Affymetrix's (Santa Clara, USA) GeneChip [13.27–29] Nanosphere's (Northbrook, USA) [13.29] technology, and Nanogen's (San Diego, USA) electronically active Nanochip [13.8–24, 30–43] technologies. Many assay techniques have been developed to carry out genotyping, gene expression analysis, forensics analysis and for a variety of other assay procedures.

Nanogen, Inc. has developed electronic microarray technology that utilizes electric fields to accelerate and manipulate biomolecules such as DNA, RNA, and proteins on a microarray surface. Each test site or microlocation on the microarray has an underlying platinum microelectrode which can be activated independently. The original 100 test site microarray with 80 μm-diameter platinum microelectrodes is fabricated on a silicon substrate. In this device, each of the 100 test site microelectrodes has a separate wire contact with

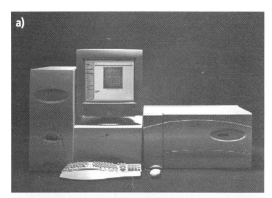

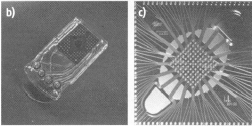

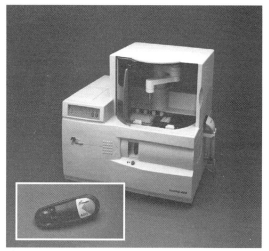

Fig. 13.1a–c First commercialized version of the Nanogen Molecular Biology Workstation. This includes the controller and fluorescent detection component ((a) *left*) and the loader system ((b) *right*) which can be used to address four 100-test site cartridges with DNA samples or DNA probes. The cartridge component containing the a 100-test site chip is shown in the *lower left* (b), and the 100-test site chip is show in the *lower right* (c)

only the microelectrode surface exposed to the sample solution. The newer 400 test site microarray with 50 μm-diameter platinum microelectrodes has CMOS control elements fabricated into the underlying silicon. This integrated CMOS is used to independently regulate the currents and voltages to each of the 400 test sites on the microarray surface. Both the 100 test site and 400 test site CMOS microarray chips are embedded within a disposable plastic fluidic cartridge that provides for automated control of sample or reagents injection onto the electronic microarray.

The first generation platform, the Molecular Biology Workstation, was developed for a 100 test site microarray cartridge. Figure 13.1a shows the Molecular Biology Workstation, Fig. 13.1b shows the 100 test site microarray cartridge, and Fig. 13.1c the 100 test site microarray itself. The Workstation platform consists of two separate instruments, a loader that is capable of

Fig. 13.2 The NanoChip 400 (NC400) is a fully integrated system capable of electronically loading samples onto a microarray and interrogating samples using the built-in fluorescent reader. The system is fully automated and has the capacity to analyze up to 364 samples in a single run. The 400 site electronic microarray, which is contained within a plastic cartridge, can be used up to ten times, allowing for greater user flexibility. A permlayer, into which streptavidin has been embedded, is molded on top of the microarray. The *insert* shows a NC 400 cartridge with 400 array sites or electrodes

processing (addressing) up to four cartridges, and a single cartridge fluorescent reader. The newer Nanochip 400 System integrates both the loader and the reader into a single instrument, and the unique 400 test site CMOS microarray has been embedded within a new cartridge design (Fig. 13.2). The Nanochip 400 System and 400 CMOS electronic microarray provide a tremendous amount of flexibility and control; each of the 400 test sites on the microarray can be easily configured and modified for a range of electronic assay formats.

13.1.1 400 Test Site CMOS Microarray

The external control of different voltages and currents via individual wires to a large number of microelectrodes can be a cumbersome process. Thus, for higher density microarrays (> 100 test sites) it is advantageous to integrate the microelectrode bias control circuitry directly into the microchip silicon structure itself. In the new 400 test site microarray, standard CMOS circuitry has been used to integrate digital

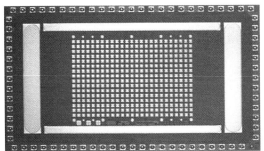

Fig. 13.3 Photograph of the 400-site CMOS ACV400-chip array. Four counter-electrodes, two positioned longitudinally and two horizontally, surround the active working electrode array

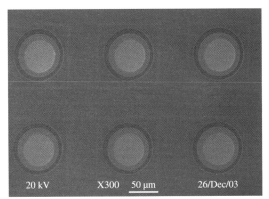

Fig. 13.4 Close-up of the current 400-site chip array. The CMOS chip has an array of 16×25 (400) sites; each electrode is 50 μm in diameter with a 150 μm center-to-center distance

communication, memory, temperature sensing, voltage/current sourcing and measuring circuits *on-chip*. After standard CMOS fabrication within the underlying silicon, thin film deposition and patterning techniques were used to fabricate the platinum microelectrodes on the surface of the standard CMOS chip [13.30]. The CMOS microelectrode array chip developed for the Nanochip 400 system consists of a 16 by 25 array of 50 μm diameter microelectrodes spaced 150 μm center to center. Figure 13.3 shows the 400 test site CMOS microelectronic array device which is only 5

by 7 mm in size. Figure 13.4 shows a close up of the 400 site microarray with several of the 50 μm diameter platinum microelectrodes. Underneath each of the 400 microelectrodes is an analog sample and hold circuit which maintains a predefined voltage on the electrode. A digital to analog converter sequentially interrogates the digitally stored bias value for each of the microelectrodes and refreshes each sample and hold cir-

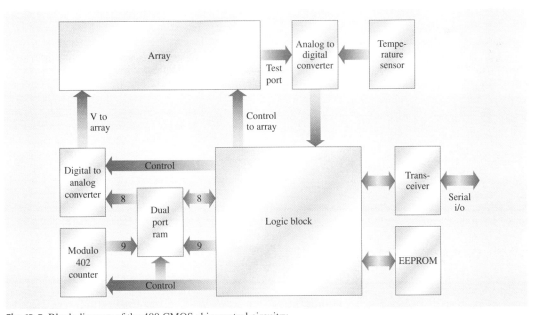

Fig. 13.5 Block diagram of the 400 CMOS chip control circuitry

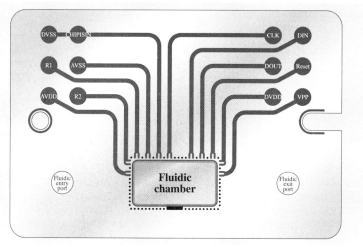

Fig. 13.6 Ceramic substrate and electrical connections to the chip

cuit accordingly. A separate loop sequentially measures the voltage and current at each of the microelectrodes. Microelectrodes can be operated at a fixed voltage, or by means of a feed back loop, at constant current or at a fixed voltage offset from a reference electrode. The microchip also contains a p-n junction temperature sensor and EEPROM memory to store thermal calibration coefficients, serial numbers, and assay-related data. Figure 13.5 shows the Nanochip 400 CMOS circuitry block diagram. The chip has only 12 external electrical connections, +5 V and ground for the digital circuits, +5 V and ground for the analog circuits, digital signal in, digital signal out, clock signal, reset, and two terminals for an external current sampling resistor. For structural support, 76 flip-chip solder bonds are used between the chip and a ceramic substrate which doubles as both a fluidic chamber and as an electrical contact interface (Fig. 13.6). In order prevent any compromise in the quality of either the circuitry or the microelectrode array, the CMOS is fabricated at one foundry while the platinum microelectrode array is fabricated at a second foundry. The CMOS process requires 16 masking step followed by an additional three masking steps for the platinum electrodes. Figure 13.7 shows a high-magnification cross section of an individual microelectrode and the underlying CMOS circuitry. After the finished wafers have been inspected for defects, the wafers are diced into individual chips and then flip-chip bonded onto the ceramic substrates. The flip-chip on substrates (FCOS) are thermally calibrated and then finally assembled into the NanoChip 400 plastic cartridge housing which provides the fluidic delivery system and outside electrical connections (Fig. 13.2, inset).

13.1.2 Electric Field Technology Description

Nanogen's microarray technology is unique among DNA microarrays due to the use of electrophoretically driven or active transport of the DNA target or DNA probe molecules on the microarray surface. This active transport over the microarray is electronically controlled by biasing different microelectrodes

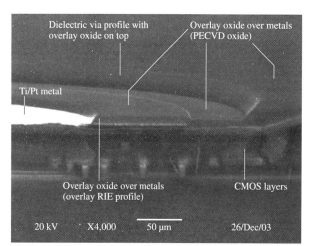

Fig. 13.7 Cross section of an electrode and accompanying CMOS circuitry

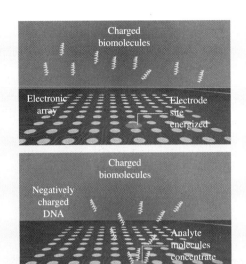

Fig. 13.8 Diagram of negatively charged DNA molecules being transported to a positively charged microelectrode

on the microarray surface. Depending on the charge of the molecule, they will be rapidly transported to the oppositely biased microelectrode. Figure 13.8 shows a diagram of negatively charged DNA molecules being transported to a positively charged microelectrode. The electronic addressing of DNA and other biomolecules onto the microarray test sites can accelerate hybridization and other molecular binding processes by up to 1000 times compared to traditional passive methods (such as those of Affymetrix and GeneChip) (Table 13.2). By way of example, hybridization on a passive microarray may take up to several hours for the low concentrations of target DNA frequently found in many clinical samples. During the operation of the electronic microarray, the bias potential is sufficiently high ($> 1.2\,\text{V}$) to cause the electrolysis of water. Oxidation occurs on the positively biased microelectrode and reduction occurs on the negatively biased microelectrode surface (see (13.1) and (13.2) below)

Oxidation $\quad H_2O \rightarrow 2H^+ + \frac{1}{2}O_2 + 2e^-$, (13.1)

Reduction $\quad 2e^- + 2H_2O \rightarrow 2OH^- + H_2$. (13.2)

Important to the operation of the device is a thin $1-2\,\mu m$ hydrogel permeation layer which covers the platinum microelectrode surface. This hydrogel permeation layer is designed to protect the more sensitive DNA and other biomolecules from the electrolysis reactions and products that occur on the platinum microarray surface. The permeation layer is usually made of either agarose or polyacrylamide. The layer also contains streptavidin which facilitates the capture and binding of biotinylated DNA probes or DNA target molecules onto the hydrogel surface.

The electric field microarray technology also takes advantage of the H^+ ions (low pH) generated at the positive electrode to carry out a unique *electronic hybridization* process. Electronic hybridization is carried out under low salt and low conductance conditions and it allows denatured DNA molecules to be hybridized only in the microscopic area around the activated microelectrode test site. A second electronic process called *electronic stringency* is achieved by reversing the polarity at the microelectrode (negative activation) causing nonspecifically bound (negatively charged) DNA molecules to be driven away from the test site. Electronic stringency can also aid in the differentiation of DNA binding strengths, allowing better single base discrimination for the determination of single nucleotide polymorphisms (SNPs).

13.1.3 Electronic DNA Hybridization and Assay Design

The electronic hybridization technology and the Nanochip 400 System provide the researcher with

Table 13.2 Comparison of electronic hybridization speed with that of conventional passive hybridization on microarrays

	Hybridization time	Concentration of targets	Concentration factor at a site	Stringency control
NanoChip	10–100 s	Directed and localized	> 1000 times	Electronic thermal chemical
Active hybridization		At the array sites; ability to control individual sites		
Passive hybridization technologies	1–2 h	Undirected; sites cannot be controlled independently	Low, diffusion-	Thermal chemical dependent

a completely flexible and open platform for performing DNA hybridizations. It offers users the option to either generate or create user-defined microarrays specific for their own targets using electronic addressing of the biotinylated probes on the array, or to perform target- or single-base-specific analysis of DNA or RNA molecules. Both processes are performed very rapidly because the electronic addressing enables hybridization of the target DNA to occur within 30 to 60 s, as compared to competitive microarray technologies where hybridization is performed within one to several hours. Electronic microarrays provide an open platform and offer flexibility in the assay design [13.31–36]. This includes either the *PCR amplicon down format* to screen one or more patients for one or more SNPs, or the *capture probe down format* to screen many patients for one or more SNPs. In addition, *sandwich-type* assays are easy to perform where oligonucleotide discriminators and/or fluorescent probe labeled oligonucleotides are hybridized, electronically or passively, to captured PCR amplicons or probes. These assay design tools enable the users to increase the discrimination at single-base resolution as well as to minimize the nonspecific binding. Multiple probes or discriminators can be attached at a single site which facilitates the detection of multiple targets on a single electrode site. This enables the worker to use over 1000 characteristic genes or single nucleotide polymorphisms (SNPs) on a single array. In multiple DNA target detection the user can use blocker nucleotides that block SNPs which are not reported at a particular location. This allows the same universal reporter and specific discriminators to be used to report one or more SNPs on another site. Two lasers are used to recognize fluorescent signal from two reporter dyes (green and red). All of the steps are extremely fast, so the users can design and generate their own assays and incorporate the preparation of the array with oligonucleotides specific to multiple targets and samples as a part of the assay.

Other flexible electronic hybridization format designs allow several types of *multiplexed DNA analyses* to be carried out, such as the determination of multiple genes in one sample, multiple samples with one gene, or multiple samples with multiple genes. The ability to control individual test sites permits genetically unrelated DNA molecules to be used simultaneously on the same microchip. In contrast, sites on a conventional DNA array cannot be controlled separately, and all process steps must be performed on an entire array. Types of multiplexed analysis include:

1. Determination of multiple genes in one sample addressed on the chip
2. Determination of multiple samples with one gene of interest addressed on the chip
3. Determination of multiple samples with multiple genes of interest addressed on the chip
4. Single site multiplexing where several targets are discriminated on the same site using different fluorescent probes.

13.1.4 DNA Genotyping Applications

Electronic hybridization technology has been developed and commercialized and is now used for a number of practical applications including clinical diagnostic DNA genotyping [13.31–40]. In all applications, the DNA probe detection step is accomplished with fluorophore reporters and laser-based fluorescence detection. In addition, due to its open platform character, Nanogen's users have developed about 200 additional assays using the electronic microarray system and technology. The DNA assay areas developed by the users include diagnostics related to the following diseases and applications: coronary artery disease, cardiovascular disease, hypertension, cardiac function, venous thrombotic disease, metabolism, drug metabolism/cancer, cancer, cytokine, transcription factor, bacterial ID, Rett syndrome, thrombophilia, thalassemia and deafness.

13.1.5 On-Chip Strand Displacement Amplification

It has also been demonstrated that a complex DNA amplification technique can be performed on separate test sites of the microarray chip. This approach significantly reduces the time for DNA analysis because it incorporates DNA amplification and detection into a single platform. The assays were demonstrated using an isothermal strand displacement amplification (SDA is licensed technology from Becton, Dickinson and Co., Franklin Lakes, USA). In the SDA amplification, DNA polymerase recognizes the nicked strand of DNA and initiates resynthesis of that strand, displacing the original strand. The released amplicons then travel in solution to primers of the complementary strand which are either in solution or anchored on the test site. Oligonucleotide primers without nicking sites, called *bumper primers*, are synthesized in the regions flanking the amplicons that were just produced, and assist in strand displacement and initial template replication [13.41].

13.1.6 Cell Separation on Microelectronic Arrays

Microelectronic arrays have also used been for cell separation applications. Disease diagnostics frequently involve identifying a small number of specific bacteria or viruses in a blood sample (infectious disease), fetal cells in maternal blood (genetic diseases) or tumor cells among a background of normal cells (early cancer detection). One powerful electric field technique used for cell separation is called dielectrophoresis (DEP). The DEP process involves the application of an asymmetric alternating current (AC) electric field to the cell population. Active microelectronic arrays have been used to achieve the separation of bacteria from whole blood [13.42], for the separation of cervical carcinoma cells from blood [13.43], and for gene expression analysis [13.44]. Microelectronic array devices utilizing high frequency ac fields have been used to carry out the DEP separation of *Listeria* bacterial cells ($\approx 1\,\mu m$) from whole blood cells ($\approx 10\,\mu m$) in a highly parallel manner. At an ac frequency of about 10 kHz the *Listeria* bacterial cells can be positioned on specific microlocations at high-field regions and the blood cells can be positioned in the low-field regions between the microelectrodes. The relative positioning of the cells between the high- and low-field regions is based on dielectric differences between the cell types. While maintaining the ac field, the microarray can be washed with a buffer solution that removes the blood cells (low-field regions) from the more firmly bound bacteria (high-field regions) near the microelectrodes. The bacteria can then be released and collected or electronically lysed to release the genomic DNA or RNA for further manipulation and analysis [13.42]. DEP represents a particularly useful process that allows difficult cell separation applications to be carried out rapidly and with high selectivity. The DEP process may also be useful for nanofabrication purposes [13.23, 24].

13.2 Electric Field-Assisted Nanofabrication Processes

Many examples of individual molecular and nanoscale components with basic electronic and photonic properties exist, including such entities as metallic nanoparticles, quantum dots, carbon nanotubes, nanowires and various organic molecules with electronic switching capabilities. However, the larger issue with enabling self-assembly-based nanoelectronics and nanophotonics is more likely to be the development of *viable* processes that will allow billions of molecular and/or nanoscale components to be assembled and interconnected into useful materials and devices. In addition to the electronic and photonic applications, nanostructures, nanomaterials and nanodevices with higher order mechanical/mechanistic, chemical/catalytic, biosensory and therapeutic properties are also envisioned [13.1–4]. The biggest challenges in enabling such devices and systems will most likely come from the stage of organizing components for higher level functioning, rather than the availability of the molecular components. Thus, a key problem with mimicking this type of nanotechnology is the lack of a viable *bottom-up* nanofabrication process to carry out the precision integration of diverse molecular and nanoscale components into viable higher order structures.

13.2.1 Electric Field-Assisted Self-Assembly Nanofabrication

As was described earlier, microelectronic array devices have been developed for applications in DNA genotyping diagnostics. These active microarray devices are able to produce reconfigurable electric field geometries on the surface of the device. The resulting electric fields are able to transport any type of charged molecule or structure, including DNA, RNA, proteins, antibodies, enzymes, nanostructures, cells or microscale devices to or from any of the sites on the array surface. When DNA hybridization reactions are carried out using the device, the electric fields are actually assisting in the self-assembly of DNA molecules at the specified test site. In principle, these active devices serve as a *motherboard or hostboard* for the assisted assembly of DNA molecules into higher order or more complex structures. Since DNA molecules have intrinsic programmable self-assembly properties and can be derivatized with electronic or photonic groups or attached to larger nanostructures (quantum dots, metallic nanoparticles and nanotubes), we have the basis for a unique bottom-up nanofabrication process. Active microelectronic arrays serving as motherboards allow one to carry out a highly

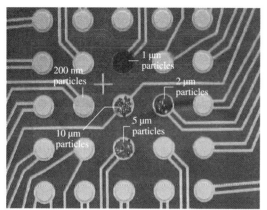

Fig. 13.9 Electronic addressing of five different types of microspheres and nanospheres to the microelectronic array test sites

with the specific complementary oligonucleotide sequences [13.8, 23, 24]. Microelectronic array devices have also been used for selective transport and addressing of larger nanoparticles and microspheres, and even objects as large as 20 μm light emitting diode structures [13.21–24]. In this context, Fig. 13.9 shows the electric field addressing of five differently sized negatively charged polystyrene microspheres and nanospheres (100 nm) to selectively activated microlocations on a 25-test site microelectronic array. The rate of transport is related to the strength of the electric field and the charge/mass ratio of the molecule or structure. Figure 13.10 shows the results for the parallel transport and positioning of two different types of microspheres onto the microelectronic

parallel electric field *pick and place* process for the heterogeneous integration of molecular, nanoscale and microscale components into complex three-dimensional structures. If desired, this process can be used to assemble molecules and/or nanocomponents within the defined perimeters of larger silicon or other semiconductor structures. Electric field-assisted self-assembly technology is based on three key physical principles:

1. The use of functionalized DNA or other high-fidelity recognition components as *molecular Lego* blocks for nanofabrication
2. The use of DNA or other high-fidelity recognition components as a *selective glue* that provides intrinsic self-assembly properties to other molecular, nanoscale or microscale components (metallic nanoparticles, quantum dots, carbon nanotubes, organic molecular electronic switches, micrometer and submicrometer silicon lift-off devices and components)
3. The use of active microelectronic array devices to provide electric field assistance or control of the intrinsic self-assembly of any modified electronic/photonic components and structures [13.18–24].

Microelectronic arrays have been used to direct the binding of derivatized nanospheres and microspheres onto selected locations on the microarray surface. In this case, fluorescent and nonfluorescent polystyrene nanospheres and microspheres derivatized with specific DNA oligonucleotides are transported and bound to selected test sites or microlocations derivatized

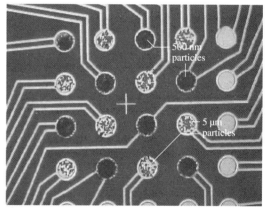

Fig. 13.10 Parallel electronic addressing of two different types of microspheres to the microelectronic array test sites

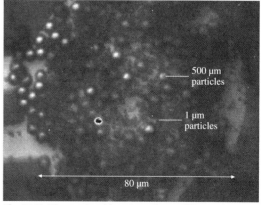

Fig. 13.11 Electronic addressing of two different layers of microspheres to the microelectronic array test sites

array surface. Finally, Fig. 13.11 now shows the results for the initial addressing of derivatized negatively charged 1 μm polystyrene microspheres to a selectively activated microlocation on a 25-site microelectronic array, and the subsequent covering of the layer of 1 μm polystyrene microspheres by larger 5 μm microspheres. Thus, it is possible to use electric field transport and addressing to form multiple layers of particles and other materials, allowing fabrication in the third dimension.

Present nanofabrication methods do not allow most nanostructures to be modified in a controlled or precise manner. For example, it would be extremely difficult to attach different DNA sequences or different kinds of protein molecules in precise locations around quantum dots or other nanoparticles (Fig. 13.12a,b). Unfortunately, without this first-order property it becomes even more difficult to then assemble these nanostructures into higher order heterogeneous 3-D structures, even though the core structure is derivatized with high-fidelity recognition components (Fig. 13.12c). Microelectronic array devices may offer the opportunity to develop processes that will allow core nanostructures to be selectively modified in a precise fashion [13.24]. The proposed electric field microarray techniques may provide the ability to carry out the precision functionalization of nanostructures by processes which involve transporting and orienting the nanostructures onto surfaces containing the selected ligand molecules which are then reacted only with a selected portion of the nanostructure. By repeating the process and reorienting

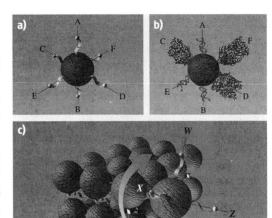

Fig. 13.12 (a,b) Precision nanosphere functionalization scheme. **(c)** Type of heterogeneous 3-D higher order structure that can only be obtained using precision nanostructures

the nanostructures, it will be possible to functionalize the core structure selectively with most biological and/or chemical groups. Such devices and processes allow one to design and create functionalized nanostructures with binding groups arranged in tetrahedral, hexagonal or other coordinate positions around the core nanostructure.

13.3 Conclusions

Active microelectronic array technology provides a number of advantages for carrying out DNA hybridization diagnostics and other affinity-based assays for molecular biology research and clinical diagnostic applications. The technology has also demonstrated the potential for assisted self-assembly and other nanofabrication applications. Microelectronic arrays have been designed and fabricated with 25 to 10 000 microscopic test sites, and devices with 100 and 400 test sites have been commercialized. The newer 400-test site devices have CMOS elements incorporated into the underlying silicon structure that provide on-board control of current and voltage to each of the test sites on the device. Microelectronic chips are incorporated into a cartridge-type device so that they can be conveniently used with a probe loading station and fluorescent detection system. Active microelectronic arrays are fundamentally different from other DNA chip or microarray devices, which are essentially passive. Active microelectronic arrays allow DNA molecules, RNA, oligonucleotide probes, PCR amplicons, proteins, nanostructures, cells and even microscale devices to be rapidly transported and selectively addressed to any of the test sites on the microelectronic array surface. Active microarray devices have considerable potential for nanofabrication by directed self-assembly of molecular, nanoscale and microscale components into higher order mechanisms, structures, and devices. This electric field technology makes possible a type of *pick and place* process for the heterogeneous integration of diverse molecular and nanoscale components into higher order structures within defined perimeters of larger silicon or semiconductor structures. The technology provides the best aspects of a top-down and bottom-up process, and has the inherent hierarchical logic of allowing one to control the organization and assembly of components from the molecular level to the nanoscale level to microscale three-dimensional integrated structures and devices.

References

13.1 National Research Council: *Small Wonders, Endless Frontiers: Review of the Nanotional Nanotechnology Initiative* (National Research Council, Washington 2002)

13.2 M.P. Hughes (Ed.): *Nanoelectromechanics in Engineering and Biology* (CRC, Boca Raton 2003)

13.3 W.A. Goddard, D. Brenner, S. Lyshevski, G. Lafrate (Eds.): *Handbook of Nanoscience, Engineering and Technology* (CRC, Boca Raton 2003)

13.4 V. Balzani, M. Venturi, A. Credi (Eds.): *Molecular Devices and Mechanics – Journey into the Nanoworld* (Wiley-VCH, Weinheim 2003)

13.5 R. Bashir: Biologically mediated assembly of artificial nanostructures and microstructures. In: *Handbook of Nanoscience, Engineering and Technology*, ed. by W.A. Goddard, D. Brenner, S. Lyshevski, G. Lafrate (CRC, Boca Raton 2003) pp. 15-1–15-31, Chap. 15

13.6 M.J. Heller, R.H. Tullis: Self-organizing molecular photonic structures based on functionalized synthetic DNA polymers, Nanotechnology **2**, 165–171 (1991)

13.7 D.M. Hartmann, D. Schwartz, G. Tu, M. Hellerand, S.C. Esener: Selective DNA attachment of particles to substrates, J. Mater. Res. **17**, 473–478 (2002)

13.8 M.J. Heller: An active microelectronics device for multiplex DNA analysis, IEEE Eng. Med. Biol. **15**, 100–103 (1996)

13.9 R.G. Sosnowski, E. Tu, W.F. Butler, J.P. O'Connell, M.J. Heller: Rapid determination of single base mismatch in DNA hybrids by direct electric field control, Proc. Natl. Acad. Sci. USA **94**, 1119–1123 (1997)

13.10 C.F. Edman, D.E. Raymond, D.J. Wu, E. Tu, R.G. Sosnowski, W.F. Butler, M. Nerenberg, M.J. Heller: Electric field directed nucleic acid hybridization on microchips, Nucl. Acids Res. **25**, 4907–4914 (1997)

13.11 M.J. Heller: An integrated microelectronic hybridization system for genomic research and diagnostic applications. In: *Micro Total Analysis Systems*, ed. by D.J. Harrison, A. van den Berg (Kluwer Academic, Dordrecht 1998) pp. 221–224

13.12 M.J. Heller, E. Tu, A. Holmsen, R.G. Sosnowski, J.P. O'Connell: Active microelectronic arrays for DNA hybridization analysis. In: *DNA Microarrays: A Practical Approach*, ed. by M. Schena (Univ. Press, Oxford 1999) pp. 167–185

13.13 M.J. Heller, A.H. Forster, E. Tu: Active microelectronic chip devices which utilize controlled electrophoretic fields for multiplex DNA hybridization and genomic applications, Electrophoresis **21**, 157–164 (2000)

13.14 C. Gurtner, E. Tu, N. Jamshidi, R. Haigis, T. Onofrey, C.F. Edman, R. Sosnowski, B. Wallace, M.J. Heller: Microelectronic array devices and techniques for electric field enhanced DNA hybridization in low-conductance buffers, Electrophoresis **23**, 1543–1550 (2002)

13.15 M.J. Heller: DNA microarray technology: devices, systems and applications, Ann. Rev. Biomed. Eng. **4**, 129–153 (2002)

13.16 M.J. Heller, E. Tu, R. Martinsons, R.R. Anderson, C. Gurtner, A. Forster, R. Sosnowski: Active microelectronic array systems for DNA hybridization, genotyping, pharmacogenomics and nanofabrication applications. In: *Integrated Microfabricated Devices*, ed. by M.J. Heller, A. Guttman (Marcel Dekker, New York 2002) pp. 223–270, Chap. 10

13.17 S.K. Kassengne, H. Reese, D. Hodko, J.M. Yang, K. Sarkar, P. Swanson, D.E. Raymond, M.J. Heller, M.J. Madou: Numerical modeling of transport and accumulation of DNA on electronically active biochips, Sens. Actuators B **94**, 81–98 (2003)

13.18 S.C. Esener, D. Hartmann, M.J. Heller, J.M. Cable: DNA assisted micro-assembly: A heterogeneous integration technology for optoelectronics, Proc. SPIE **70**, 113–140 (1998)

13.19 C. Gurtner, C.F. Edman, R.E. Formosa, M.J. Heller: Photoelectrophoretic transport and hybridization of DNA on unpatterned silicon substrates, J. Am. Chem. Soc. **122**(36), 8589–8594 (2000)

13.20 Y. Huang, K.L. Ewalt, M. Tirado, R. Haigis, A. Forster, D. Ackley, M.J. Heller, J.P. O'Connell, M. Krihak: Electric manipulation of bioparticles and macromolecules on microfabricated electrodes, Anal. Chem. **73**, 1549–1559 (2001)

13.21 C.F. Edman, C. Gurtner, R.E. Formosa, J.J. Coleman, M.J. Heller: Electric-field-directed pick-and-place assembly, HDI **3**(10), 30–35 (2000)

13.22 C.F. Edman, R.B. Swint, C. Gurthner, R.E. Formosa, S.D. Roh, K.E. Lee, P.D. Swanson, D.E. Ackley, J.J. Colman, M.J. Heller: Electric field directed assembly of an InGaAs LED onto silicon circuitry, IEEE Photon. Tech. Lett. **12**(9), 1198–1200 (2000)

13.23 C.F. Edman, M.J. Heller, R. Formosa, C. Gurtner: Methods and apparatus for the electronic homogeneous assembly and fabrication of devices, US Patent 6569382 (2003)

13.24 M.J. Heller, J.M. Cable, S.C. Esener: Methods for the electronic assembly and fabrication of devices, US Patent 6652808 (2003)

13.25 C.F. Edman, M.J. Heller, C. Gurtner, R. Formosa: Systems and devices for the photoelectrophoretic transport and hybridization of oligonucleotides, US Patent 6706473 (2004)

13.26 A. Taton, C. Mirkin, R. Letsinger: Scanometric DNA array detection with nanoparticle probes, Science **289**, 1757–1760 (2000)

13.27 M. Chee, R. Yang, E. Hubbell, A. Berno, X. Huang, D. Stern, J. Winkler, D. Lockhart, M. Morris, S. Fodor: Accessing genetic information with high-density DNA arrays, Science **274**, 610–614 (1996)

13.28 A. Pease, D. Solas, E. Sullivan, M. Cronin, C. Holmes, S. Fodor: Light-generated oligonucleotide arrays for rapid DNA sequence analysis, Proc. Natl. Acad. Sci. USA **99**, 5022–5026 (1994)

13.29 R.J. Lipshutz, D. Morris, M. Chee, E. Hubbell, M.J. Kozal, N. Shah, N. Shen, R. Yang, S.P. Fodor: Using oligonucleotide probe arrays to access genetic diversity, Biotechniques **19**(3), 442–447 (1995)

13.30 P. Swanson, R. Gelbart, E. Atlas, L. Yang, T. Grogan, W.F. Butler, D.E. Ackley, E. Sheldon: A fully multiplexed CMOS biochip for DNA analysis, Sens. Actuators B **64**, 22–30 (2000)

13.31 P.N. Gilles, D.J. Wu, C.B. Foster, P.J. Dillion, S.J. Channock: Single nucleotide polymorphic discrimination by an electronic dot blot assay on semiconductor microchips, Nat. Biotechnol. **17**(4), 365–370 (1999)

13.32 N. Narasimhan, D. O'Kane: Validation of SNP genotyping for human serum paraoxonase gene, Clin. Chem. **34**(7), 589–592 (2001)

13.33 R. Sosnowski, M.J. Heller, E. Tu, A. Forster, R. Radtkey: Active microelectronic array system for DNA hybridization, genotyping and pharmacogenomic applications, Psychiatr. Genet. **12**, 181–192 (2002)

13.34 Y.R. Sohni, J.R. Cerhan, D.J. O'Kane: Microarray and microfluidic methodology for genotyping cytokine gene polymorphisms, Hum. Immunol. **64**, 990–997 (2003)

13.35 E.S. Pollak, L. Feng, H. Ahadian, P. Fortina: Microarray-based genetic analysis for studying susceptibility to arterial and venous thrombotic disorders, Ital. Heart J. **2**, 569–572 (2001)

13.36 W.A. Thistlethwaite, L.M. Moses, K.C. Hoffbuhr, J.M. Devaney, E.P. Hoffman: Rapid genotyping of common MeCP2 mutations with an electronic DNA microchip using serial differential hybridization, J. Mol. Diagn. **5**(2), 121–126 (2003)

13.37 V.R. Mas, R.A. Fisher, D.G. Maluf, D.S. Wilkinson, T.G. Carleton, A. Ferreira-Gonzalez: Hepatic artery thrombosis after liver transplantation and genetic factors: Prothrombin G20210A polymorphism, Transplantation **76**(1), 247–249 (2003)

13.38 R. Santacroce, A. Ratti, F. Caroli, B. Foglieni, A. Ferraris, L. Cremonesi, M. Margaglione, M. Seri, R. Ravazzolo, G. Restagno, B. Dallapiccola, E. Rappaport, E.S. Pollak, S. Surrey, M. Ferrari, P. Fortina: Analysis of clinically relevant single-nucleotide polymorphisms by use of microelectric array technology, Clin. Chem. **48**(12), 2124–2130 (2002)

13.39 A. Åsberg, K. Thorstensen, K. Hveem, K. Bjerve: Hereditary hemochromatosis: The clinical significance of the S64C mutation, Genet. Test. **6**(1), 59–62 (2002)

13.40 J.G. Evans, C. Lee-Tataseo: Determination of the factor V Leiden single-nucleotide polymorphism in a commercial clinical laboratory by use of NanoChip microelectric array technology, Clin. Chem. **48**(9), 1406–1411 (2002)

13.41 T. Walker, J. Nadeau, P. Spears, J. Schram, C. Nycz, D. Shank: Multiplex strand displacement amplification (SDA) and detection of DNA sequences from *Mycobacterium tuberculosis* and other mycobacteria, Nucl. Acids Res. **22**(13), 2670–2677 (1994)

13.42 J. Cheng, E.L. Sheldon, L. Wu, A. Uribe, L.O. Gerrue, J. Carrino, M.J. Heller, J.P. O'Connell: Electric field controlled preparation and hybridization analysis of DNA/RNA from *E. coli* on microfabricated bioelectronic chips, Nat. Biotechnol. **16**, 541–546 (1998)

13.43 J. Cheng, E.L. Sheldon, L. Wu, M.J. Heller, J. O'Connell: Isolation of cultured cervical carcinoma cells mixed with peripheral blood cells on a bioelectronic chip, Anal. Chem. **70**, 2321–2326 (1998)

13.44 Y. Huang, J. Sunghae, M. Duhon, M.J. Heller, B. Wallace, X. Xu: Dielectrophoretic separation and gene expression profiling on microelectronic chip arrays, Anal. Chem. **74**, 3362–3371 (2002)

14. Single-Walled Carbon Nanotube Sensor Concepts

Cosmin Roman, Thomas Helbling, Christofer Hierold

Carbon nanotubes are nanocomponents par excellence that offer unique properties to be exploited in next-generation devices. Sensing applications are perhaps the class that has most to gain from single-walled carbon nanotubes (SWNTs); virtually any property of SWNTs (e.g., electronic, electrical, mechanical, and optical) can result or has already resulted in sensor concept demonstrators. The basic questions that this chapter will attempt to address are: *why* use SWNTs, and *how* can SWNTs be used in sensing applications? A tour through the gallery of basic nanotube properties is used to reveal the richness and uniqueness of this material's intrinsic properties. Together with examples from the literature showing performance of SWNT-based sensors at least comparable to (and sometimes surpassing) that of state-of-the-art micro- or macrodevices, these nanotube properties should explain *why* so much effort is currently being invested in this field. Because nanotubes, like any other nanoobject, are not easy to probe, a versatile strategy for accessing their properties, via the carbon nanotube field-effect transistor (CNFET) concept, will be described in this chapter. Fabricating CNFET devices, together with examples of SWNT sensor demonstrators utilizing the CNFET principle, will outline a proposal for *how* nanotubes can be utilized in sensors.

In Sect. 14.1 design considerations for SWNT sensors are brought into attention, starting with

14.1	Design Considerations for SWNT Sensors	404
	14.1.1 CNT Properties for Sensing	405
	14.1.2 Carbon Nanotube FET Structures	409
	14.1.3 Sensor Characterization	411
14.2	Fabrication of SWNT Sensors	412
	14.2.1 Methods for SWNT Production	412
	14.2.2 Strategies for SWNT Assembly into Devices	413
14.3	Example State-of-the-Art Applications	416
	14.3.1 Chemical and Biochemical Sensors	416
	14.3.2 Piezoresistive Sensors	418
	14.3.3 Resonant Sensors	420
14.4	Concluding Remarks	421
	References	421

a brief survey of SWNT properties useful for sensing. The CNFET is introduced in Sect. 14.1.2 as a platform enabling access to individual SWNT properties during the sensing process. The current status of CNFET-based sensor characterization is captured in Sect. 14.1.3. Methods for fabricating, or supporting the fabrication of, SWNT FETs are reviewed in Sect. 14.2. Finally, Sect. 14.3 will be devoted to examples of CNT-based sensors, encompassing three main case studies, namely (bio)chemical, piezoresistive, and resonator sensors.

Sensors are only one possible application of SWNTs. Other notable applications include field-emission devices, energy storage, composites, and nanoelectronics [14.1]. For example, in nanoelectronics, CNTs have been assessed by the International Technology Roadmap for Semiconductors 2007 (ITRS), edited by a group of scientists from all major semiconductor manufacturers and academic institutions, to have greater potential for post-complementary metal–oxide–semiconductor (CMOS) device concepts than any other on the horizon (e.g., molecular electronic devices, ferromagnetic logic devices, and spin transistors). For sensing devices, carbon nanotubes present several key advantages, including:

1. *Nanometer feature size*: useful for building highly localized sensing units (active spots) and for large-scale integration (sensor arrays)
2. *High sensitivity to stimuli*: from their unique structural and electronic properties (Sect. 14.1.1) and high surface-to-volume ratio
3. *Low power consumption*: whether operating as transistors or mechanical resonators, the power used to excite or probe a SWNT is on the order of 10 nW.

Because of these and other advantages, at relatively short time after their discovery, SWNTs have resulted in sensor device demonstrators fueling optimism worldwide. Many of these investigations have been published in prestigious research journals. There are, however, still some challenges to overcome before broader acceptance in industrial product development activities will be observed. In fact, carbon nanotubes are today at the crossroads between basic science and engineering. CNT device demonstrators and theoretical extrapolations surpass in performance state-of-the-art devices, more than motivating any future attempts to solve the remaining issues. Based on the steep evolution slope experienced so far, it may not take long until carbon-nanotube-based sensors will appear on device and product roadmaps.

14.1 Design Considerations for SWNT Sensors

The range of sensing schemes involving carbon nanotubes is already impressive considering the recentness of this material. A rough classification of CNT-based sensors can be made according to: *material*, *input*, and *output*. The carbon nanotube *material* utilized in sensors can vary from individual SWNTs, multi-walled nanotubes (MWNTs) and bundles, to CNT networks and composites, and even to bulk (forests) CNT material (Table 14.1). Sensors based on nanotubes can respond to a wide range of *inputs*, including (bio)chemical (molecules), mechanical (deformation), optical (radiation), and electrical (fields due to charges). Also different transduction mechanisms have been employed in CNT sensors to generate *outputs* such as electrical (conductive, capacitive), mechanical (resonance frequency), and optical (luminescence). A nonexhaustive CNT sensor catalog with references is given in Table 14.1.

For reasons of clarity and concreteness, in this chapter the discussion will be restricted to a particular class of CNT sensors, namely carbon nanotube field-effect transistor (CNFET) sensors. This class refers to FET configurations with *an individual single-walled carbon nanotube* channel that transforms *input stimuli of different origin* into *electrical signals at the output*. The CNFET is one of the simplest means to probe the properties of an individual SWNT, and at the same time perhaps the most versatile building

Table 14.1 Nonexhaustive catalog of CNT-based sensors with references

Sensor type	Sensor input	Sensor output	CNT material
Chemical sensors [14.2–4]	Gas molecules (i.e., NO_2, NH_3, O_2); organic vapor	Conductance change; CNFETs: shift of gate threshold	Individual SWNT
Chemical sensors [14.5, 6]	Gas molecules, organic vapor	Conductance change; CNFETs: shift of gate threshold	CNT networks; functionalization
Biochemical sensors [14.7]	Biomolecules (liquid phase)	Conductance change; CNFETs: shift of gate threshold	Individual SWNT, CNT networks
Electromechanical sensors [14.8–12]	Pressure, displacement, strain	Conductance change	Individual SWNT
Pressure sensors [14.13]	Pressure	Conductance change	CNT block
Resonant cantilevers [14.14, 15]	Molecules (mass loading)	Resonance frequency change, readout via field emission	MWNT, DWNT
Doubly clamped resonators [14.16–18]	Molecules (mass loading), strain	Resonance frequency change; readout via conductance change	Individual SWNT
Optical sensors [14.19, 20]	Photons, light	Photocurrent	Individual SWNT

block available for engineering sensing devices. Furthermore, most of the knowledge gained in studying the rich variety of SWNT FET sensors is transferable to more complex devices involving MWNTs, bundles or networks.

In the next subsection, some of the most important SWNT properties and property modulation mechanisms useful for sensing are listed and exemplified. The operation and particularities of CNFETs are reviewed in Sect. 14.1.2, whereas in Sect. 14.1.3 the current status of sensor characterization is briefly discussed.

14.1.1 CNT Properties for Sensing

SWNTs have many interesting properties for nanodevices in general, and nanosensors in particular. The properties of SWNTs are discussed in many textbooks [14.22–24] and are also surveyed in Chap. 3 of this book. In the following, we condense the most relevant SWNT properties, supported by examples of how these have been employed in sensor devices.

SWNTs as Nanometer-Thin Semiconducting or Metallic Wires

Structurally, SWNTs are molecular cylinders with monoatomic-thick walls, resembling a honeycomb lattice of carbon atoms (graphene) rolled into a tube (Fig. 14.1). The structure of a SWNT is uniquely identified by its chiral indices (n, m) that give the so-called chiral vector $C_h = na_1 + ma_2$ determining the circumference of the nanotube upon rolling [$a_{1,2}$ are the two-dimensional (2-D) graphene lattice vectors as in Fig. 14.1a]. The (n, m) indices can be interchanged with (d_t, θ), where d_t is the nanotube diameter given by $d_t \approx (a/\pi)\sqrt{n^2 + m^2 + nm}$ [nm] and θ is the chiral angle defined by $\tan\theta = \sqrt{3}m/(2n+m)$ (where

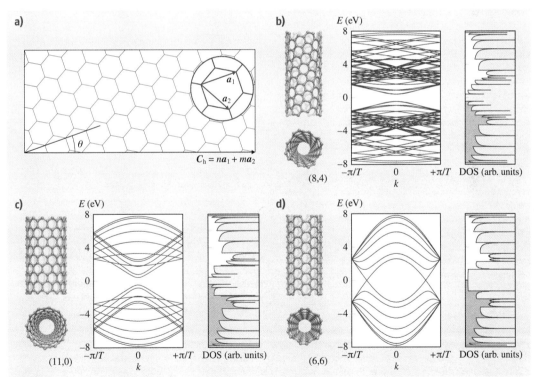

Fig. 14.1a–d Structural and electronic properties of SWNTs. (a) The *unfolded* unit cell of a chiral (8,4) CNT showing chiral vector C_h, chiral angle θ, and 2-D graphene lattice vectors $a_{1,2}$ (in *inset*). (b) The *folded* (8,4) structure, electronic band structure, and density of states (DOS). Bands are obtained via the zone-folding procedure applied to tight-binding dispersion relations [14.21]. (c,d) Band structure and DOS for zigzag (11,0) and armchair (6,6) nanotubes, respectively

$a = 0.249$ nm is the graphene lattice constant). SWNTs can be classified with the help of θ into zigzag ($\theta = 0$; $m = 0$), armchair ($\theta = \frac{\pi}{6}$; $n = m$) or chiral ($0 < \theta < \frac{\pi}{6}$; $n \neq m \neq 0$). In practice, SWNTs have diameters ranging from 0.4 to 3 nm, and lengths of around a few micrometers, although tubes almost a centimeter in length have been produced [14.25].

From the *electronic* point of view, depending on the chiral indices (n, m), SWNTs can be either semiconducting or metallic (hereon labeled s-SWNT or m-SWNT). A simple model for the electronic structure of SWNTs (zone folding of graphene tight-binding π bands) [14.21] predicts that those tubes for which $p \equiv (n - m) \mod 3 = 0$ are metallic (a third of all SWNTs), the rest ($p = \pm 1$) being semiconducting (two-thirds of all SWNTs). Structure, bands, and densities of states for three selected SWNTs (a chiral, a zigzag, and an armchair tube) are displayed in Fig. 14.1b–d. As for any one-dimensional (1-D) structure, the density of states of SWNTs is singular at energies corresponding to subband extrema (van Hove singularities). A rough estimation for the electronic band gap of s-SWNTs is $E_g \approx (2at_0)/(\sqrt{3}d_t)$ [eV] [14.21] (where $t_0 = 2.6$ eV is the so-called hopping tight-binding parameter for π orbitals). More accurate electronic structure calculations, taking into account the surface curvature of nanotubes, revealed that in fact only armchair ($n = m$) SWNTs are truly metallic, whereas other tubes with $p = 0$ actually have a small bandgap $E_g \approx 40/d_t^2$ [meV] [14.26]; these tubes are labeled small-gap semiconducting (SGS)-SWNTs.

The mentioned structural and electronic properties have resulted in a few SWNT sensor concepts. For example, FETs based on s-SWNTs have been utilized as charge detectors in flow meters [14.27] or (bio)chemical sensors [14.2, 7] (see Sect. 14.3.1 for more details). The high curvature ($d_t \approx 1$ nm) of carbon nanotubes (both tips and bodies) has been exploited, for example, in gas ionization sensors [14.28] and capacitance gas sensors [14.29].

SWNTs as Diamond-Stiff, Ultralight Strings

The basic *inertial* and mechanical properties of SWNTs are: linear mass density $\rho_L = 2.33 d_t$ [zg/nm] (1 zg = 10^{-21} g, and d_t is the tube diameter in nm, as given above), Young's modulus E in the range of ≈ 1.25 TPa [14.30], maximum tensile strain of 6%, and strength of ≈ 45 GPa [14.31, 32]. These properties promote carbon nanotubes as ideal nanosized beams for mechanical sensors. For example, consider a straight, doubly clamped SWNT. Assuming that the SWNT can

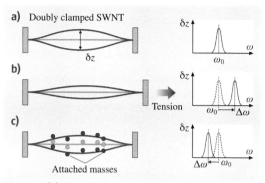

Fig. 14.2 (a) Sketch illustrating a doubly clamped SWNT beam and a resonance in the frequency response spectrum corresponding to the fundamental bending mode ω_0. (b) With applied tension the resonance frequency ω shifts upwards, whereas (c) with attached particles (mass loading) ω shifts downwards

be treated as a continuum elastic beam, the resonance frequency of the fundamental flexural mode is given by [14.33]

$$\omega = \frac{4\pi^2}{L^2} \sqrt{\frac{EI}{3\rho A} \left(1 + \frac{L^2 \sigma}{4\pi^2 EI}\right)}, \quad (14.1)$$

where L is the tube length, I is the moment of inertia, A is the cross-sectional area, ρ is the mass density ($\rho_L = \rho A$), and σ is the initial tension in the beam.

The dependence of ω on tension σ has been utilized by *Sazonova* et al. [14.16] and others [14.17, 18] to demonstrate *tunable* CNT resonators. These devices can thus be employed as sensitive strain/stress sensors. On the other hand, the attachment of a small mass to the tube increases the effective mass of the beam and as a result downshifts ω (via the ρ term in (14.1)). This principle, known as resonant inertial balance, has been utilized [14.14, 15, 17] to detect minute mass loading on a CNT. Sensing mechanisms for strain sensors and inertial balances are sketched in Fig. 14.2, and are discussed in more detail in Sect. 14.3.3.

Mechanically Tunable Electronic Properties of SWNTs

In analogy to crystalline semiconductors such as silicon, straining carbon nanotubes modifies their electronic bandgap. With an extended version of the model that was used to obtain the bands in Fig. 14.1, *Yang* et al. [14.34] calculated the modulation of bandgaps E_g of SWNTs of different chiralities under tensile (axial) and shear (torsional) strain (Fig. 14.3c,d). A linear approximation of the bandgap change ΔE_g with axial ε and torsional γ strain, was derived by the same group as [14.35]

$$\Delta E_g(\varepsilon, \gamma) \approx 3t_0 \operatorname{sgn}(2p+1)$$
$$\times [(1+\nu)\varepsilon \cos 3\theta + \gamma \sin 3\theta] \quad [\text{eV}],$$

with hopping parameter t_0, family p, and chirality θ introduced earlier, and $\nu = 0.2$ being the Poisson's ratio. The sign of ΔE_g depends on p, being positive for $p \in \{0, 1\}$ (bandgap increases with strain), and negative for $p = -1$ (bandgap decreases with strain). Also, the sensitivity with respect to strain depends on the chiral angle; as such, armchair tubes ($\theta = \frac{\pi}{6}$) are insensitive to axial strain but maximally sensitive to torsional strain. Zigzag tubes ($\theta = 0$) are just the opposite, whereas chiral tubes are sensitive to both types of strain.

For ordinary semiconductors such as silicon the conductance varies exponentially with bandgap, i.e., $G \propto \exp(-E_g/k_B T)$. This insight has led to many sensor concepts (strain, torsion, force, pressure, etc.) that involve SWNTs as strain gauges [14.8–12]. More details about piezoresistive properties of SWNT in CN-FET configurations will follow in Sect. 14.3.2.

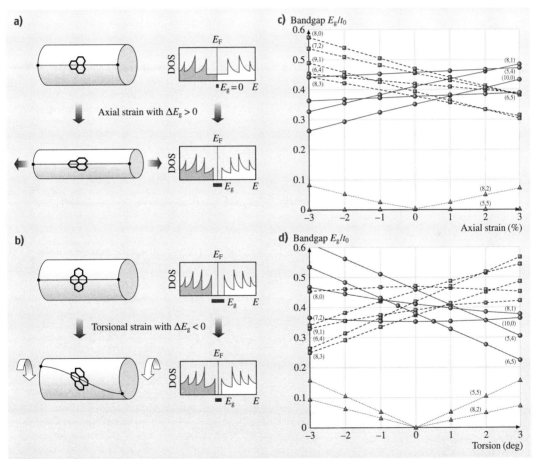

Fig. 14.3a–d Influence of strain on electronic properties of SWNTs. (**a,b**) Density of states modulation by strain. The example in (**a**) corresponds to tensile strain applied to a metallic zigzag CNT with $p = 0$ (bandgap opening), and the one in (**b**) to torsional strain applied to a chiral CNT with $p = 1$ (bandgap closing). (**c,d**) Calculated bandgaps as a function of tensile (**c**) and torsional (**d**) strain, for several chiralities belonging to each of the three families ($p = 0$ *dotted*, $p = 1$ *dashed*, and $p = -1$ *solid lines*). ((**c,d**) after [14.34])

SWNTs as Optically Active, Direct-Gap Materials

Inspecting Fig. 14.1 reveals that subband extrema come in pairs at the same lattice wavenumber k (electron–hole symmetry). SWNTs are therefore optically active materials. The absorption of photons generates electron–hole (e^-–h^+) pairs, and conversely, e^-–h^+ recombination occurs over the nanotube bandgap via photon emission. Resonant absorption of photons happens at energies corresponding to transitions between symmetric (with respect to E_F) van Hove singularities. These transition energies are labeled $E_{nn}^{s/m}$, where "s/m" denotes semiconducting or metallic and $n = 1, 2, \ldots$ is the index of the van Hove singularity in increasing energy order ($E_{n+1n+1}^{s/m} > E_{nn}^{s/m}$, with $E_{11}^s \equiv E_g$).

Excited e^-–h^+ pairs with energies $E_{nn}^{s/m} > E_{11}^{s/m}$ (i.e., $n > 1$) deexcite nonradiatively into lower-energy pairs whenever there is a continuum of states below $E_{nn}^{s/m}$, until an energy gap is reached. Therefore no photoluminescent emission is possible for m-SWNTs, whereas s-SWNT emit mainly over their bandgap. Transition energies $E_{nn}^{s/m}$ cover a wide spectrum ($\subset (-9, 9)\,\text{eV}$), thus SWNTs absorb light in the ultraviolet (UV) ($> 3.1\,\text{eV}$), visible, and infrared (IR) ($< 1.8\,\text{eV}$) domains, but they emit mostly in IR ($d_t > 0.5\,\text{nm}$ corresponds to $E_g < 1.5\,\text{eV}$). The basic optical properties of SWNTs are summarized in Fig. 14.4. In addition, because of their 1-D character, SWNTs only absorb and emit light linearly polarized along their axes. Optical properties of SWNTs have been exploited in devices such as polarized photodetectors [14.19, 20].

SWNT Sensitivity to Molecule Adsorption

SWNTs are hollow structures, made of surface only. Molecular adsorption is therefore expected to have a huge impact on CNT electronic properties. Theoretically, the interaction of molecules with carbon nanotubes is a complex topic that can only be treated ab initio within quantum mechanics. Generally speaking, molecules either physisorb or chemisorb at the CNT surface. Bond breaking is rare, as the sp^2 hybridized carbon network is very stable. However, local sp^3 hybridization is possible, and this adsorption mechanism is enhanced by curvature in small-diameter carbon nanotubes [14.36]. Defect sites in the nanotube wall greatly enhance adsorption as well [14.37].

At low molecular coverage, doping is the main effect, which is equivalent to a simple charge-neutrality level shift E_F with respect to the mid-gap (Fig. 14.5a). However, at large coverage densities, the bands of the tube itself can be distorted (by mixing with molecular orbitals). In addition to the charge-neutrality level shift, in this regime, new bands may appear inside the nanotube gap, or new gaps may open within the conduction or valence bands [14.38] (Fig. 14.5b). Even the electron affinity/work function of the nanotube can be modified in the process [14.36]. In m-SWNTs (as well

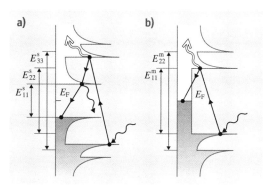

Fig. 14.4a,b Optical-related processes taking place in a SWNT after photon absorption. Wiggly *single lines* represent photons, *solid lines* represent electron transitions (excitation/deexcitation), and *wiggly white arrows* represent nonradiative processes (e.g., phonons). (**a**) A semiconducting tube can reemit photons over the bandgap, whereas in (**b**) metallic tubes, deexcitation is mostly nonradiative

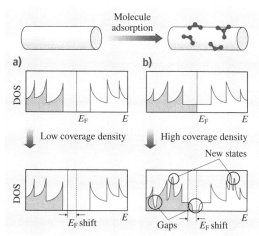

Fig. 14.5a,b The effect of molecular adsorbates on SWNT electronic properties. (**a**) An s-SWNT in the low-coverage limit suffers only a shift in the Fermi level E_F. (**b**) In the high-coverage limit, new gaps and states, E_F shifts, and even work function changes are possible

as in the valence or conduction band of s-SWNTs) adsorbing molecules may create a disorder potential which results in bandgap opening at the Fermi level.

Chemical and biochemical sensors based on the modulation of SWNT electronic properties upon exposure to target molecules have been demonstrated [14.2, 7]. The most widely used method for reading out these modulations is via transport measurements (CNFETs). However, capacitive [14.29] and optical [14.39] detection have also been employed. Chemical sensors will be treated in Sect. 14.3.1.

14.1.2 Carbon Nanotube FET Structures

The preceding section focused on the properties of SWNTs and property modulation subject to mechanical, chemical or optical stimuli. However, these properties are intrinsic to isolated SWNTs, and in order to become useful they need to be probed from the macroscopic world. A simple way to probe CNT properties is to contact the nanotube electrically with metallic leads. In combination with a nearby gate electrode this configuration is known as the carbon nanotube field-effect transistor (CNFET). A basic CNFET is sketched in Fig. 14.6, consisting of an individual SWNT (the channel) placed on an insulator (typically SiO$_2$), contacted by source and drain electrodes, and gated by the conducting substrate (typically highly doped Si). The fabrication of such a structure (and of more advanced CNFETs) is covered in Sect. 14.2.2. Before giving examples of sensor demonstrators in Sect. 14.3 we review the basic operation and particularities of CNFETs. Understanding CNFET operation is important in identifying to what extent contacting SWNTs by metal leads affects their properties.

The right frameset to study transport in carbon nanotubes is mesoscopic physics, since nanotubes are crystalline, low-defect-density conductors for which quantum-mechanical effects are expected to play a role. In a general conductor with length below the elastic mean free path (ℓ_e) and phase relaxation length (L_φ), at low temperature and low bias (quasiequilibrium), each mode (Bloch wave) carries a current of $2e/h$. Therefore, the total source and drain current I_{sd}, can be obtained by summing the average number of modes M, going from source to drain, with energies in the range (eV_s, eV_d) centered at the Fermi level E_F. This yields $I_{sd} = (2e^2/h)MV_{sd}$ [14.40], where e and h are the usual electron charge and Planck's constant, respectively, and $V_{sd} = V_s - V_d$ is the applied source–drain (low bias). In other words, the sample conductance $G = I_{sd}/V_{sd}$ is simply MG_0, where the quantum conductance $G_0 = (2e^2/h) = (1/12.9)\,\text{k}\Omega^{-1}$ is introduced.

For a m-SWNT, at any energy close to E_F (to within ± 1 eV), there are precisely two subbands (Fig. 14.1d), meaning that $M = 2$ and the predicted conductance of a metallic tube is $G = 2G_0$ ($= 1/6.5\,\text{k}\Omega^{-1}$). Conductance values approaching $2G_0$ have indeed been measured experimentally [14.41, 42] for m-SWNTs. However, in most situations, G is well below $2G_0$. One reason for this is that the metallic leads used to contact the tube differ in geometry, lattice, and electronic properties (most notably work function and effective mass) from SWNTs. The heterojunction between a metal and the SWNT incorporates interface barriers that introduce scattering and reduce the electron transfer rates. Another reason for reduced conductance $G < 2G_0$ is disorder in the nanotube, caused by structural defects, adsorbed molecules or simply the random potential profile of the underlying oxide (trapped charges, dangling bonds, etc.). Accordingly, in order to include interface and disorder, the conductance is reduced by a multiplicative factor $\mathcal{T}$ ($G \to \mathcal{T} 2G_0$), describing the average transmission probability for an electron to propagate from source to drain. At room temperature, for long enough nanotubes, phonon scattering leads to incoherent, classically diffusive transport, respecting Ohm's law ($\mathcal{T} \propto \frac{1}{L}$).

As mentioned previously, for m-SWNTs the transmission $\mathcal{T}$ can approach 1, which corresponds to thin contact barriers through which electrons can easily tunnel, and clean SWNT surfaces. For s-SWNTs, the situation is different. The physics of heterostructures predicts that, at the interface between a metal and a semiconductor, Schottky barriers decaying slowly within the semiconductor may arise. Figure 14.7a depicts schematically the band diagram at the interface between a metal of work function ϕ_m, and an intrinsic

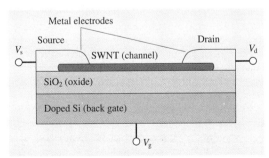

Fig. 14.6 Schematic representation of a simple SWNT field-effect transistor in cross-sectional view

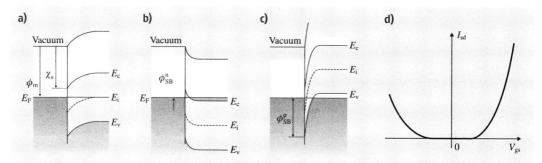

Fig. 14.7a–d Band-structure diagram at the interface between a metal and an s-SWNT. (a) No gate bias applied $V_{gs} = 0$, (b) $V_{gs} > 0$ just above the threshold for electron conduction, and (c) $V_{gs} < 0$ just below the threshold for hole conduction. (d) The $I_{sd}(V_{gs})$ characteristic of the transistor, emphasizing the asymmetric electron versus hole conductivity, due to $\phi_{SB}^p > \phi_{SB}^n$

s-SWNT of electron affinity χ_s ($\chi_s < \phi_m < \chi_s + E_g/2$). There are two such barriers, one at the source and the other at the drain. For simplicity, assume low V_{sd} bias (quasithermal equilibrium), which allows a flat Fermi level to be defined throughout the structure. Assume as well that the conductance of the device can be written as $1/G = 1/G_s + 1/G_t + 1/G_d$, where $G_{s,t,d}$ is the conductance of the source barrier, nanotube channel, and drain barrier, respectively. The latter assumption ignores interference effects and holds only if the length L of the nanotube segment exceeds L_φ [14.40]. Although still debated, this seems to be the case for SWNTs with $L \geq 1\,\mu m$ at room temperature [14.43]. Within this picture, $G_d = G_s$ and we can focus on just one interface.

At finite temperature ($T \neq 0$), the charge flow across any metal–semiconductor junction, characterized by some interface barrier ϕ_B, has two components: tunneling through ϕ_B and thermionic emission over ϕ_B. At $T = 0$ only direct tunneling at E_F is possible. There are two such barriers, one for electrons ϕ_B^n and one for holes ϕ_B^p. In Fig. 14.7a, corresponding to $V_{gs} = 0$ and an intrinsic s-SWNT, $\phi_B^n = E_g/2$, $\phi_B^p = E_g + \chi_s - \phi_m$, and only thermionic emission is possible. In this situation I_{sd} is practically 0. A positive voltage $V_{gs} > 0$ applied to the gate (Fig. 14.7b) downshifts the bands in the bulk of the nanotube, but not the position of the band edges at the interface with respect to E_F. Thermionic emission increases for electrons and decreases for holes. At some point, when the nanotube conduction band E_C becomes aligned to E_F, $\phi_B^n \equiv \phi_{SB}^n = \phi_m - \chi_s$ and $\phi_B^p = E_g$. The barrier ϕ_B^n is also a lot thinner than at $V_{gs} = 0$ because the electron density inside the nanotube is higher, more effectively screening any interface potential. Tunneling of electrons increases and will eventually exceed thermionic emission (see later). Above this threshold, increasing V_{gs} further thins down the barrier, increasing I_{sd}. The barrier width depends, besides on the position of E_F, on the geometry of the leads and oxide thickness [14.44]. Similarly, for negative, large enough gate voltage, the bulk valence band E_V becomes aligned to E_F (Fig. 14.7c). In this case $\phi_B^p \equiv \phi_{SB}^p = E_g + \chi_s - \phi_m$, $\phi_B^n = E_g$, and hole tunneling dominates. The inequality between the electron and hole barriers (in this example $\phi_{SB}^p > \phi_{SB}^n$) translates into asymmetric electron versus hole conductivity [the two branches of the $I_{sd}(V_{gs})$ characteristic], sketched qualitatively in Fig. 14.7d.

As opposed to normal, planar junctions (e.g., metal–silicon), Fermi-level pinning is proposed not to exist in metal–CNT junctions, because of the particular tubular geometry of the CNTs [14.45]. As a result, different metals lead to barriers of different heights, which should also depend on the CNT diameter d_t (d_t determines the bandgap E_g as discussed in Sect. 14.1.1). This fact, evidenced experimentally [14.46], is a strong confirmation of the Schottky barrier (SB) model of CNFETs. Consequently, it seems possible to eliminate the Schottky barrier completely by either choosing metals with large work function such as Au, Pt or Pd ($\phi_m \approx 5.1\,eV$) to cancel ϕ_{SB}^p, or metals with low work function such as Al ($\phi_m \approx 4.2\,eV$) to cancel ϕ_{SB}^n, knowing that the midgap work function of SWNTs is around 4.5 eV. Based on this principle *Javey* et al. [14.47] have demonstrated G within 50% of the ideal $2G_0$. More experimental evidence in support of the SB-CNFET model has been supplied by *Appenzeller* et al. [14.48, 49]. By measuring the temperature dependence of the $I_{sd}(V_{gs})$ characteristic [14.49], they concluded that tunneling is the main injection mechanism in SB-CNFETs and not thermionic emission. However, the work function alone is not enough, as for example Pt ($\phi_m \approx 5.6\,\text{eV}$) yields poor contacts, likely due to poor wetting or native-oxide tunnel barriers [14.50].

For certain CNFET sensor devices, the Schottky barrier FET model is essential in understanding the device operation and sensing mechanism. This is the case of (bio)chemical sensors to be discussed in Sect. 14.3.1. In other devices, Schottky barriers, if present at all, contribute only to the contact resistance, which to a first approximation is constant during the sensing process, and can be thus factored out from the sensing mechanism. This is the case with piezoresistive gauges and resonators as presented in Sects. 14.3.2 and 14.3.3, respectively. For these devices, the transport picture introduced in this section, in terms of thermionic emission and tunneling, is still useful for explaining their operation. The only difference is that potential barriers for electrons and holes will be found on the nanotube and not only at the interface (Fig. 14.14c,d). Further insight into the physics of CNFETs in different regimes (large-bias, low-temperature phenomena such as Luttinger liquid, Coulomb blockade or orbital Kondo effects, etc.) have also been described [14.51–53].

14.1.3 Sensor Characterization

Sensor characterization refers to assessing the performance metrics for a certain sensing technology. Some of the most important performance metrics are sensitivity, signal-to-noise ratio (SNR), limit-of-detection (LoD), cross-sensitivity/selectivity, signal rise/fall time (speed), repeatability, offset/sensitivity, drift, hysteresis, and lifetime/robustness. So far, mainly *sensitivity* and *dynamic properties*, e.g., *signal rise/fall time*, have been targeted by the CNT sensor community (Sect. 14.3). This is a natural situation, considering that the technology for fabricating individual SWNT FETs is not yet scalable (Sect. 14.2). Efforts have then been focused on demonstrating that CNT nanosensors are superior to existing devices, to motivate and justify future investment. On the other hand, the lack of batch fabrication processes has hindered gathering statistical information about the characteristics of CNT sensors in operation; most publications on individual SWNT sensors (in particular FET devices) refer to data obtained from just a few fabricated samples. Therefore the *reproducibility* of sensor characteristics has been ignored, or marginally addressed.

To better illustrate matters, consider the story of chemical CNFET sensors. In 2000 *Kong* et al. [14.2] first reported detection of NO_2 and NH_3 with individual SWNT FETs. In 2008 *Cho* et al. (in collaboration with *Kong*) published the first results for a hybrid CNT chemical sensor with a CMOS interface chip [14.54]. The fabrication of CNT sensor arrays is still far from scalable, yet by this approach the interface captures signals from up to 24 CNFETs in parallel, further multiplexed and sent to analysis. The authors manage to measure 414 devices, which results in a histogram of the distribution of resistance, spread over six orders of magnitude. This large dispersion is the result of varying number and type of CNTs in the devices, which is not controlled during fabrication. Dispersion is also visible in the recorded sensor response on the measurand. This kind of platform is a first step towards capturing device performance statistics, a prerequisite for process control and optimization.

Acquiring sensor signal-to-noise ratio is currently in progress. Electronic noise in CNFETs has received a lot of attention lately. *Collins* et al. [14.55] were the first to measure $1/f$ noise in SWNTs, both individual tubes and mats. The noise spectrum was found to agree with the classical noise power formula $S_V(f) = (A/f^\beta)V^2$, where A is the noise amplitude, f is the frequency, $\beta \approx 1$, and V is the bias voltage. Furthermore, the noise amplitude seems to obey the empirical Hooge law $A = \alpha_H/N$, where α_H is the Hooge constant and N is the number of carriers in the channel, because on average A decreases with the number of tubes in the sample [14.55]. Interestingly, the $S_V(f)$ formula is expected to hold for classically diffusive transport, which is questionable for CNFETs. However, *Appenzeller* et al. [14.56] have calculated the average number of carriers inside SWNT ballistic FETs, for different contact metals, and found the Hooge law to be valid even in this regime, with a fitted α_H value of $\approx 2 \times 10^{-3}$ (the same as for bulk silicon). Upon closer examination, it has been observed by *Tersoff* [14.57] that the Hooge law is not accurate in the subthreshold region of a CNFET. He proposed adding a phenomenological term to the noise power, proportional to $(dI_{sd}/dV_{gs})^2$, to account for gating of the CNFET by fluctuating charges in the vicinity of the tube. This model has recently been confirmed experimentally [14.58]. Overall, regardless of the model utilized, the noise in CNFETs is found to be significant, which limits the SNR of CNT sensors. The fact that noise is proposed to be mainly extrinsic [14.56], i.e., caused by external fluctuations, is however encouraging since it sets clear technical objectives in achieving better control of the CNT environment.

14.2 Fabrication of SWNT Sensors

Fabrication of CNT devices involves many different aspects that can be grouped into two main tasks, namely synthesis of SWNTs and assembly of nanotubes into devices. *Synthesis* is concerned with the production of SWNTs with controlled properties such as diameter, chirality, length, and defect densities. Sometimes synthesis is followed by postsynthesis methods for CNT purification, sorting, and most importantly functionalization. On the other hand, *assembly* refers to techniques and methods for placing SWNTs at predefined locations on a substrate with controlled number of nanotubes, orientation, and slack (straightness). Post assembly, other processing steps such as nanotube electrical contacting or device encapsulation may follow. The development of complete processes for the fabrication of micro- and nanosystems that integrate nanotube devices/sensors with acceptable yield is currently one of the key topics in CNT research. This section attempts to survey some of the available methods and processes aimed at controlled *diameter*, *chirality*, *location*, and *orientation*.

As discussed in Sect. 14.1, most of the intrinsic properties and property modulations of SWNTs depend on their *diameter* d_t. For example, the bandgap E_g of an s-SWNT is inversely proportional to d_t, as are the Schottky barriers $\phi_{SB}^{n/p}$ in CNFETs (Sect. 14.1.2). Other properties, most notably piezoresistance, depend as well on *chirality* θ. Since d_t and θ of a SWNT are defined during synthesis (production), methods for nanotube production are briefly surveyed in Sect. 14.2.1. *Location* and *orientation* control during assembly are important for building complex structures, for developing batch fabrication processes, and for large-scale device integration (sensor arrays). Section 14.2.2 will thus review processes for the assembly of SWNTs into devices, together with some general strategies for achieving location and orientation control.

14.2.1 Methods for SWNT Production

Carbon nanotube production methods can be classified into two categories, namely high temperature (*arc discharge* and *laser ablation*; $T = 1200–3000\,°C$) and medium temperature (*catalytic chemical vapor deposition*, CCVD; $T = 400–1100\,°C$). Details about each method can be obtained from textbooks [14.22, 24] and also Chap. 3 of this book. Here, only those aspects which are relevant for CNT devices and integration are discussed.

Because of higher production temperature (potential defect annealing), arc discharge and laser ablation are believed to produce better crystalline SWNTs. However, with these methods it is not possible to produce SWNTs directly on the target substrate for device fabrication; only postsynthesis assembly is possible. Typically as-produced nanotubes are first dispersed in a liquid that can be subsequently utilized to deposit SWNTs on the target substrate (Sect. 14.2.2). Also, the distribution of diameters is difficult to control with these methods during SWNT production. Nevertheless, after dispersing nanotubes in liquid, different techniques can be employed to separate (semiconducting from metallic) and sort (by diameter) CNTs, as explained below.

In catalytic chemical vapor deposition, SWNTs are synthesized from metallic catalyst particles from a carbon-containing gas feedstock. Since CCVD is a catalytic process, patterning of a catalyst-containing layer can be utilized to grow nanotubes at selected locations directly on preprocessed silicon chips (see Sect. 14.2.2 for more details). The key observation for CCVD is that *the size of the* catalyst particle *correlates with the SWNT diameter* [14.59], and the *catalyst particle density determines* the final *CNT density* [14.60]. In [14.59] *Cheung* et al. prepared Fe nanoparticles, with narrow diameter distributions centered around 3, 9, and 13 nm, on SiO_2. The grown CNT diameter distribution mirrors the initial particle size distribution, with a standard deviation of roughly 30% of the average. However, 3 nm particles produce SWNTs with double-walled CNTs (DWNT), 9 nm particles produce D/MWNT with just a few SWNTs, and 13 nm particles produce only MWNTs, showing that a particle size < 3 nm is required for CCVD synthesis of SWNTs. For SWNT-based FETs, a good compromise between low contact resistance (low $\phi_{SB}^{n/p}$) and large gap E_g (large on/off ratio) would correspond to diameters in the range 1.5–2 nm [14.46]. Obtaining such small catalyst nanoparticles, while keeping a narrow size distribution, is challenging. *Li* et al. [14.61] have demonstrated SWNT diameter distributions of either (1.5 ± 0.4) nm or (3.0 ± 0.9) nm (again standard deviation/average $\approx 30\%$). Their method involved loading the hollow cavity of the iron-storage protein ferritin (internal cavity diameter 8 nm) with Fe, followed by deposition on a substrate and calcination of the protein shell. Particle size was controlled by the iron loading time.

Currently, much activity is going into trying to achieve SWNTs with acceptable diameter distributions. For CCVD-grown tubes, the most promising approach seems to remain improving particle size distribution. Post synthesis, for SWNT dispersed in solution, regardless of their origin (arc discharge, laser ablation or CCVD), several methods have been proposed to narrow down the diameter distribution. For example, *Tromp* et al. [14.62] have recently demonstrated diameter separation of SWNTs by noncovalent functionalization with anchor molecules optimally attaching to specific tube diameters (1.2 nm). This functionalization improves solubility of the selected tubes, which are subsequently separated from the insoluble bulk via sonication and centrifugation. By using DNA wrapping and density gradient ultracentrifugation (DGU), *Arnold* et al. [14.63] have shown good separation by diameter, for SWNTs with diameters below 1 nm. Soon afterwards, the same group [14.64] extended the technique to surfactants, and managed to isolate narrow distributions of SWNTs with > 97% within a 0.02 nm-diameter range, with, e.g., 84% of the tubes being of (6,5) chiral indices. Furthermore, by using mixtures of surfactants, they were able to produce predominantly semiconducting or metallic SWNTs. Combined with the recently proposed *continued growth* [14.65] the DGU method might one day achieve macroquantities of essentially single-chirality SWNTs. However, to date, achieving SWNTs with a single chirality remains the key challenge in CNT production.

14.2.2 Strategies for SWNT Assembly into Devices

Perhaps the simplest way to survey SWNT assembly strategies is to split the discussion into two parts by considering the nanotube source: *liquid suspension* (laser, arc or CCVD SWNTs dispersed in liquid) or in situ growth (CCVD SWNTs grown directly on the target substrate). In-depth reviews of the various assembly strategies can be found elsewhere [14.66, 67]. Here, we focus only on those techniques resulting directly in CNFET device structures or that have some relevance to this topic. Regarding liquid suspension or in situ growth, there is an ongoing debate as to which one produces best results. Table 14.2 compares the two in several aspects, showing a good balance of advantages and disadvantages. In general, fluid-dispersed tubes offer better SWNTs (because of laser or arc production and separation techniques), but assembly is more complex than for in situ growth.

Liquid Suspension Assembly Methods

Historically liquid suspension assembly methods were the first to appear. The very first CNFETs, demonstrated in 1998 [14.68, 69], were built through a similar, simple process. In [14.69] laser-ablation SWNTs were dispersed by sonication in dichloroethane and then spread over Au electrodes predefined using electron-beam lithography (EBL) on a doped Si wafer substrate covered by a thick gate oxide film (SiO_2). The Si substrate served as a back gate to all devices. A schematic cross-

Table 14.2 A comparison between liquid suspension and in situ growth assembly methods

SWNT source	Liquid suspension[a]	In situ growth[b]
CNT quality[c]	+	+/−[d]
Average diameter control[e]	−[f]	+[g]
Diameter standard deviation[h]	+	−
Chirality control[i]	−	−
FET performance[j]	−[k]	+
Alignment precision	−/+[l]	+
Ease of integration	+/−[m]	+[n]

[a] CNTs as produced by arc discharge or laser ablation
[b] In situ growth mainly by CCVD (Sect. 14.2.1)
[c] SWNT quality is attributed to crystallinity and defect density: Arc discharge or laser ablation SWNTs are still considered to produce better CNTs (Sect. 14.2.1)
[d] Progress in CCVD leads to constant increase in CNT quality. Measures include post annealing, H_2 treatment
[e] Average SWNT diameters determine Schottky barrier heights in CNFETs (Sect. 14.1.2)
[f] The average diameter of arc discharge and laser ablation SWNTs is between 1.2 and 1.5 nm and is difficult to control
[g] The average diameter of SWNTs correlates with catalyst particle size and growth conditions and can be tuned (Sect. 14.2.1)
[h] A small diameter standard deviation should reduce performance variations in CNFET sensor devices
[i] Functionalization of SWNTs, sonication and centrifugation post synthesis narrows diameter distribution. Individual chiralities may be selected (Sect. 14.2.1). Only applicable to SWNTs assembled from liquid suspension
[j] FET performance for sensing applications refers to low contact resistance, low noise, and high on/off ratio
[k] Liquid suspension involves CNT surface treatment by (strongly binding) surfactants, sonication, and centrifugation. These steps may degrade the CNFET performance
[l] Localized surface functionalization and dielectrophoresis increases the alignment precision. However, bundling of individual CNTs is still an issue in these processes
[m] Wet processing is more compatible with CMOS integrated circuit (IC) substrates, but wet processing may influence the SWNTs' electronic properties
[n] In situ growth is compatible with MEMS substrates including suspended MEMS and nanoelectromechanical systems (NEMS) structures

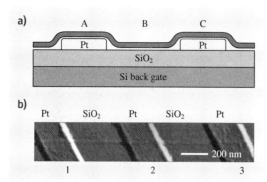

Fig. 14.8a,b First CNFET: (**a**) schematic view, and (**b**) AFM image (after [14.68])

section of a CNFET stack, together with an atomic force microscopy (AFM) image of a fabricated device, is shown in Fig. 14.8 [14.68]. The same process, extended with a patterned poly(methyl methacrylate) (PMMA) top layer, has been used to demonstrate one of the first CNT logic inverters [14.70]. The only controlled parameter in this process is tube density, which can be adjusted – taking into account the electrode design – to obtain devices with only one nanotube bridging the correct electrodes, albeit at very low yield.

It was soon realized that surface functionalization can significantly help in guiding SWNT deposition. In [14.71] *Liu* et al. observed that SWNTs preferentially adsorb on amino-functionalized surfaces. Subsequently, they developed a process to define amino-functionalized regions, where tubes are adsorbed, on a tube-repelling surface: a trimethylsilyl (TMS) self-assembled monolayer (SAM). The process, as summarized in Fig. 14.9a, was utilized to place a SWNT between two contacts, as shown by the AFM image in Fig. 14.9b. Since then, other groups have utilized the same principle but with different nanotube surface chemistries, for example, the hydroxamic acid group (Al_2O_3) [14.72], patterned aminosilane monolayer [14.3], and polar–nonpolar interfaces [14.73].

Another important process family for SWNT assembly from solution is based on electric fields. *Krupke* et al. [14.74] used alternating-current (AC) dielectrophoresis to place SWNT bundles from suspension between predefined electrodes. Electrode arrays were defined using EBL, and connected to a frequency generator. On top of the electrodes a droplet of the CNT suspension was applied and evaporated. In general $\approx 70\%$ of the electrodes were bridged by at least one bundle, and even individual SWNTs bridged occasionally. The same group [14.75] proposed that, in D_2O (heavy water) with dielectric constant $\varepsilon_{D_2O} = 80$, s-SWNTs ($\varepsilon_s < 5$) will experience negative dielectrophoresis (repulsion from electrodes), whereas m-SWNTs ($\varepsilon_m \to \infty$) will experience positive dielectrophoresis (attraction to electrodes). They showed that under these conditions m-SWNT are indeed deposited on electrodes, while s-SWNTs remain in solution, thus

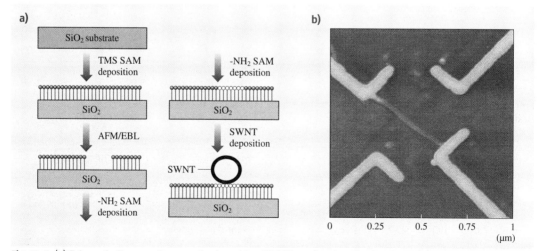

Fig. 14.9 (**a**) Schematic diagram of process for depositing SWNTs on functionalized surfaces. (**b**) AFM image of a device made on Au electrodes (after [14.71])

providing a basic method to separate metallic from semiconducting tubes.

In Situ Growth Assembly Methods

The second major class of assembly methods is based on in situ growth (via CCVD) of SWNTs directly onto silicon chips/wafers. CCVD has one major advantage over liquid suspension deposition regarding location control, namely that the catalyst can be patterned [14.77]. A first process for fabricating CNFETs using CCVD has been proposed by *Soh* et al. [14.78]. EBL was first used to pattern wells, into which catalyst was deposited. A resist lift-off left isolated catalyst islands, from which SWNTs were grown by CCVD. Metal electrodes are defined on top of the catalyst islands by EBL, contacting some of the grown tubes. The location of tubes is thus approximately controlled, but not the orientation nor the number of nanotubes between two electrodes, although by changing the catalyst particle density and tuning the growth parameters the yield can be improved. Nevertheless, considerable progress has been made lately in catalyst deposition and patterning. *Javey* and *Dai* [14.79] have developed a method that allows positioning of *individual* catalyst particles with EBL resolution. Furthermore tube-to-particle number ratio approaching 1/1, i.e., close to 100% SWNT growth yield, was achieved.

Since nanotube location control can be achieved by catalyst patterning, a lot of work has focused subsequently on orienting CCVD growth. Field-directed growth has been pioneered by *Zhang* et al. [14.76], exploiting the large anisotropic polarizability of SWNTs. Elevated polysilicon structures were first defined onto a quartz substrate by optical lithography, with a catalyst layer on top, transferred by contact printing. Electric fields were applied during growth via outer poly-Si pads contacted by metal leads (electrodes). Figure 14.10 shows both the process flow and growth results, with and without applied fields. Field-directed growth was subsequently demonstrated on flat surfaces as well [14.80], and catalyst patterning and field-directed growth have been combined to build CN-FETs by *Dittmer* et al. [14.81]. Another approach for orienting CNT growth was proposed in [14.82], where a strong correlation between nanotube orientation and feedstock *gas flow direction* was observed. Finally, numerous recent studies show that *surface-directed growth* is also possible on substrates such as A- and R-planes of sapphire [14.83] or miscut C-plane sapphire [14.84]. The oriented SWNTs on sapphire have been contacted into FET configurations by *Liu* et al. [14.85].

As discussed in Sect. 14.1.1, SWNTs have interesting electromechanical properties, which often require free-standing (suspended) nanotube segments in order to manifest. Assembly of SWNTs into suspended microstructures tends to be easier with in situ growth than *liquid suspension*, because of complications arising from capillary forces. Directed growth of SWNTs from predefined silicon towers was first shown by *Cassell* et al. [14.86, 87], via a process that preceded the field-directed growth shown in Fig. 14.10. Basic electromechanical structures were presented later by the same group, with a process involving patterned CCVD growth directly from the surface of Mo, a refractory metal capable of withstanding CCVD temperatures ($\approx 900\,^\circ\text{C}$) [14.88] and inhibiting catalyst particle diffusion. Integration of CCVD nanotube growth into

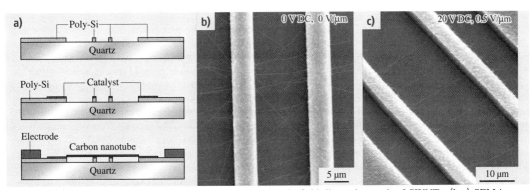

Fig. 14.10 (a) Schematic diagram of the process flow for electric-field-directed growth of SWNTs. (b,c) SEM images of SWNTs, grown in various fields. At zero field, tubes grow randomly (b), whereas alignment is seen in 0.5 V/µm (c) (after [14.76])

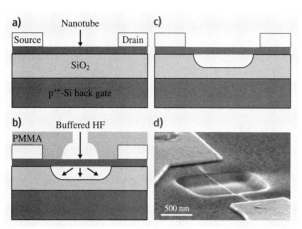

Fig. 14.11 (a–c) Schematic overview of an EBL fabrication process for SWNT resonators. (**d**) SEM image of a device (after [14.18])

We end this survey of CNT assembly methods with one of the oldest processes for building SWNT devices, yet still the *Swiss army knife* for rapid demonstration (proof of concepts) and basic investigations. The method proceeds from SWNTs already present in low density on the substrate, whether from liquid suspension or grown in situ. Location of nanotubes is recorded by AFM imaging, and electrodes are defined on nanotubes by e-beam lithography, metal evaporation, and lift-off. Since the location and orientation of the tube is known, some basic yet precise micromachining is also possible in this approach. For example, *Witkamp* et al. [14.18] used a PMMA layer to define an etch mask over a top-contacted CNFET. The SiO_2 underneath the SWNT is etched away in HF, releasing the central part of the tube (Fig. 14.11). The resulting suspended nanotube device is an electromechanical resonator, as discussed in Sect. 14.3. The process is time consuming (because of AFM imaging and EBL), area inefficient (because of nonaligned sparse tubes), and therefore nonscalable. However, currently it is the only process that can guarantee individual SWNT devices with optimum alignment precision with respect to electrodes and other postdefined structures.

a state-of-the-art fabrication process for microelectromechanical systems (MEMS) has been demonstrated recently by *Jungen* et al. [14.89]. Both location and orientation control were achieved by confining the growth of the tubes geometrically between sharp poly-Si tips.

14.3 Example State-of-the-Art Applications

In Sect. 14.1 the intrinsic properties of SWNTs and the operation principle of CNFETs were introduced. Fabrication of CNFET devices, including a wide range of different SWNT production methods and various assembly strategies, was covered in the preceding Sect. 14.2. Here we finally show, by way of examples, how various properties of SWNTs assembled in CNFET configurations can be turned into sensor functions. The main focus of this section is therefore placed on explaining the concept and operation, and on discussing the sensing mechanisms, for a few selected device demonstrators, rather than giving a thorough state-of-the-art review. References to review articles are specified when available. An attempt will be made to discuss the operation of the selected SWNT sensor examples through the perspective of the CNFET model laid out in Sect. 14.1.2. In support, whenever possible, band diagrams (Schottky barriers, nanotube body bands) and their modulation by stimuli and/or bias conditions will be provided. The following subsections will cover *chemical and biochemical sensors* (Sect. 14.3.1) and *physical sensors* (strain, pressure, force, and mass), either piezoresistive (Sect. 14.3.2) or resonating (Sect. 14.3.3).

14.3.1 Chemical and Biochemical Sensors

Shortly after the first CNFETs were demonstrated, it was noticed that the electronic properties of these devices are very sensitive to environmental conditions. Specifically, *Collins* et al. [14.4] have shown that the resistance of SWNTs changes reversibly by up to 15% under cycling the environment from air to vacuum, a phenomenon attributed to O_2 doping. Since then, the number and diversity of proposed (bio)chemical SWNT sensors have grown tremendously; perhaps, this class of sensors is the most researched today. Excellent reviews on chemical and biochemical sensors based on carbon nanotubes are available [14.90–92].

The first gas sensor based on an individual SWNT FET was demonstrated by *Kong* et al. [14.2]. The device is a back-gated FET, fabricated with a process as

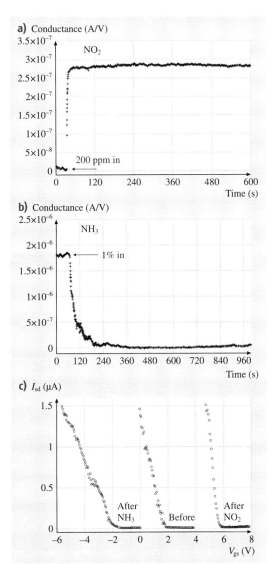

Fig. 14.12a–c SWNT FET chemical sensor measurements. (**a,b**) Time response of the same device upon exposure to 200 ppm NO_2 (**a**) and after recovery to 1% NH_3 (**b**). (**c**) $I_{sd}(V_{gs})$ characteristics, before and after exposure to gas, showing a gate threshold shift (after [14.2])

described in [14.78]. Such CNFET devices were exposed to NO_2 and NH_3 diluted in Ar or air, resulting in a significant change in their conductance. Figure 14.12a shows the response of a sensor to 200 ppm of NO_2. The conductance increased over three orders of magnitude in the time range of 2–10 s. The same device was exposed, after recovery, to 1% NH_3 (Fig. 14.12b), resulting in a 100-fold conductance decrease within 1–2 min. Upon measuring the $I_{sd}(V_{gs})$ characteristic (Fig. 14.12c) the authors conclude that NH_3 depletes the initially p-type SWNT of holes, shifting E_F away from the valence band, whereas NO_2 does the contrary, enhancing the hole concentration and pushing E_F closer to the valence band.

To date, the exact mechanism responsible for the change in the $I_{sd}(V_{gs})$ characteristic, attributed to doping by *Kong* et al. in their paper [14.2], remains a controversial issue. *Heinze* et al. [14.44] have analyzed the adsorption of molecules either on the nanotube body or onto the metal contacts. Transport in the two situations is expected to have different signatures. In the former case, molecules dope the nanotube, shifting its charge-neutrality level E_F. If this process does not change the electron affinity of the SWNT χ_s (valid at least at low doping concentrations), then the electron and hole Schottky barriers $\phi_{SB}^{n/p}$ remain unchanged; just the bulk bands of the CNT shift up or down depending on the doping polarity. The result is simply a *rigid shift* of the $I_{sd}(V_{gs})$ characteristic, as sketched in Fig. 14.13b. On the other hand, molecules attaching mainly to the leads modify the metal work function ϕ_m. This modifies the relative height of the two Schottky barriers $\phi_{SB}^{n/p}$, which translates into changed polarity of the device, as illustrated by the $I_{sd}(V_{gs})$ characteristic in Fig. 14.13a. These signatures have been observed experimentally [14.44] and assigned to one of the two mechanisms with the help of device simulations. Exposing a CNFET to potassium showed the *tube doping* signature, whereas oxygen showed the *contact adsorption* signature (Fig. 14.13c,d).

In attempting to identify the sensing mechanisms, other groups have devised means to expose just parts of the CNFET via selective passivation [14.93–95], with so far inconclusive results. The sensing mechanism gets further convoluted by the presence of the gate oxide. *Helbling* et al. [14.96] showed that etching away the underlying SiO_2 from a CNFET results in chemically sensitive devices. On the other hand, *Auvray* et al. [14.3] observed that SWNTs on aminosilane are 4–5 times more sensitive to triethylamine than are SWNTs on SiO_2.

For CNFETs to work as biochemical sensors, some modifications are required. First, because of operation in aqueous solutions that contain ions, the back gate is changed to a reference electrode known as the liquid

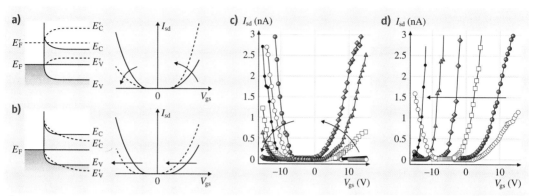

Fig. 14.13a–d Mechanisms encountered in CNT chemical sensors. (**a,b**) Band diagrams and corresponding $I_{sd}(V_{gs})$ characteristics, before (*solid line*) and after (*dashed line*) adsorption of molecules, for the cases of adsorption at the contacts (**a**) and nanotube doping (**b**). In (**a**) the metal work function ϕ_m is modified, whereas in (**b**) the charge-neutrality level E_F changes in the CNT. (**c,d**) Measured $I_{sd}(V_{gs})$ characteristics subject to oxygen exposure (**c**), assigned to mechanism (**a**), and potassium doping (**d**), assigned to mechanism (**b**). *Arrows* in (**a–d**) indicate the change trend in $I_{sd}(V_{gs})$ characteristics. ((**c,d**) after [14.44])

gate [14.97]. With this modification, CNFETs exhibit similar $I_{sd}(V_{lg})$ characteristics to devices in normal, gaseous environment ($V_{lg} \Leftrightarrow V_{gs}$). Second, to facilitate adsorption of biomolecules onto SWNTs, functionalization is often required [14.98]. The first biochemical CNFET sensor was demonstrated by *Besteman* et al. [14.7]. Glucose oxidase was attached to a SWNT, which subsequently showed response to both pH and glucose concentration (0.1 M). In this study functionalization is an enabler; unfunctionalized devices simply do not respond. More generally, functionalization is essential both in liquid- and gas-phase sensors for increasing sensitivity and selectivity [14.99]. For biosensors, the debate concerning the sensing mechanism mirrors the one for chemical sensors. Recently, combining simulation and measurements, *Heller* et al. [14.100] found that both electrostatic gating and Schottky barriers are contributing, and that the substrate also influences the response.

Concerning the sensing performance of (bio)chemical CNFETs, *Kong* et al. [14.2] report concentration limits of detection of ≈ 2 ppm for NO_2 and $\approx 0.1\%$ for NH_3, *Auvray* et al. [14.3] report ≈ 10 ppb for triethylamine, whereas *Besteman* et al. [14.7] speculate that CNFET sensors have the potential to measure the enzymatic activity of even a single redox enzyme. These sensitivities, combined with ≈ 10 nW power consumption and fast response times, give SWNT FET (bio)chemical sensors key advantages over competing technologies.

14.3.2 Piezoresistive Sensors

The electronic structure and thus the electrical conductance of all three classes of SWNTs (metallic, semiconducting, and small-bandgap semiconducting) depend strongly on the mechanical deformation of the SWNT structure, as theoretically predicted in [14.26, 35] and briefly outlined in Sect. 14.1.1 for axial and torsional strain.

This theoretical expectation was first corroborated experimentally in pioneering work by *Tombler* et al. [14.8], for a suspended m-SWNT clamped to and contacted by two metal electrodes. Upon deforming and straining the nanotube by using an AFM tip, the overall conductance of the device decreased reversibly by two orders of magnitude (strain $\varepsilon \approx 3\%$). The authors have attributed this large conductance change to a mechanically induced modification of the atomic structure of the SWNT region immediately below the AFM tip, which could raise a tunnel barrier in the path of free carriers.

A suspended doubly clamped CNFET configuration to measure the piezoresistance (i. e., the change of the bandgap as a function of axial strain $dE_g/d\varepsilon$) was first proposed by *Minot* et al. [14.9]. The experimental setup was similar to that of Tombler. The CNFETs were strained by an AFM tip while the tip was simultaneously used as a local gate in the center region of the suspended SWNT (Fig. 14.14a). The overall CNFET conductance was tuned by the gate voltage at the tip (V_{tip}). Figure 14.14b shows a series of sweeps of V_{tip} for an intrinsic m-SWNT when no strain is applied. By gradually straining the suspended doubly clamped CNFET a small bandgap opens ($dE_g/d\varepsilon > 0$), according to the theory of SGS-SWNTs ($p = 0$), and a conductance dip

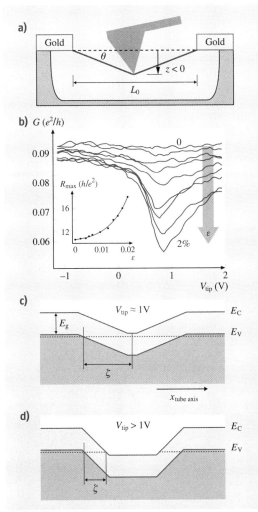

Fig. 14.14a–d Strain-dependent conductance measurement on a doubly clamped fully suspended CNFET. (**a**) Schematic of the experimental setup; an AFM tip strains the nanotube. (**b**) $G(V_{tip})$ measurements for an intrinsic metallic SWNT for 0–2% strain (*arrow* indicates increasing strain). *Inset* shows $R_{max}(\varepsilon)$ for $V_{tip} \approx 1$ V. (**c,d**) Band diagrams at the center of the SWNT at $V_{tip} \approx 1$ V (**c**) and $V_{tip} > 1$ V (**d**) displaying a barrier of width ζ below the AFM tip (after [14.9])

forms in the gate characteristic at $V_{tip} \approx 1$ V. In this configuration it is assumed that the tip gate is acting locally in the middle part of the SWNT, while the sections close to the contacts are p-type (this is because, for $V_{tip} = 0$ V, the CNFET is in the on-state for hole conductance). The strained CNFET characteristic can be explained starting from this initial situation. For a small positive gate voltage (0 V $\leq V_{tip} < 1$ V) a barrier for holes is created, leading to a reduction of the total current (Fig. 14.14c). The hole current reaches its minimum (I_{min}) when the barrier formed by V_{tip} reaches E_g. Up to this point only thermionic emission is accounted for, since no tunneling is expected due to the width of the barrier formed by the tip gate (> 20 nm). For $V_{tip} > 1$ V the CNFET section of the SWNT reaches inversion and transport increases due to tunneling through the thinned barriers ζ (Fig. 14.14d). Since the barrier height for holes at I_{min} is approximately equal to the bandgap E_g of the CNFET, and since thermionic emission dominates, the overall change in the bandgap of the CNFET due to strain can be extracted by employing the thermally activated conductance formula $G \propto \exp(-E_g(\varepsilon)/k_B T)$. As a result an overall change of the bandgap of $dE_g/d\varepsilon \approx 35$ meV/% was extracted. This value is below the theoretical prediction of a maximum bandgap change of $dE_{g\,max}/d\varepsilon \approx 94$ meV/% [14.35].

In contrast to doubly clamped suspended SWNTs the electromechanical properties of CNFETs adhering to thin-film membranes were investigated in two studies [14.10, 11], confirming piezoresistivity also in this situation. *Grow* et al. [14.10] apply the gate voltage at the metalized backside of a SiN$_x$ membrane. For SGS-SWNTs FETs the $I_{sd}(V_{gs})$ characteristic shows that the unstrained device is p-type, with no or only small Schottky barriers for the holes (small on-state resistance). The applied strain increases the bandgap ($dE_g/d\varepsilon > 0$). By increasing the back-gate voltage V_{gs}, a small but wide hole barrier (zero tunneling) is formed along the length of the CNFET channel, reducing the hole current and reaching I_{min} when the barrier is close to E_g. The largest change in channel conductance at various strain levels therefore happens at I_{min}.

The sensitivity of the devices can be expressed as $|dE_g/d\varepsilon|$ in meV/% or alternatively by the gauge factor (GF) which is usually used for classical piezoresistors. The GF is defined by the relative change of resistance divided by the strain of the SWNT and is only constant for small strains ($d\varepsilon$) around a working point ε_0. *Minot* et al. [14.9] reported a maximum sensitivity of 54 meV/% for the suspended CNFET, while *Grow* et al. [14.10] extracted 180 meV/% for a SGS-SWNT adhering to a substrate. For a metallic SWNT a sensitivity of 340–430 meV/% was reported [14.11]. Such high sensitivities of CNFETs in contact with the substrate may be explained by additional mechanical distortions of the SWNTs due to interaction with the dielectric substrate. The maximum reported GF for a membrane-based unstrained CNFET is 856 [14.10], and in a prestrained suspended metallic SWNT with $\varepsilon_0 \approx 0.4\%$ a GF of 2900 was shown [14.12].

In general the piezoresistive effect of SWNTs has been experimentally manifested in CNFETs by loading the tubes with axial strain. Piezoresistive CNFETs excel in terms of their very high sensitivity even at small strain ($< 0.2\%$), nanoscale size, and low power consumption (nW), in stark contrast to the mW of classical piezoresistors.

14.3.3 Resonant Sensors

Along with piezoresistive sensors, resonant sensors are useful for measuring mechanical stimuli such as strain and stress. In addition, resonators can be utilized to measure inertial properties (mainly mass), in which case they are known as inertial balances. The core principle, briefly explained in Sect. 14.1.1, exploits the dependence of the resonant frequency (typically the fundamental flexural mode) of a SWNT beam on quantities such as tension σ and mass density ρ, $\omega = \omega(\sigma, \rho)$.

For measuring mass via $\omega(\rho)$ either a cantilever or doubly clamped beam configuration can be employed. Indeed, the cantilever configuration has been utilized from the beginning to acquire the Young's moduli of CNT [14.101] or to demonstrate mass sensing. *Poncharal* et al. [14.14] were the first to show that, with a MWNT cantilever, a mass as small as 22 ± 6 fg (1 fg = 10^{-15} g) can be measured. Recently, *Jensen* et al. [14.15] claim to have detected only 51 Au atoms, corresponding to a mass of roughly 17 zg (1 zg = 10^{-21} g), with a DWNT cantilever inside a field-emitting device.

On the other hand, the doubly clamped beam configuration is useful for measuring stress and strain via $\omega(\sigma)$, hardly possible with cantilevers which have a free end. Moreover, for doubly clamped beams, biasing and measuring currents through the nanotube is straightforward, which is a prerequisite for electronic readout. The first electromechanical resonator based on a doubly clamped carbon nanotube was proposed by *Sazonova* et al. [14.16]. Structurally, the device is a standard backgate CNFET with a SWNT channel doubly clamped by top electrodes and partly or totally suspended over a trench (Figs. 14.11d and 14.15a [14.18]). The suspended region is the vibrating part of the nanotube.

The mechanical actuation of the nanotube is electrostatic. An applied gate voltage V_{gs} induces a capacitive Coulomb force F, deflecting the suspended nanotube segment. V_{gs} contains a direct-current (DC) compo-

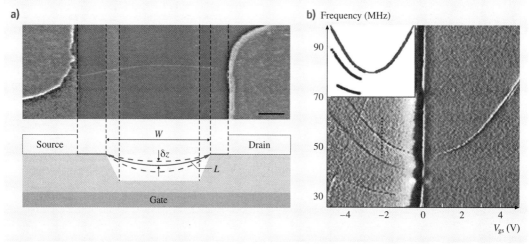

Fig. 14.15a,b Suspended SWNT resonator. (**a**) SEM image and cross-section schematic (*scale bar* 300 nm). (**b**) Detected current as a function of gate voltage and frequency. *Inset* shows the extracted positions of the resonance frequency ω_0 (after [14.16])

nent to prestrain the nanotube ($\sigma \neq 0$), and a harmonic at frequency ω to excite the nanotube beam vibration. At frequencies close to the fundamental flexural mode ω_0, the nanotube behaves like a forced classical beam [14.102], oscillating with an amplitude δz that follows a classical Lorentzian centered at ω_0 (Fig. 14.2).

Readout of the nanotube displacement is done via the channel conductance. For a CNFET the conductance G depends on both the gate voltage V_{gs} and the distance of the nanotube from the gate z through the gate capacity C_{gs}. This is so because the conductance is proportional to the free-carrier concentration in the channel ($G \propto Q = C_{gs}V_{gs}$). At frequencies close to ω_0, the oscillation amplitude δz becomes important, resulting in a detectable conductance oscillation $\delta G(t)$. Theoretically, $\delta G(t)$ is as easy to read as applying a source–drain voltage V_{sd} and measuring the current I_{sd}. However, because $f_0 = \omega_0/(2\pi) > 10\,\text{MHz}$, Sazonova et al. [14.16] have proposed using the CNFET as a mixer to downconvert $\delta G(t)$ from ω to $\Delta\omega$, conveniently chosen at around 10 kHz. In Fig. 14.15b, an intensity map of the current amplitude $\delta I_{sd}(V_{gs}, \omega)$ is shown [14.16]. As expected, ω_0 increases with V_{gs}, since the static gate voltage changes the tension σ in the nanotube. Branches visible below the fundamental ω_0 in this map are due to initial slack. Indeed, Witkamp et al. [14.18] have shown that, for straight tubes (no slack), the modes below the fundamental are suppressed.

As for sensing capabilities of CNT resonators, Sazonova et al. [14.16] have estimated the force sensitivity to be around $1\,\text{fN}/\sqrt{\text{Hz}}$ ($1\,\text{fN} = 10^{-15}\,\text{N}$), within a factor of ten of the best measured sensitivities at room temperature. In [14.17], Peng et al. have loaded a CNFET resonator with thermally evaporated Fe. Assuming a uniform coating of 2 nm of Fe ($\approx 3.5 \times 10^{-17}\,\text{g}$), they extrapolate a minimum detectable mass on the order of 1 ag ($10^{-18}\,\text{g}$). One of the limiting factors so far for both inertial balances and force sensors has been the quality factor Q, which in all reports has not exceeded 300 [14.16–18]. However Q optimization has not been addressed so far.

14.4 Concluding Remarks

Since CNTs are at the crossroads between fundamental science and engineering, transdisciplinary development in this field is highly demanded. Concerning device fabrication, synthetic chemists and process engineers still have challenges ahead of them in trying to develop strategies for controlling nanotube electronic properties and local integration into functional systems. Even if chirality control is not yet here, this does not mean that technology transfer in the midterm is not possible. In fact, the sensor concepts presented in Sect. 14.3 (excepting piezoresistive gauges) do not require chirality control, but bandgap control (diameter and metallic versus semiconducting separation). With state-of-the-art methods for SWNT synthesis, separation, and sorting, and CCVD, catalyst control, and in situ growth, bandgap control is almost here. Actually, one should not rule out even piezoresistive gauges, since calibration, tuning, and performance characterization, as usually done for sensors, may result in acceptable product yield. Nanosystem technology may become the first user of SWNTs because, in contrast to mainstream technologies (CMOS logic, memories) for which ultralarge-scale integration (ULSI) is the goal, sensors will require the integration of individual and *just a few* structures on the wafer level, only. An ideal platform for exploring new sensing devices is the CNFET sensor tool box as presented in this chapter.

References

14.1   R.H. Baughman, A.A. Zakhidov, W.A. de Heer: Carbon nanotubes – The route toward applications, Science **297**, 787–792 (2002)

14.2   J. Kong, N.R. Franklin, C. Zhou, M.G. Chapline, S. Peng, K. Cho, H. Dai: Nanotube molecular wires as chemical sensors, Science **287**, 622–625 (2000)

14.3   S. Auvray, V. Derycke, M. Goffman, A. Filoramo, O. Jost, J.-P. Bourgoin: Chemical optimization of self-assembled carbon nanotube transistors, Nano Lett. **5**, 451–455 (2005)

14.4   P.G. Collins, K. Bradley, M. Ishigami, A. Zettl: Extreme oxygen sensitivity of electronic properties of carbon nanotubes, Science **287**, 1801–1804 (2000)

14.5 J. Li, Y. Lu, Q. Ye, M. Cinke, J. Han, M. Meyyappan: Carbon nanotube sensors for gas and organic vapor detection, Nano Lett. **3**, 929–933 (2003)

14.6 Q.F. Pengfei, O. Vermesh, M. Grecu, A. Javey, Q. Wang, H. Dai, S. Peng, K.J. Cho: Toward large arrays of multiplex functionalized carbon nanotube sensors for highly sensitive and selective molecular detection, Nano Lett. **3**, 347–351 (2003)

14.7 K. Besteman, J.O. Lee, F.G.M. Wiertz, H.A. Heering, C. Dekker: Enzyme-coated carbon nanotubes as single-molecule biosensors, Nano Lett. **3**, 727–730 (2003)

14.8 T.W. Tombler, C. Zhou, L. Alexeyev, J. Kong, H. Dai, W. Liu, C.S. Jayanthi, M. Tang, S.Y. Wu: Reversible electromechanical characteristics of carbon nanotubes under local-probe manipulation, Nature **405**, 769–772 (2000)

14.9 E.D. Minot, Y. Yaish, V. Sazonova, J.Y. Park, M. Brink, P.L. McEuen: Tuning carbon nanotube band gaps with strain, Phys. Rev. Lett. **90**, 156401-1–156401-4 (2003)

14.10 R.J. Grow, Q. Wang, J. Cao, D. Wang, H. Dai: Piezoresistance of carbon nanotubes on deformable thin-film membranes, Appl. Phys. Lett. **86**, 093104-1–093104-3 (2005)

14.11 C. Stampfer, T. Helbling, D. Obergfell, B. Schoberle, M.K. Tripp, A. Jungen, S. Roth, V.M. Bright, C. Hierold: Fabrication of single-walled carbon-nanotube-based pressure sensors, Nano Lett. **6**, 233–237 (2006)

14.12 C. Stampfer, A. Jungen, R. Linderman, D. Obergfell, S. Roth, C. Hierold: Nano-electromechanical displacement sensing based on single-walled carbon nanotubes, Nano Lett. **6**, 1449–1453 (2006)

14.13 V.L. Pushparaj, L. Ci, S. Sreekala, A. Kumar, S. Kesapragada, D. Gall, O. Nalamasu, A.M. Pulickel, J. Suhr: Effects of compressive strains on electrical conductivities of a macroscale carbon nanotube block, Appl. Phys. Lett. **91**, 153116-3 (2007)

14.14 P. Poncharal, Z.L. Wang, D. Ugarte, W.A. de Heer: Electrostatic deflections and electromechanical resonances of carbon nanotubes, Science **283**, 1513–1516 (1999)

14.15 K. Jensen, K. Kim, A. Zettl: An atomic-resolution nanomechanical mass sensor, Nat. Nanotechnol. **3**, 533–537 (2008)

14.16 V. Sazonova, Y. Yaish, H. Üstünel, D. Roundy, T.A. Arias, P.L. McEuen: A tunable carbon nanotube electromechanical oscillator, Nature **431**, 284–287 (2004)

14.17 H.B. Peng, C.W. Chang, S. Aloni, T.D. Yuzvinsky, A. Zettl: Ultrahigh frequency nanotube resonators, Phys. Rev. Lett. **97**, 087203-1–087203-4 (2006)

14.18 B. Witkamp, M. Poot, H.S.J. van der Zant: Bending-mode vibration of a suspended nanotube resonator, Nano Lett. **6**, 2904–2908 (2006)

14.19 M. Freitag, Y. Martin, J.A. Misewich, R. Martel, P. Avouris: Photoconductivity of single carbon nanotubes, Nano Lett. **3**, 1067–1071 (2003)

14.20 X. Qiu, M. Freitag, V. Perebeinos, P. Avouris: Photoconductivity spectra of single-carbon nanotubes: Implications on the nature of their excited states, Nano Lett. **5**, 749–752 (2005)

14.21 C.T. White, J.W. Mintmire: Density of states reflects diameter in nanotubes, Nature **394**, 29–30 (1998)

14.22 M.S. Dresselhaus, G. Dresselhaus, P. Avouris: *Carbon Nanotubes: Synthesis, Structure, Properties and Applications* (Springer, Berlin, Heidelberg 2001)

14.23 R. Saito, G. Dresselhaus, M.S. Dresselhaus: *Physical Properties of Carbon Nanotubes* (Imperial College Press, London 2001)

14.24 A. Jorio, G. Dresselhaus, M.S. Dresselhaus: *Carbon Nanotubes: Advanced Topics in the Synthesis, Structure, Properties and Applications* (Springer, Berlin, Heidelberg 2008)

14.25 S. Li, Z. Yu, C. Rutherglen, P.J. Burke: Electrical properties of 0.4 cm long single-walled carbon nanotubes, Nano Lett. **4**, 2003–2007 (2004)

14.26 A. Kleiner, S. Eggert: Curvature, hybridization, and STM images of carbon nanotubes, Phys. Rev. B **64**, 113402-1–113402-4 (2001)

14.27 B. Bourlon, J. Wong, C. Mikó, L. Forró, M. Bockrath: A nanoscale probe for fluidic and ionic transport, Nat. Nanotechnol. **2**, 104–107 (2006)

14.28 A. Modi, N. Koratkar, E. Lass, B. Wei, P.M. Ajayan: Miniaturized gas ionization sensors using carbon nanotubes, Nature **424**, 171–174 (2003)

14.29 E.S. Snow, F.K. Perkins, E.J. Houser, S.C. Badescu, T.L. Reinecke: Chemical detection with a single-walled carbon nanotube capacitor, Science **307**, 1942–1945 (2005)

14.30 A. Krishnan, E. Dujardin, T.W. Ebbesen, P.N. Yianilos, M.M.J. Treacy: Young's modulus of single-walled nanotubes, Phys. Rev. B **58**, 14013 (1998)

14.31 D.A. Walters, L.M. Ericson, M.J. Casavant, J. Liu, D.T. Colbert, K.A. Smith, R.E. Smalley: Elastic strain of freely suspended single-wall carbon nanotube ropes, Appl. Phys. Lett. **74**, 3803–3805 (1999)

14.32 M.-F. Yu, B.S. Files, S. Arepalli, R.S. Ruoff: Tensile loading of ropes of single wall carbon nanotubes and their mechanical properties, Phys. Rev. Lett. **84**, 5552–5555 (2000)

14.33 H.W.C. Postma, I. Kozinsky, A. Husain, M.L. Roukes: Dynamic range of nanotube- and nanowire-based electromechanical systems, Appl. Phys. Lett. **86**, 223105-1–223105-3 (2005)

14.34 L. Yang, M.P. Anantram, J. Han, J.P. Lu: Band-gap change of carbon nanotubes: Effect of small uniaxial and torsional strain, Phys. Rev. B **60**, 13874–13878 (1999)

14.35 L. Yang, J. Han: Electronic structure of deformed carbon nanotubes, Phys. Rev. Lett. **85**, 154–157 (2000)

14.36 P. Ruffieux, O. Gröning, M. Bielmann, P. Mauron, L. Schlapbach, P. Gröning: Hydrogen adsorption on sp^2-bonded carbon: Influence of the local curvature, Phys. Rev. B **66**, 245416-1–245416-8 (2002)

14.37 S. Picozzi, S. Santucci, L. Lozzi, L. Valentini, B. Delley: Ozone adsorption on carbon nanotubes: The role of Stone–Wales defects, J. Chem. Phys. **120**, 7147–7152 (2004)

14.38 D.L. Carroll, P. Redlich, X. Blase, J.-C. Charlier, S. Curran, P.M. Ajayan, S. Roth, M. Rühle: Effects of nanodomain formation on the electronic structure of doped carbon nanotubes, Phys. Rev. Lett. **81**, 2332–2335 (1998)

14.39 R.J. Chen, N.R. Franklin, J. Kong, J. Cao, T.W. Tombler, Y. Zhang, H. Dai: Molecular photodesorption from single-walled carbon nanotubes, Appl. Phys. Lett. **79**, 2258–2260 (2001)

14.40 S. Datta: *Electronic Transport in Mesoscopic Systems* (Cambridge Univ. Press, Cambridge 1997)

14.41 J. Kong, E. Yenilmez, T.W. Tombler, W. Kim, H. Dai, R.B. Laughlin, L. Liu, C.S. Jayanthi, S.Y. Wu: Quantum interference and ballistic transmission in nanotube electron waveguides, Phys. Rev. Lett. **87**, 106801-1–106801-4 (2001)

14.42 W. Liang, M. Bockrath, D. Bozovic, J.H. Hafner, M. Tinkham, H. Park: Fabry–Perot interference in a nanotube electron waveguide, Nature **411**, 665–669 (2001)

14.43 J.-C. Charlier, X. Blase, S. Roche: Electronic and transport properties of nanotubes, Rev. Mod. Phys. **79**, 677–732 (2007)

14.44 S. Heinze, J. Tersoff, R. Martel, V. Derycke, J. Appenzeller, P. Avouris: Carbon nanotubes as Schottky barrier transistors, Phys. Rev. Lett. **89**, 106801-1–106801-4 (2002)

14.45 F. Leonard, J. Tersoff: Role of Fermi-level pinning in nanotube Schottky diodes, Phys. Rev. Lett. **84**, 4693–4696 (2000)

14.46 Z.H. Chen, J. Appenzeller, J. Knoch, Y.M. Lin, P. Avouris: The role of metal-nanotube contact in the performance of carbon nanotube field-effect transistors, Nano Lett. **5**, 1497–1502 (2005)

14.47 A. Javey, J. Guo, Q. Wang, M. Lundstrom, H. Dai: Ballistic carbon nanotube field-effect transistors, Nature **424**, 654–657 (2003)

14.48 J. Appenzeller, J. Knoch, V. Derycke, R. Martel, S. Wind, P. Avouris: Field-modulated carrier transport in carbon nanotube transistors, Phys. Rev. Lett. **89**, 126801-1–126801-4 (2002)

14.49 J. Appenzeller, M. Radosavljević, J. Knoch, P. Avouris: Tunneling versus thermionic emission in one-dimensional semiconductors, Phys. Rev. Lett. **92**, 048301-1–048301-4 (2004)

14.50 D. Mann, A. Javey, J. Kong, Q. Wang, H. Dai: Ballistic transport in metallic nanotubes with reliable Pd ohmic contacts, Nano Lett. **3**, 1541–1544 (2003)

14.51 M.P. Anantram, F. Léonard: Physics of carbon nanotube electronic devices, Rep. Prog. Phys. **69**, 507–561 (2006)

14.52 M.J. Biercuk, S. Ilani, C.M. Marcus, P.L. McEuen: Electrical transport in single-wall carbon nanotubes. In: *Carbon Nanotubes*, ed. by A. Jorio, G. Dresselhaus, M.S. Dresselhaus (Springer, Berlin, Heidelberg 2008) pp. 455–493

14.53 S. Roche, E. Akkermans, O. Chauvet, F. Hekking, J.-P. Issi, R. Martel, G. Montambaux, P. Poncharal: Transport properties. In: *Understanding Carbon Nanotubes*, ed. by A. Loiseau, P. Launois, P. Petit, S. Roche, J.-P. Salvetat (Springer, Berlin, Heidelberg 2006) pp. 335–437

14.54 T.S. Cho, K.-J. Lee, J. Kong, A.P. Chandrakasan: The design of a low power carbon nanotube chemical sensor system, Des. Autom. Conf. 2008 (DAC 2008), Anaheim (2008) pp. 84–89

14.55 P.G. Collins, M.S. Fuhrer, A. Zettl: $1/f$ noise in carbon nanotubes, Appl. Phys. Lett. **76**, 894–896 (2000)

14.56 J. Appenzeller, Y.M. Lin, J. Knoch, Z.H. Chen, P. Avouris: $1/f$ noise in carbon nanotube devices – On the impact of contacts and device geometry, IEEE Trans. Nanotechnol. **6**, 368–373 (2007)

14.57 J. Tersoff: Low-frequency noise in nanoscale ballistic transistors, Nano Lett. **7**, 194–198 (2007)

14.58 J. Männik, I. Heller, A.M. Janssens, S.G. Lemay, C. Dekker: Charge noise in liquid-gated single-wall carbon nanotube transistors, Nano Lett. **8**, 685–688 (2007)

14.59 C.L. Cheung, A. Kurtz, H. Park, C.M. Lieber: Diameter-controlled synthesis of carbon nanotubes, J. Phys. Chem. B **106**, 2429–2433 (2002)

14.60 Y. Tu, Z.P. Huang, D.Z. Wang, J.G. Wen, Z.F. Ren: Growth of aligned carbon nanotubes with controlled site density, Appl. Phys. Lett. **80**, 4018–4020 (2002)

14.61 Y. Li, W. Kim, Y. Zhang, M. Rolandi, D. Wang, H. Dai: Growth of single-walled carbon nanotubes from discrete catalytic nanoparticles of various sizes, J. Phys. Chem. B **105**, 11424–11431 (2001)

14.62 R.M. Tromp, A. Afzali, M. Freitag, D.B. Mitzi, Z. Chen: Novel strategy for diameter-selective separation and functionalization of single-walled carbon nanotubes, Nano Lett. **8**, 469–472 (2008)

14.63 M.S. Arnold, S.I. Stupp, M.C. Hersam: Enrichment of single-walled carbon nanotubes by diameter in density gradients, Nano Lett. **5**, 713–718 (2005)

14.64 M.S. Arnold, A.A. Green, J.F. Hulvat, S.I. Stupp, M.C. Hersam: Sorting carbon nanotubes by electronic structure using density differentiation, Nat. Nanotechnol. **1**, 60–65 (2006)

14.65 R.E. Smalley, Y. Li, V.C. Moore, B.K. Price, R. Colorado, H.K. Schmidt, R.H. Hauge, A.R. Barron, J.M. Tour: Single wall carbon nanotube amplification: en route to a type-specific growth mechanism, J. Am. Chem. Soc. **128**, 15824–15829 (2006)

14.66 E. Joselevich, H. Dai, J. Liu, K. Hata, A.H. Windle: Carbon nanotube synthesis and organization. In: *Carbon Nanotubes*, ed. by A. Jorio, G. Dresselhaus, M.S. Dresselhaus (Springer, Berlin, Heidelberg 2008) pp. 101–164

14.67 Y. Yan, M.B. Chan-Park, Q. Zhang: Advances in carbon-nanotube assembly, Small **3**, 24–42 (2007)

14.68 S.J. Tans, A.R.M. Verschueren, C. Dekker: Room-temperature transistor based on a single carbon nanotube, Nature **393**, 49–52 (1998)

14.69 R. Martel, T. Schmidt, H.R. Shea, T. Hertel, P. Avouris: Single- and multi-wall carbon nanotube field-effect transistors, Appl. Phys. Lett. **73**, 2447–2449 (1998)

14.70 V. Derycke, R. Martel, J. Appenzeller, P. Avouris: Carbon nanotube inter- and intramolecular logic gates, Nano Lett. **1**, 453–456 (2001)

14.71 J. Liu, M.J. Casavant, M. Cox, D.A. Walters, P. Boul, W. Lu, A.J. Rimberg, K.A. Smith, D.T. Colbert, R.E. Smalley: Controlled deposition of individual single-walled carbon nanotubes on chemically functionalized templates, Chem. Phys. Lett. **303**, 125–129 (1999)

14.72 C. Klinke, J.B. Hannon, A. Afzali, P. Avouris: Field-effect transistors assembled from functionalized carbon nanotubes, Nano Lett. **6**, 906–910 (2006)

14.73 S.G. Rao, L. Huang, W. Setyawan, S. Hong: Nanotube electronics: Large-scale assembly of carbon nanotubes, Nature **425**, 36–37 (2003)

14.74 R. Krupke, F. Hennrich, H.B. Weber, M.M. Kappes, H. von Löhneysen: Simultaneous deposition of metallic bundles of single-walled carbon nanotubes using AC-dielectrophoresis, Nano Lett. **3**, 1019–1023 (2003)

14.75 R. Krupke, F. Hennrich, H. von Löhneysen, M.M. Kappes: Separation of metallic from semiconducting single-walled carbon nanotubes, Science **301**, 344–347 (2003)

14.76 Y. Zhang, A. Chang, J. Cao, Q. Wang, W. Kim, Y. Li, N. Morris, E. Yenilmez, J. Kong, H. Dai: Electric-field-directed growth of aligned single-walled carbon nanotubes, Appl. Phys. Lett. **79**, 3155–3157 (2001)

14.77 J. Kong, H.T. Soh, A.M. Cassell, C. Quate, H. Dai: Synthesis of individual single-walled carbon nanotubes on patterned silicon wafers, Nature **395**, 878–881 (1998)

14.78 H.T. Soh, A. Morpurgo, J. Kong, C. Marcus, C. Quate, H. Dai: Integrated nanotube circuits: Controlled growth and ohmic contacting of single-walled carbon nanotubes, Appl. Phys. Lett. **75**, 627–629 (1999)

14.79 A. Javey, H. Dai: Regular arrays of 2 nm metal nanoparticles for deterministic synthesis of nanomaterials, J. Am. Chem. Soc. **127**, 11942–11943 (2005)

14.80 E. Joselevich, C.M. Lieber: Vectorial growth of metallic and semiconducting single-wall carbon nanotubes, Nano Lett. **2**, 1137–1141 (2002)

14.81 S. Dittmer, J. Svensson, E.E.B. Campbell: Electric field aligned growth of single-walled carbon nanotubes, Curr. Appl. Phys. **4**, 595–598 (2004)

14.82 S. Huang, X. Cai, J. Liu: Growth of millimeter-long and horizontally aligned single-walled carbon nanotubes on flat substrates, J. Am. Chem. Soc. **125**, 5636–5637 (2003)

14.83 S. Han, X. Liu, C. Zhou: Template-free directional growth of single-walled carbon nanotubes on a- and r-plane sapphire, J. Am. Chem. Soc. **127**, 5294–5295 (2005)

14.84 A. Ismach, L. Segev, E. Wachtel, E. Joselevich: Atomic-step-templated formation of single wall carbon nanotube patterns, Angew. Chem. Int. Ed. **43**, 6140–6143 (2004)

14.85 X. Liu, S. Han, C. Zhou: Novel nanotube-on-insulator (NOI) approach toward single-walled carbon nanotube devices, Nano Lett. **6**, 34–39 (2005)

14.86 A.M. Cassell, N.R. Franklin, T.W. Tombler, E.M. Chan, J. Han, H. Dai: Directed growth of free-standing single-walled carbon nanotubes, J. Am. Chem. Soc. **121**, 7975–7976 (1999)

14.87 N.R. Franklin, H. Dai: An enhanced CVD approach to extensive nanotube networks with directionality, Adv. Mater. **12**, 890–894 (2000)

14.88 N.R. Franklin, Q. Wang, T.W. Tombler, A. Javey, M. Shim, H. Dai: Integration of suspended carbon nanotube arrays into electronic devices and electromechanical systems, Appl. Phys. Lett. **81**, 913–915 (2002)

14.89 A. Jungen, S. Hofmann, J.C. Meyer, C. Stampfer, S. Roth, J. Robertson, C. Hierold: Synthesis of individual single-walled carbon nanotube bridges controlled by support micromachining, J. Micromech. Microeng. **17**, 603–608 (2007)

14.90 T. Zhang, S. Mubeen, N.V. Myung, M.A. Deshusses: Recent progress in carbon nanotube-based gas sensors, Nanotechnology **19**, 332001 (2008)

14.91 S.N. Kim, J.F. Rusling, F. Papadimitrakopoulos: Carbon nanotubes for electronic and electrochemical detection of biomolecules, Adv. Mater. **19**, 3214–3228 (2007)

14.92 N. Sinha, J. Ma, J.T.W. Yeow: Carbon nanotube-based sensors, J. Nanosci. Nanotechnol. **6**, 573–590 (2006)

14.93 K. Bradley, J.C.P. Gabriel, A. Star, G. Gruner: Short-channel effects in contact-passivated nanotube chemical sensors, Appl. Phys. Lett. **83**, 3821–3823 (2003)

14.94 J. Zhang, A. Boyd, A. Tselev, M. Paranjape, P. Barbara: Mechanism of NO_2 detection in carbon nanotube field effect transistor chemical sensors, Appl. Phys. Lett. **88**, 123112-1–123112-3 (2006)

14.95   X.L. Liu, Z.C. Luo, S. Han, T. Tang, D.H. Zhang, C.W. Zhou: Band engineering of carbon nanotube field-effect transistors via selected area chemical gating, Appl. Phys. Lett. **86**, 243501-1–243501-3 (2005)

14.96   T. Helbling, R. Pohle, L. Durrer, C. Stampfer, C. Roman, A. Jungen, M. Fleischer, C. Hierold: Sensing NO_2 with individual suspended single-walled carbon nanotubes, Sens. Actuators B **132**, 491–497 (2008)

14.97   M. Krüger, M.R. Buitelaar, T. Nussbaumer, C. Schönenberger, L. Forró: Electrochemical carbon nanotube field-effect transistor, Appl. Phys. Lett. **78**, 1291–1293 (2001)

14.98   R.J. Chen, Y. Zhang, D. Wang, H. Dai: Noncovalent sidewall functionalization of single-walled carbon nanotubes for protein immobilization, J. Am. Chem. Soc. **123**, 3838–3839 (2001)

14.99   K. Balasubramanian, M. Burghard: Chemically functionalized carbon nanotubes, Small **1**, 180–192 (2005)

14.100 I. Heller, A.M. Janssens, J. Männik, E.D. Minot, S.G. Lemay, C. Dekker: Identifying the mechanism of biosensing with carbon nanotube transistors, Nano Lett. **8**, 591–595 (2008)

14.101 M.M.J. Treacy, T.W. Ebbesen, J.M. Gibson: Exceptionally high Young's modulus observed for individual carbon nanotubes, Nature **381**, 678–680 (1996)

14.102 D. Garcia-Sanchez, A. San Paulo, M.J. Esplandiu, F. Perez-Murano, L. Forró, A. Aguasca, A. Bachtold: Mechanical detection of carbon nanotube resonator vibrations, Phys. Rev. Lett. **99**, 085501-1–085501-4 (2007)

15. Nanomechanical Cantilever Array Sensors

Hans Peter Lang, Martin Hegner, Christoph Gerber

Microfabricated cantilever sensors have attracted much interest in recent years as devices for the fast and reliable detection of small concentrations of molecules in air and solution. In addition to application of such sensors for gas and chemical-vapor sensing, for example as an artificial nose, they have also been employed to measure physical properties of tiny amounts of materials in miniaturized versions of conventional standard techniques such as calorimetry, thermogravimetry, weighing, photothermal spectroscopy, as well as for monitoring chemical reactions such as catalysis on small surfaces. In the past few years, the cantilever-sensor concept has been extended to biochemical applications and as an analytical device for measurements of biomaterials. Because of the label-free detection principle of cantilever sensors, their small size and scalability, this kind of device is advantageous for diagnostic applications and disease monitoring, as well as for genomics or proteomics purposes. The use of microcantilever arrays enables detection of several analytes simultaneously and solves the inherent problem of thermal drift often present when using single microcantilever sensors, as some of the cantilevers can be used as sensor cantilevers for detection, and other cantilevers serve as passivated reference cantilevers that do not exhibit affinity to the molecules to be detected.

15.1	**Technique**	427
	15.1.1 Cantilevers	428
	15.1.2 History of Cantilever Sensors	428
15.2	**Cantilever Array Sensors**	429
	15.2.1 Concept	429
	15.2.2 Compressive and Tensile Stress	429
	15.2.3 Disadvantages of Single Microcantilevers	429
	15.2.4 Reference and Sensor Cantilevers in an Array	430
15.3	**Modes of Operation**	430
	15.3.1 Static Mode	430
	15.3.2 Dynamic Mode	432
	15.3.3 Heat Mode	433
	15.3.4 Further Operation Modes	434
15.4	**Microfabrication**	434
15.5	**Measurement Setup**	434
	15.5.1 Measurements in Gaseous or Liquid Environments	434
	15.5.2 Readout Principles	436
15.6	**Functionalization Techniques**	438
	15.6.1 General Strategy	438
	15.6.2 Functionalization Methods	438
15.7	**Applications**	439
	15.7.1 Chemical Detection	439
	15.7.2 Biochemical Environment	442
	15.7.3 Microcantilever Sensors to Measure Physical Properties	444
15.8	**Conclusions and Outlook**	445
	References	446

15.1 Technique

Sensors are devices that detect, or sense, a signal. Moreover, a sensor is also a transducer, i.e. it transforms one form of energy into another or responds to a physical parameter. Most people will associate sensors with electrical or electronic devices that produce a change in response when an external physical parameter is changed. However, many more types of transducers exist, such as electrochemical (pH probe), electromechanical (piezoelectric actuator, quartz, strain gauge), electroacoustic (gramophone pick-up, microphone), photoelectric (photodiode, solar cell), electromagnetic (antenna), magnetic (Hall-effect sensor, tape or hard-disk head for storage applications), electrostatic (electrometer), thermoelectric (thermocouple, thermoresistors), and electrical (capacitor, resistor). Here we want to concentrate on a further type of sensor not yet mentioned: the mechanical sensor. It responds to changes of an external parameter, such as temperature changes or molecule adsorption, by a mechanical response, e.g. by bending or deflection.

15.1.1 Cantilevers

Mechanical sensors consist of a fixed and a movable part. The movable part can be a thin membrane, a plate or a beam, fixed at one or both ends. The structures described here are called cantilevers. A cantilever is regarded here as a microfabricated rectangular bar-shaped structure that is longer than it is wide and has a thickness that is much smaller than its length or width. It is a horizontal structural element supported only at one end on a chip body; the other end is free (Fig. 15.1). Most often it is used as a mechanical probe to image the topography of a sample using a technique called atomic force microscopy (AFM) or scanning force microscopy (SFM) [15.1], invented by *Binnig* et al. in the mid 1980s [15.1]. For AFM a microfabricated sharp tip is attached to the apex of the cantilever and serves as a local probe to scan the sample surface. The distance between tip and surface is controlled via sensitive measurement of interatomic forces in the piconewton range.

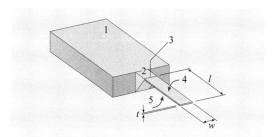

Fig. 15.1 Schematic of a cantilever: (1) rigid chip body, (2) solid cantilever-support structure, (3) hinge of cantilever, (4) upper surface of the cantilever, which is usually functionalized with a sensor layer for detection of molecules, (5) lower surface of the cantilever, usually passivated in order not to show affinity to the molecules to be detected. The geometrical dimensions, length l, width w and thickness t, are indicated

By scanning the tip across a conductive or nonconductive surface using an *x*-*y*-*z* actuator system (e.g. a piezoelectric scanner), an image of the topography is obtained by recording the correction signal that has to be applied to the *z*-actuation drive to keep the interaction between tip and sample surface constant. SFM methods are nowadays well established in scientific research, education and, to a certain extent, also in industry. Beyond imaging of surfaces, cantilevers have been used for many other purposes. However, here we focus on their application as sensor devices.

15.1.2 History of Cantilever Sensors

The idea of using beams of silicon as sensors to measure deflections or changes in resonance frequency is actually quite old. First reports go back to 1968, when *Wilfinger* et al. [15.2] investigated silicon cantilever structures of $50 \times 30 \times 8 \text{ mm}^3$, i.e. quite large structures, for detecting resonances. On the one hand, they used localized thermal expansion in diffused resistors (piezoresistors) located near the cantilever support to create a temperature gradient for actuating the cantilever at its resonance frequency. On the other hand, the piezoresistors could also be used to sense mechanical deflection of the cantilever. This early report already contains concepts for sensing and actuation of cantilevers. In the following years only a few reports are available on the use of cantilevers as sensors, e.g. *Heng* [15.3], who fabricated gold cantilevers capacitively coupled to microstrip lines in 1971 to mechanically trim high-frequency oscillator circuits. In 1979, *Petersen* [15.4] constructed cantilever-type micromechanical membrane switches in silicon that should have filled the gap between silicon transistors and mechanical electromagnetic relays. *Kolesar* [15.5] suggested the use of cantilever structures as electronic nerve-agent detectors in 1985.

Only with the availability of microfabricated cantilevers for AFM [15.1] did reports on the use of cantilevers as sensors become more frequent. In 1994, *Itoh* and *Suga* [15.6] presented a cantilever coated with a thin film of zinc oxide and proposed piezoresistive deflection readout as an alternative to optical beam-deflection readout. *Cleveland* et al. [15.7] reported the tracking of cantilever resonance frequency to detect nanogram changes in mass loading when small particles are deposited onto AFM probe tips. *Thundat* et al. [15.8] showed that the resonance frequency as well as static bending of microcantilevers are influenced by ambient conditions, such as moisture adsorption, and that deflection of metal-coated cantilevers can be further influenced by thermal effects (bimetallic effect). The first chemical sensing applications were presented by *Gimzewski* et al. [15.9], who used static cantilever bending to detect chemical reactions with very high sensitivity. Later *Thundat* et al. [15.10] observed changes in the resonance frequency of microcantilevers due to adsorption of analyte vapor on exposed surfaces. Frequency changes have been found to be caused by mass loading or adsorption-induced changes in the cantilever spring constant. By coating cantilever surfaces with hygroscopic materials, such as phosphoric acid or gelatin, the cantilever can sense water vapor with picogram mass resolution.

The deflection of individual cantilevers can easily be determined using AFM-like optical beam-deflection electronics. However, single cantilever responses can be prone to artifacts such as thermal drift or unspecific adsorption. For this reason the use of passivated reference cantilevers is desirable. The first use of cantilever arrays with sensor and reference cantilevers was reported in 1998 [15.11], and represented significant progress for the understanding of true (difference) cantilever responses.

15.2 Cantilever Array Sensors

15.2.1 Concept

For the use of a cantilever as a sensor, neither a sharp tip at the cantilever apex nor a sample surface is required. The cantilever surfaces serve as sensor surfaces and allow the processes taking place on the surface of the beam to be monitored with unprecedented accuracy, in particular the adsorption of molecules. The formation of molecule layers on the cantilever surface will generate surface stress, eventually resulting in a bending of the cantilever, provided the adsorption preferentially occurs on one surface of the cantilever. Adsorption is controlled by coating one surface (typically the upper surface) of a cantilever with a thin layer of a material that exhibits affinity to molecules in the environment (sensor surface). This surface of the cantilever is referred to as the functionalized surface. The other surface of the cantilever (typically the lower surface) may be left uncoated or be coated with a passivation layer, i.e. a chemical surface that does not exhibit significant affinity to the molecules in the environment to be detected. To enable functionalized surfaces to be established, often a metal layer is evaporated onto the surface designed as sensor surface. Metal surfaces, e.g. gold, may be used to covalently bind a monolayer that represents the chemical surface sensitive to the molecules to be detected from environment. Frequently, a monolayer of thiol molecules covalently bound to a gold surface is used. The gold layer is also favorable for use as a reflection layer if the bending of the cantilever is read out via an optical beam-deflection method.

15.2.2 Compressive and Tensile Stress

Given a cantilever coated with gold on its upper surface for adsorption of alkanethiol molecules and left uncoated on its lower surface (consisting of silicon and silicon oxide), the adsorption of thiol molecules will take place on the upper surface of the cantilever, resulting in a downward bending of the cantilever due to the formation of surface stress. We will call this process development of compressive surface stress, because the forming self-assembled monolayer produces a downward bending of the cantilever (away from the gold coating). In the opposite situation, i.e. when the cantilever bends upwards, we would speak of tensile stress. If both the upper and lower surfaces of the cantilevers are involved in the reaction, then the situation will be much more complex, as a predominant compressive stress formation on the lower cantilever surface might appear like tensile stress on the upper surface. For this reason, it is of utmost importance that the lower cantilever surface is passivated in order that ideally no processes take place on the lower surface of the cantilever.

15.2.3 Disadvantages of Single Microcantilevers

Single microcantilevers are susceptible to parasitic deflections that may be caused by thermal drift or chemical interaction of a cantilever with its environment, in particular if the cantilever is operated in a liquid.

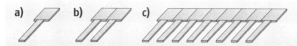

Fig. 15.2 (a) Single cantilever; **(b)** a pair of cantilevers, one to be used as a sensor cantilever, the other as a reference cantilever, and **(c)** an array of cantilevers with several sensor and reference cantilevers

Often, a baseline drift is observed during static-mode measurements. Moreover, nonspecific physisorption of molecules on the cantilever surface or nonspecific binding to receptor molecules during measurements may contribute to the drift.

15.2.4 Reference and Sensor Cantilevers in an Array

To exclude such influences, simultaneous measurement of reference cantilevers aligned in the same array as the sensing cantilevers is crucial [15.11]. As the difference in signals from the reference and sensor cantilevers shows the net cantilever response, even small sensor responses can be extracted from large cantilever deflections without being dominated by undesired effects. When only single microcantilevers are used, no thermal-drift compensation is possible. To obtain useful data under these circumstances, both microcantilever surfaces have to be chemically well defined. One of the surfaces, typically the lower one, has to be passivated; otherwise the cantilever response will be convoluted with undesired effects originating from uncontrolled reactions taking place on the lower surface (Fig. 15.2a). With a pair of cantilevers, reliable measurements are obtained. One cantilever is used as the sensor cantilever (typically coated on the upper side with a molecule layer exhibiting affinity to the molecules to be detected), whereas the other cantilever serves as the reference cantilever. It should be coated with a passivation layer on the upper surface so as not to exhibit affinity to the molecules to be detected. Thermal drifts are canceled out if difference responses, i.e. difference in deflections of sensor and reference cantilevers, are taken. Alternatively, both cantilevers are used as sensor cantilevers (sensor layer on the upper surfaces), and the lower surface has to be passivated (Fig. 15.2b). It is best to use a cantilever array (Fig. 15.2c), in which several cantilevers are used either as sensor or as reference cantilevers so that multiple difference signals can be evaluated simultaneously. Thermal drift is canceled out as one surface of all cantilevers, typically the lower one, is left uncoated or coated with the same passivation layer.

15.3 Modes of Operation

In analogy to AFM, various operating modes for cantilevers are described in the literature. The measurement of static deflection upon the formation of surface stress during adsorption of a molecular layer is termed the *static mode*. Ibach used cantileverlike structures to study adsorbate-induced surface stress [15.12] in 1994. Surface-stress-induced bending of cantilevers during the adsorption of alkanethiols on gold was reported by *Berger* et al. in 1997 [15.13]. The mode corresponding to noncontact AFM, termed the *dynamic mode*, in which a cantilever is oscillated at its resonance frequency, was described by *Cleveland* et al. [15.7]. They calculated mass changes from shifts in the cantilever resonance frequency upon the mounting of tiny tungsten particle spheres at the apex of the cantilever. The so-called *heat mode* was pioneered by *Gimzewski* et al. [15.9], who took advantage of the bimetallic effect that produces a bending of a metal-coated cantilever when heat is produced on its surface. Therewith they constructed a miniaturized calorimeter with picojoule sensitivity. Further operating modes exploit other physical effects such as the production of heat from the absorption of light by materials deposited on the cantilever (photothermal spectroscopy) [15.14], or cantilever bending caused by electric or magnetic forces.

15.3.1 Static Mode

The continuous bending of a cantilever with increasing coverage by molecules is referred to as operation in the static mode (Fig. 15.3a). Adsorption of molecules onto the functional layer produces stress at the interface between the functional layer and the molecular layer forming. Because the forces within the functional layer try to keep the distance between molecules constant, the cantilever beam responds by bending because of its extreme flexibility. This property is described by the spring constant k of the cantilever, which for a rectangu-

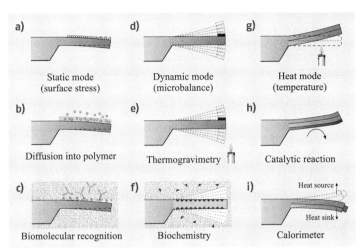

Fig. 15.3a–i Basic cantilever operation modes: (**a**) static bending of a cantilever on adsorption of a molecular layer. (**b**) Diffusion of molecules into a polymer layer leads to swelling of the polymer and eventually to a bending of the cantilever. (**c**) Highly specific molecular recognition of biomolecules by receptors changes the surface stress on the upper surface of the cantilever and results in bending. (**d**) Oscillation of a cantilever at its resonance frequency (dynamic mode) allows information on mass changes taking place on the cantilever surface to be obtained (application as a microbalance). (**e**) Changing the temperature while a sample is attached to the apex of the cantilever allows information to be gathered on decomposition or oxidation process. (**f**) Dynamic-mode measurements in liquids yield details on mass changes during biochemical processes. (**g**) In the heat mode, a bimetallic cantilever is employed. Here bending is due to the difference in the thermal expansion coefficients of the two materials. (**h**) A bimetallic cantilever with a catalytically active surface bends due to heat production during a catalytic reaction. (**i**) A tiny sample attached to the apex of the cantilever is investigated, taking advantage of the bimetallic effect. Tracking the deflection as a function of temperature allows the observation of phase transitions in the sample in a calorimeter mode

lar microcantilever of length l, thickness t and width w is calculated as

$$k = \frac{Ewt^3}{4l^3}, \qquad (15.1)$$

where E is the Young's modulus [$E_{Si} = 1.3 \times 10^{11}$ N/m^2 for Si(100)].

As a response to surface stress, e.g. owing to adsorption of a molecular layer, the microcantilever bends, and its shape can be approximated as part of a circle with radius R. This radius of curvature is given by [15.15, 16]

$$\frac{1}{R} = \frac{6(1-\nu)}{Et^2}. \qquad (15.2)$$

The resulting surface stress change is described using *Stoney's* formula [15.15]

$$\Delta\sigma = \frac{Et^2}{6R(1-\nu)}, \qquad (15.3)$$

where E is Young's modulus, t the thickness of the cantilever, ν the Poisson's ratio ($\nu_{Si} = 0.24$), and R the bending radius of the cantilever.

Static-mode operation has been reported in various environments. In its simplest configuration, molecules from the gaseous environment adsorb on the functionalized sensing surface and form a molecular layer (Fig. 15.3a), provided the molecules exhibit some affinity to the surface. In the case of alkanethiol covalently binding to gold, the affinity is very high, resulting in a fast bending response within minutes [15.13]. Polymer sensing layers only exhibit a partial sensitivity, i.e. polymer-coated cantilevers always respond to the presence of volatile molecules, but the magnitude and temporal behavior are specific to the chemistry of the polymer. Molecules from the environment diffuse into the polymer layer at different rates, mainly depending on the size and solubility of the molecules in the polymer layer (Fig. 15.3b). A wide range of hydrophilic/hydrophobic polymers can be selected, differing in their affinity to polar/unpolar molecules. Thus, the polymers can be chosen according to what an application requires.

Static-mode operation in liquids, however, usually requires rather specific sensing layers, based on molecular recognition, such as DNA hybridization [15.17] or antigen–antibody recognition (Fig. 15.3c). Cantilevers functionalized by coating with biochemical sensing layers respond very specifically using biomolecular key–lock principles of molecular recognition. However, whether molecular recognition will actually lead to a bending of the cantilever depends on the efficiency of transduction, because the surface stress has to be generated very close to the cantilever surface to produce bending. By just scaling down standard gene-chip strategies to cantilever geometry utilizing long spacer molecules so that DNA molecules become more accessible for hybridization, the hybridization takes place at a distance of several nanometers from the cantilever surface. In such experiments, no cantilever bending was observed [15.18].

15.3.2 Dynamic Mode

Mass changes can be determined accurately by using a cantilever actuated at its eigenfrequency. The eigenfrequency is equal to the resonance frequency of an oscillating cantilever if the elastic properties of the cantilever remain unchanged during the molecule-adsorption process and if damping effects are insignificant. This mode of operation is called the dynamic mode (e.g., the use as a microbalance, Fig. 15.3d). Owing to mass addition on the cantilever surface, the cantilever's eigenfrequency will shift to a lower value. The frequency change per mass change on a rectangular cantilever is calculated [15.19] according to

$$\Delta f/\Delta m = \frac{1}{4\pi n_l l^3 w} \times \sqrt{\frac{E}{\rho^3}}, \qquad (15.4)$$

where $\rho = m/(lwt)$ is the mass density of the microcantilever and the deposited mass, and $n_l \approx 1$ is a geometrical factor.

The mass change is calculated [15.8] from the frequency shift using

$$\Delta m = \frac{k}{4\pi^2} \times \left(\frac{1}{f_1^2} - \frac{1}{f_0^2}\right), \qquad (15.5)$$

where f_0 is the eigenfrequency before the mass change occurs, and f_1 the eigenfrequency after the mass change.

Mass-change determination can be combined with varying environment temperature conditions (Fig. 15.3e) to obtain a method introduced in the literature as *micromechanical thermogravimetry* [15.20]. A tiny piece of sample to be investigated has to be mounted at the apex of the cantilever. Its mass should not exceed several hundred nanograms. Adsorption, desorption and decomposition processes, occurring while changing the temperature, produce mass changes in the picogram range that can be observed in real time by tracking the resonance-frequency shift.

Dynamic-mode operation in a liquid environment is more difficult than in air, because of the large damping of the cantilever oscillation due to the high viscosity of the surrounding media (Fig. 15.3f). This results in a low quality factor Q of the oscillation, and thus the resonance frequency shift is difficult to track with high resolution. The quality factor is defined as

$$Q = 2\Delta f/f_0 . \qquad (15.6)$$

Whereas in air the resonance frequency can easily be determined with a resolution of below 1 Hz, only a frequency resolution of about 20 Hz is expected for measurements in a liquid environment.

The damping or altered elastic properties of the cantilever during the experiment, e.g. by a stiffening or softening of the spring constant caused by the adsorption of a molecule layer, result in the fact that the measured resonance frequency will not be exactly equal to the eigenfrequency of the cantilever, and therefore the mass derived from the frequency shift will be inaccurate. In a medium, the vibration of a cantilever is described by the model of a driven damped harmonic oscillator

$$m^* \frac{d^2 x}{dt^2} + \gamma \frac{dx}{dt} + kx = F\cos(2\pi ft), \qquad (15.7)$$

where $m^* = \text{const}(m_c + m_l)$ is the effective mass of the cantilever (for a rectangular cantilever the constant is 0.25). Especially in liquids, the mass of the comoved liquid m_l adds significantly to the mass of the cantilever m_c. The term $\gamma \frac{dx}{dt}$ is the drag force due to damping, $F\cos(2\pi ft)$ is the driving force executed by the piezo-oscillator, and k is the spring constant of the cantilever.

If no damping is present, the eigenfrequencies of the various oscillation modes of a bar-shaped cantilever are calculated according to

$$f_n = \frac{\alpha_n^2}{2\pi}\sqrt{\frac{k}{2(m_c + m_l)}}, \qquad (15.8)$$

where f_n are the eigenfrequencies of the n-th mode, α_n are constants depending on the mode: $\alpha_1 = 1.8751, \alpha_2 = 4.6941, \alpha_n = \pi(n - 0.5)$; k is the spring constant of the cantilever, m_c the mass of the cantilever, and m_l the mass of the medium surrounding the cantilever, e.g. liquid [15.21].

Addition of mass to the cantilever due to adsorption will change the effective mass as follows

$$m^* = \text{const}(m_c + m_l + \Delta m), \quad (15.9)$$

where Δm is the additional mass adsorbed. Typically, the comoved mass of the liquid is much larger than the adsorbed mass.

Figure 15.4 clearly shows that the resonance frequency is only equal to the eigenfrequency if no damping is present. With damping, the frequency at which the peak of the resonance curve occurs is no longer identical to that at which the turning point of the phase curve occurs. For example, resonance curve 2 with damping γ_2 has its maximum amplitude at frequency f_2. The corresponding phase would be $\varphi_{\text{res}}(\gamma_2)$, which is not equal to $\pi/2$, as would be expected in the undamped case. If direct resonance-frequency tracking or a phase-locked loop is used to determine the frequency of the oscillating cantilever, then only its resonance frequency is detected, but not its eigenfrequency. Remember that the eigenfrequency, and not the resonance frequency, is required to determine mass changes.

15.3.3 Heat Mode

If a cantilever is coated with metal layers, thermal expansion differences in the cantilever and the coating layer will further influence cantilever bending as a function of temperature. This mode of operation is referred to as the *heat mode* and causes cantilever bending because of differing thermal expansion coefficients in the sensor layer and cantilever materials [15.9] (Fig. 15.3g).

$$\Delta z = \frac{5}{4}(\alpha_1 - \alpha_2)\frac{t_1 + t_2}{t_2^2 \kappa}\frac{l^3}{(\lambda_1 t_1 + \lambda_2 t_2)w}P. \quad (15.10)$$

Here α_1, α_2 are the thermal expansion coefficients of the cantilever and coating materials, respectively, λ_1, λ_2 their thermal conductivities, t_1, t_2 the material thicknesses, P is the total power generated on the cantilever, and κ is a geometry parameter of the cantilever device.

Heat changes are either caused by external influences (change in temperature, Fig. 15.3g), occur directly on the surface by exothermal, e.g. catalytic,

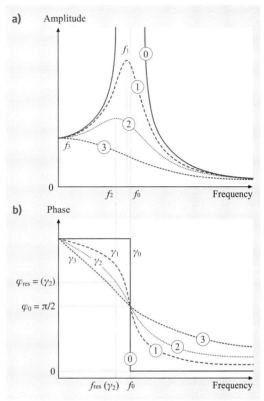

Fig. 15.4 (a) Resonance curve with no damping (0), and increasing damping (1)–(3). The undamped curve with resonance frequency f_0 exhibits a very high amplitude, whereas the resonance peak amplitude decreases with damping. This also involves a shift in resonance frequencies from f_1 to f_3 to lower values. (b) Corresponding phase curves showing no damping (0), and increasing damping (1)–(3). The steplike phase jump at resonance of the undamped resonance gradually broadens with increasing damping

reactions (Fig. 15.3h), or are due to material properties of a sample attached to the apex of the cantilever (micromechanical calorimetry, Fig. 15.3i). The sensitivity of the cantilever heat mode is orders of magnitude higher than that of traditional calorimetric methods performed on milligram samples, as it only requires nanogram amounts of sample and achieves nanojoule [15.20], picojoule [15.22] and femtojoule [15.23] sensitivity.

These three measurement modes have established cantilevers as versatile tools to perform experiments in nanoscale science with very small amounts of material.

15.3.4 Further Operation Modes

Photothermal Spectroscopy

When a material adsorbs photons, a fraction of the energy is converted into heat. This photothermal heating can be measured as a function of the light wavelength to provide optical absorption data of the material. The interaction of light with a bimetallic microcantilever creates heat on the cantilever surface, resulting in a bending of the cantilever [15.14]. Such bimetallic-cantilever devices are capable of detecting heat flows due to an optical heating power of 100 pW, which is two orders of magnitude better than in conventional photothermal spectroscopy.

Electrochemistry

A cantilever coated with a metallic layer (measurement electrode) on one side is placed in an electrolytic medium, e.g. a salt solution, together with a metallic reference electrode, usually made of a noble metal. If the voltage between the measurement and the reference electrode is changed, electrochemical processes on the measurement electrode (cantilever) are induced, such as adsorption or desorption of ions from the electrolyte solution onto the measurement electrode. These processes lead to a bending of the cantilever due to changes in surface stress and in the electrostatic forces [15.24].

Detection of Electrostatic and Magnetic Forces

The detection of electrostatic and magnetic forces is possible if charged or magnetic particles are deposited on the cantilever [15.25, 26]. If the cantilever is placed in the vicinity of electrostatic charges or magnetic particles, attractive or repulsion forces occur according to the polarity of the charges or magnetic particles present on the cantilever. These forces will result in an upward or a downward bending of the cantilever. The magnitude of the bending depends on the distribution of charged or magnetic particles on both the cantilever and in the surrounding environment according to the laws of electrostatics and magnetism.

15.4 Microfabrication

Silicon cantilever sensor arrays have been microfabricated using a dry-etching silicon-on-insulator (SOI) fabrication technique developed in the micro-/nanomechanics department at the IBM Zurich Research Laboratory. One chip comprises eight cantilevers, having a length of 500 μm, a width of 100 μm, and a thickness of 0.5 μm, and arranged on a pitch of 250 μm. For dynamic-mode operation, the cantilever thickness may be up to 7 μm. The resonance frequencies of the cantilevers vary by 0.5% only, demonstrating the high reproducibility and precision of cantilever fabrication. A scanning electron microscopy image of a cantilever sensor-array chip is shown in Fig. 15.5.

Fig. 15.5 Scanning electron micrograph of a cantilever-sensor array. © Viola Barwich, University of Basel, Switzerland

15.5 Measurement Setup

15.5.1 Measurements in Gaseous or Liquid Environments

A measurement set-up for cantilever arrays consists of four major parts: (1) the measurement chamber containing the cantilever array, (2) an optical or electrical system to detect the cantilever deflection [e.g. laser sources, collimation lenses and a position-sensitive detector (PSD), or piezoresistors and Wheatstone-bridge detection electronics], (3) electronics to amplify, process and acquire the signals from the detector, and (4) a gas- or liquid-handling system to inject samples

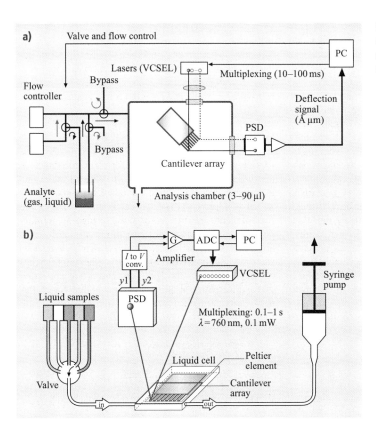

Fig. 15.6 Schematic of measurement setups for (**a**) a gaseous (artificial nose) and (**b**) a liquid environment (biochemical sensor)

reproducibly into the measurement chamber and purge the chamber.

Figure 15.6 shows the schematic set-up for experiments performed in a gaseous (Fig. 15.6a) and a liquid, biochemical (Fig. 15.6b) environment for the optical beam-deflection embodiment of the measurement set-up. The cantilever sensor array is located in an analysis chamber with a volume of 3–90 µl, which has inlet and outlet ports for gases or liquids. The cantilever deflection is determined by means of an array of eight vertical-cavity surface-emitting lasers (VCSELs) arranged at a linear pitch of 250 µm that emit at a wavelength of 760 nm into a narrow cone of 5 to 10°.

The light of each VCSEL is collimated and focused onto the apex of the corresponding cantilever by a pair of achromatic doublet lenses, 12.5 mm in diameter. This size has to be selected in such a way that all eight laser beams pass through the lens close to its center to minimize scattering, chromatic and spherical aberration artifacts. The light is then reflected off the gold-coated surface of the cantilever and hits the surface of a position-sensing detector (PSD). PSDs are light-sensitive photopotentiometer-like devices that produce photocurrents at two opposing electrodes. The magnitude of the photocurrents depends linearly on the distance of the impinging light spot from the electrodes. Thus the position of an incident light beam can easily be determined with micrometer precision. The photocurrents are transformed into voltages and amplified in a preamplifier. As only one PSD is used, the eight lasers cannot be switched on simultaneously. Therefore, a time-multiplexing procedure is used to switch the lasers on and off sequentially at typical intervals of 10–100 ms. The resulting deflection signal is digitized and stored together with time information on a personal computer (PC), which also controls the multiplexing of the VCSELs as well as the switching of the valves and mass flow controllers used for setting the composition ratio of the analyte mixture.

The measurement setup for liquids (Fig. 15.6b) consists of a polyetheretherketone (PEEK) liquid cell, which contains the cantilever array and is sealed by a viton O-ring and a glass plate. The VCSELs and the PSD are mounted on a metal frame around the liquid cell. After preprocessing the position of the deflected light beam in a current-to-voltage converter and amplifier stage, the signal is digitized in an analog-to-digital converter and stored on a PC. The liquid cell is equipped with inlet and outlet ports for liquids. They are connected via 0.18 mm inner-diameter Teflon tubing to individual thermally equilibrated glass containers, in which the biochemical liquids are stored. A six-position valve allows the inlet to the liquid chamber to be connected to each of the liquid-sample containers separately. The liquids are pulled (or pushed) through the liquid chamber by means of a syringe pump connected to the outlet of the chamber. A Peltier element is situated very close to the lumen of the chamber to allow temperature regulation within the chamber. The entire experimental setup is housed in a temperature-controlled box regulated with an accuracy of 0.01 K to the target temperature.

15.5.2 Readout Principles

This section describes various ways to determine the deflection of cantilever sensors. They differ in sensitivity, effort for alignment and setup, robustness and ease of readout as well as their potential for miniaturization.

Piezoresistive readout

Piezoresistive cantilevers [15.6, 20] are usually U-shaped, having diffused piezoresistors in both of the legs close to the hinge (Fig. 15.7a). The resistance in the piezoresistors is measured by a Wheatstone-bridge technique employing three reference resistors, one of which is adjustable. The current flowing between the two branches of the Wheatstone bridge is initially nulled by changing the resistance of the adjustable resistor. If the cantilever bends, the piezoresistor changes its value and a current will flow between the two branches of the Wheatstone bridge. This current is converted via a differential amplifier into a voltage for static-mode measurement. For dynamic-mode measurement,

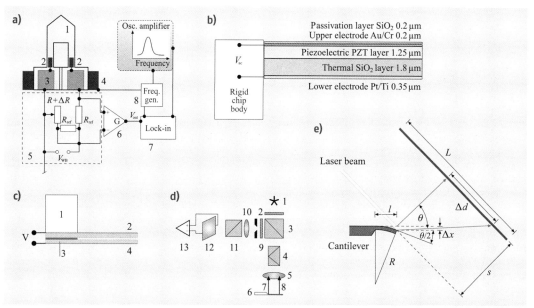

Fig. 15.7 (a) Piezoresistive readout: (1) cantilever, (2) piezoresistors, (3) Au contact pads, (4) external piezocrystal for actuation, (5) Wheatstone-bridge circuit, (6) differential amplifier, (7) lock-in amplifier, (8) function generator. **(b)** Piezoelectric readout. **(c)** Capacitive readout: (1) solid support, (2) rigid beam with counter-electrode, (3) insulation layer (SiO$_2$), (4) flexible cantilever with electrode. **(d)** Interferometric readout: (1) laser diode, (2) polarizer, (3) nonpolarizing beam splitter, (4) Wollaston prism, (5) focusing lens, (6) cantilever, (7) reference beam (near cantilever hinge), (8) object beam (near cantilever apex), (9) diaphragm and $\lambda/4$ plate, (10) focusing lens, (11) Wollaston prism, (12) quadrant photodiode, (13) differential amplifier. **(e)** Beam-deflection readout

the piezoresistive cantilever is externally actuated via a frequency generator connected to a piezocrystal. The alternating current (AC) actuation voltage is fed as reference voltage into a lock-in amplifier and compared with the response of the Wheatstone-bridge circuit. This technique allows one to sweep resonance curves and to determine shifts in resonance frequency.

Piezoelectric Readout

Piezoelectric cantilevers [15.27] are actuated by applying an electric AC voltage via the inverse piezoelectric effect (self-excitation) to the piezoelectric material (PZT or ZnO). Sensing of bending is performed by recording the piezoelectric current change due to the fact that the PZT layer may produce a sensitive field response to weak stress through the direct piezoelectric effect. Such cantilevers are multilayer structures consisting of an SiO_2 cantilever and the PZT piezoelectric layer. Two electrode layers, insulated from each other, provide electrical contact. The entire structure is protected using passivation layers (Fig. 15.7b). An identical structure is usually integrated into the rigid chip body to provide a reference for the piezoelectric signals from the cantilever.

Capacitive Readout

For capacitive readout (Fig. 15.7c), a rigid beam with an electrode mounted on the solid support and a flexible cantilever with another electrode layer are used [15.28, 29]. Both electrodes are insulated from each other. Upon bending of the flexible cantilever the capacitance between the two electrodes changes and allows the deflection of the flexible cantilever to be determined. Both static- and dynamic-mode measurements are possible.

Optical (Interferometric) Readout

Interferometric methods [15.30, 31] are most accurate for the determination of small movements. A laser beam passes through a polarizer plate (polarization 45°) and is partially transmitted by a nonpolarized beam splitter (Fig. 15.7d). The transmitted beam is divided in a Wollaston prism into a reference and an object beam. These mutually orthogonally polarized beams are then focused onto the cantilever. Both beams (the reference beam from the hinge region and the object beam from the apex region of the cantilever) are reflected back to the objective lens, pass the Wollaston prism, where they are recombined into one beam, which is then reflected into the other arm of the interferometer, where after the $\lambda/4$ plate a phase shift of a quarter wavelength between object and reference beam is established. Another Wollaston prism separates the reference and object beams again for analysis with a four-quadrant photodiode. A differential amplifier is used to obtain the cantilever deflection with high accuracy. However, the interferometric setup is quite bulky and difficult to handle.

Optical (Beam-Deflection) Readout

The most frequently used approach to read out cantilever deflections is optical beam deflection [15.32], because it is a comparatively simple method with an excellent lateral resolution. A schematic of this method is shown in Fig. 15.7e.

The actual cantilever deflection Δx scales with the cantilever dimensions; therefore the surface stress $\Delta \sigma$ in N/m is a convenient quantity to measure and compare cantilever responses. It takes into account the cantilever material properties, such as Poisson's ratio v, Young's modulus E and the cantilever thickness t. The radius of curvature R of the cantilever is a measure of bending, (15.2). As shown in the drawing in Fig. 15.7e, the actual cantilever displacement Δx is transformed into a displacement Δd on the PSD. The position of a light spot on a PSD is determined by measuring the photocurrents from the two facing electrodes. The movement of the light spot on the linear PSD is calculated from the two currents I_1 and I_2 and the size L of the PSD by

$$\Delta d = \frac{I_1 - I_2}{I_1 + I_2} \cdot \frac{L}{2} . \qquad (15.11)$$

As all angles are very small, it can be assumed that the bending angle of the cantilever is equal to half of the angle θ of the deflected laser beam, i.e. $\theta/2$. Therefore, the bending angle of the cantilever can be calculated to be

$$\frac{\theta}{2} = \frac{\Delta d}{2s} , \qquad (15.12)$$

where s is the distance between the PSD and the cantilever. The actual cantilever deflection Δx is calculated from the cantilever length l and the bending angle $\theta/2$ by

$$\Delta x = \frac{\theta/2}{2} \cdot l . \qquad (15.13)$$

Combination of (15.12) and (15.13) relates the actual cantilever deflection Δx to the PSD signal

$$\Delta x = \frac{l \Delta d}{4s} . \qquad (15.14)$$

The relation between the radius of curvature and the deflection angle is

$$\frac{\theta}{2} = \frac{l}{R},\quad(15.15)$$

and after substitution becomes

$$R = \frac{2ls}{\Delta d},\quad(15.16)$$

or $R = \frac{2\Delta x}{l^2}$.

15.6 Functionalization Techniques

15.6.1 General Strategy

To serve as sensors, cantilevers have to be coated with a sensor layer that is either highly specific, i.e. is able to recognize target molecules in a key–lock process, or partially specific, so that the sensor information from several cantilevers yields a pattern that is characteristic of the target molecules.

To provide a platform for specific functionalization, the upper surface of these cantilevers is typically coated with 2 nm of titanium and 20 nm of gold, which yields a reflective surface and an interface for attaching functional groups of probe molecules, e.g. for anchoring molecules with a thiol group to the gold surface of the cantilever. Such thin metal layers are believed not to contribute significantly to bimetallic bending, because the temperature is kept constant.

15.6.2 Functionalization Methods

There are numerous ways to coat a cantilever with material, both simple and more advanced ones. The method of choice should be fast, reproducible, reliable and allow one or both of the surfaces of a cantilever to be coated separately.

Simple Methods

Obvious methods to coat a cantilever are thermal or electron-beam-assisted evaporation of material, electrospray or other standard deposition methods. The disadvantage of these methods is that they only are suitable for coating large areas, but not individual cantilevers in an array, unless shadow masks are used. Such masks need to be accurately aligned to the cantilever structures, which is a time-consuming process.

Other methods to coat cantilevers use manual placement of particles onto the cantilever [15.9, 20, 33–35], which requires skillful handling of tiny samples. Cantilevers can also be coated by directly pipetting solutions of the probe molecules onto the cantilevers [15.36] or by employing air-brush spraying and shadow masks to coat the cantilevers separately [15.37].

All these methods have only limited reproducibility and are very time-consuming if a larger number of cantilever arrays has to be coated.

Microfluidics

Microfluidic networks (μFN) [15.38] are structures of channels and wells, etched several ten to hundred micrometer deep into silicon wafers. The wells can be filled easily using a laboratory pipette, so that the fluid with the probe molecules for coating the cantilever is guided through the channels towards openings at a pitch matched to the distance between individual cantilevers in the array (Fig. 15.8a).

The cantilever array is then introduced into the open channels of the μFN that are filled with a solution of the probe molecules. The incubation of the cantilever array in the channels of the μFN takes from a few seconds (self-assembly of alkanethiol monolayers) to several tens of minutes (coating with protein solutions). To prevent evaporation of the solutions, the channels are covered by a slice of poly(dimethylsiloxane) (PDMS). In addition, the microfluidic network may be placed in an environment filled with saturated vapor of the solvent used for the probe molecules.

Array of Dimension-Matched Capillaries

A similar approach is insertion of the cantilever array into an array of dimension-matched disposable glass capillaries. The outer diameter of the glass capillaries is 240 μm so that they can be placed neatly next to each other to accommodate the pitch of the cantilevers in the array (250 μm). Their inner diameter is 150 μm, providing sufficient room to insert the cantilevers (width: 100 μm) safely (Fig. 15.8b). This method has been successfully applied for the deposition of a variety of materials onto cantilevers, such as polymer solutions [15.37], self-assembled monolayers [15.39], thiol-functionalized single-stranded DNA oligonucleotides [15.40], and protein solutions [15.41].

Inkjet Spotting

All of the above techniques require manual alignment of the cantilever array and functionalization tool, and are therefore not ideal for coating a large number of cantilever arrays. The inkjet-spotting technique, however, allows rapid and reliable coating of cantilever arrays [15.42, 43]. An x-y-z positioning system allows a fine nozzle (capillary diameter: 70 μm) to be positioned with an accuracy of approximately 10 μm over a cantilever. Individual droplets (diameter: 60–80 μm, volume 0.1–0.3 nl) can be dispensed individually by means of a piezo-driven ejection system in the inkjet nozzle. When the droplets are spotted with a pitch smaller than 0.1 mm, they merge and form continuous films. By adjusting the number of droplets deposited on the cantilevers, the resulting film thickness can be controlled precisely. The inkjet-spotting technique allows a cantilever to be coated within seconds and yields very homogeneous, reproducibly deposited layers of well-controlled thickness. Successful coating of self-assembled alkanethiol

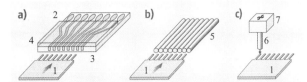

Fig. 15.8 (a) Cantilever functionalization in microfluidic networks. (b) Incubation in dimension-matched microcapillaries. (c) Coating with an inkjet spotter: (1) cantilever array, (2) reservoir wells, (3) microfluidic network with channels, (4) PDMS cover to avoid evaporation, (5) microcapillaries, (6) inkjet nozzle, (7) inkjet x-y-z positioning unit

monolayers, polymer solutions, self-assembled DNA single-stranded oligonucleotides [15.43], and protein layers has been demonstrated. In conclusion, inkjet spotting has turned out to be a very efficient and versatile method for functionalization, which can even be used to coat arbitrarily shaped sensors reproducibly and reliably [15.44, 45].

15.7 Applications

15.7.1 Chemical Detection

Hydrogen

Early reports on detection of gases such as hydrogen involve nanomechancal detection of catalytic reactions of bimetallic microcantilevers coated with aluminum and a top layer of platinum in thermal mode [15.9]. The catalytic reaction of oxygen present in a reaction chamber with hydrogen being introduced into the chamber produces oscillatory chemical reactions resulting in mechanical oscillations of the cantilever due to heat formation related to catalytic conversion of H_2 and O_2 to form H_2O. By use of an array of four platinum coated and four uncoated microcantilevers, a change of the deflection signal due to bending of the platinum coated cantilever relative to the uncoated cantilevers is observed upon hydrogen adsorption in the presence of oxygen [15.11]. Similar responses are obtained with Pd coated glass cantilevers [15.46] and with Pd coated silicon microcantilevers using dynamic mode [15.47], capacitive readout [15.48] or beam-deflection readout in static mode [15.49].

Water Vapor

First observation of microcantilever resonance frequency detuning is reported in [15.8]. A dependence on relative humidity of ZFM5 zeolites attached to resonating microcantilevers was observed in [15.50]. Relative humidity was measured with an accuracy of 1% using piezoresistive sensors embedded in polymer [15.51]. A detection limit of 10 ppm is achieved using Al_2O_3 coated microcantilevers [15.52].

Other Vapors

ZFM5 zeolites were used to detect vapor of p-nitroaniline dye in dynamic mode with picogram sensitivity [15.50]. A freon gas sensor using a piezoelectric microcantilever coated with MFI zeolite is described in [15.53]. Ethanol vapor detection in dynamic mode is described in [15.54].

Alkane Thiol Vapors

Surface stress changes and kinetics were measured in situ during the self-assembly of alkanethiols on gold by means of a micromechanical sensor, observing scaling of compressive surface stress with the length of the alkane chain [15.13, 55]. 65 ppb of 2-mercaptoethanol have been measured evaluating the response of gold-coated silicon nitride microcantilevers [15.56]. The mechanism of stress formation upon adsorption of thiol layers has been studied by exposing monolayers of alkanethiols on gold to low energy Ar ions, resulting in formation of a large tensile stress [15.57]. The influence of surface morphology and thickness of the gold coating of the cantilever is discussed in [15.58, 59]. A multiple-point deflection technique has been used to investigate stress evolution during the adsorption of dodecanethiol on microcantilever sensors, allowing to assess the cantilever bending profile [15.60]. Using gold-coated, piezoelectric-excited, millimeter-sized cantilevers exposed to 1-hexadecanethiol (HDT) in ethanol, a detection range between 1 fM to 1 mM is claimed [15.61]. The formation of alkanedithiol (HS–(CH)SH) monolayers on gold in solution is monitored using microcantilever sensors [15.62]. The nanomechanical bending of microfabricated cantilevers during the immobilization of alkanethiols of different chain lengths has been investigated in the liquid phase [15.63].

Metal Vapors

Detection of mercury vapor was one of the first applications of microcantilever sensors in dynamic mode [15.10]. 20 ppb of Hg vapor was detected using a microcantilever with an integrated piezoelectric film [15.64]. A monolayer of 1,6-hexanedithiol has been identified as an unusually specific recognition agent for CH_3Hg^+ [15.65]. 15 ppb detection limit for mercury is reported using microcantilevers that are thermally excited at the fundamental and first three higher order modes [15.66, 67]. Cs ion concentrations in the range of 10^{-11}–10^{-7} M were detected using a 1,3-alternate 25,27-bis(11-mercapto-1-undecanoxy)-26,28-calix[4]benzo-crown-6 caesium recognition agent bound to a gold coated microcantilever [15.68]. The crown cavity is highly selective to Cs, as compared to K or Na. An atomic force microscope cantilever has been used as a bending-beam sensor to measure surface stress changes which occur during electrochemical processes, such as the formation of a Pb layer on Au [15.69].

HF and HCN

Microcantilevers have been used as a gas sensor to detect hydrofluoric acid (HF) at a threshold of 0.2 ppm [15.70]. Femtomolar HF concentrations, which is also a decomposition component of nerve agents, were detected using a SiO_2 microcantilever. The high sensitivity is considered to be due to the reaction of HF with SiO_2 [15.71]. The etching rate is determined to 0.05 nm/min for SiO_2 and 0.7 nm/min for Si_3N_4 [15.72]. Sensor responses towards HCN at concentration of 150 ppm within seconds are reported [15.73].

Ion Sensing

Using microcantilevers coated with a self-assembled monolayer of triethyl-12-mercaptododecylammonium bromide on gold CrO_4^{2-} ions are detected at a concentration of 10^{-9} M. Other anions, such as Cl^-, Br^-, CO_3^{2-}, HCO_3^- and SO_4^{2-} do not deflect such modified cantilevers significantly [15.74]. Hg^{2+} has been measured at a concentration of 10^{-11} M using a microcantilever coated with gold. Almost no affinity to other cations exists, such as K^+, Na^+ Pb^{2+}, Zn^{2+}, Ni^{2+}, Cd^{2+}, Cu^{2+}, and Ca^{2+} [15.75]. Adsorption characteristics of Ca^{2+} ions as a function of concentration in aequous $CaCl_2$ solution was investigated in static and dynamic mode [15.76]. Microcantilevers functionalized with metal-binding protein AgNt84-6 are able to detect heavy metal ions like Hg^{2+} and Zn^{2+}, but are insensitive to Mn^{2+} [15.77]. Hydrogels containing benzo-18-crown-6 have been used to modify microcantilevers for measurements of the concentration of Pb^{2+} in aqueous solutions [15.78]. Using different thiolated ligands as self-assembled monolayers (SAMs) functionalized on silicon microcantilevers (MCs) coated with gold allows to detect Cs^+, Co^{2+} and Fe^{3+} [15.79]. A gold coated microcantilever is utilized as the working electrode to detect Cr(VI) [15.80]. Others use 11-undecenyltriethylammonium bromide [15.81] or sol–gel layers [15.82] for detection of Cr(VI). Based on the EDTA-Cd(II) complex and its binding capability to bovine serum albumine (BSA) and antibody-based Cd(II) sensor using microcantilevers is presented [15.83].

Volatile Organic Compounds

A microcantilever-based alcohol vapor sensor is described in [15.84] using the piezoresistive technique and polymer coating. They also present a simple evaporation model that allows determining the concentration. The detection limit found is 10 ppm for methanol, ethanol and 2-propanol. In [15.85] an integrated complementary metal oxide semiconductor (CMOS) chemical microsensor with piezoresistive detection (Wheatstone bridge configuration) using poly(etherurethane) (PEUT) as the sensor layer is presented. They are able to reversibly detect volatile organic compounds (VOCs) such as toluene, n-octane, ethyl acetate and ethanol with a sensitivity level down to 200 ppm. An improved version of that device is described in [15.86]. The sensitivity could be increased to 5 ppm for n-octane.

Later the technique has been refined by using electromagnetic rather than electrothermal actuation and transistor-based readout reducing power dissipation on the cantilever [15.87]. Piezoelectric readout in dynamic mode and electromagnetic actuation of cantilevers spray-coated with PEUT is reported in [15.88], achieving a sensitivity of 14 ppm for ethanol. In [15.89] a study is published how to prepare polyethylene glycol (PEG) coated microcantilever sensors using a microcapillary pipette assisted method. PEG coating is suitable for ethanol sensing as ethanol quickly forms hydrogen bonds with the OH groups of the PEG. Sensor operation is reported to be reversible and reproducible. In [15.90] artificial neural networks are used for analyte species and concentration identification with polymer coated optically read-out microcantilevers. The analytes detected are carbon dioxide, dichloromethane, diisopropylmethylphosphonate (DIMP), dioxane, ethanol, water, 2-propanol, methanol, trichloroethylene and trichloromethylene. In [15.91] the chemical sensing performance of a silicon reconant microcantilever sensor is investigated in dependence on the thickness of the sensitive coating. For a coating thickness of 1, 4 and 21 µm of PEUT a limit of detection of 30 ppm was found for ethanol. A new concept of parylene micromembrane array for chemical sensing is presented [15.92] using the capacitive method. The parylene membrane is suspended over a metal pad patterned on the substrate. The pad and part of the membrane that is metal-coated serve as electrodes for capacitive measurement. The top electrode located on the membrane is chemically modified by applying a gold layer and self-assembled thiol monolayers (–COOH, –CH3 and –OH) for detection of analyte molecules. Successful detection of 2-propanol and toluene is reported. In [15.93] a sensitive self-oscillating cantilever array is described for quantitative and qualitative analysis of organic vapor mixtures. The cantilevers are electromagnetically actuated and the resonance frequency is measured using a frequency counter. Sensor response is reproducible and reversible. Using PEUT coating the smallest measured concentration is 400 ppm, but the limit of detection is well below 1 ppm. In [15.94] a combination of gas chromatography with a microcantilever sensor array for enhanced selectivity is reported. Test VOC mixtures composed of acetone, ethanol and trichloroethylene in pentane, as well as methanol with acetonitrile in pentane were first separated in a gas chromatography column and then detected using micocantilevers coated with responsive phases such as 3-aminopropyltriethoxy silane, copper phtalocyanine and methyl-β-cyclodextrin. Analytes detected include pentane, methanol, acetonitrile, acetone, ethanol and trichloroethylene. In [15.95] results are presented on independent component analysis (ICA) of ethanol, propanol and DIMP using cantilever coated with molecular recognition phases (MRP), whereby ICA has proven its feature extraction ability for components in mixtures.

Toxins

Detection of the organochlorine insecticide compound dichlorodiphenyltrichloroethane (DDT) is reported using a synthetic hapten of the pesticide as recognition site conjugated with bovine serum albumin (BSA) covalently immobilised on the gold-coated side of the cantilever by using thiol self assembled monolayers [15.96].

Explosives, Chemical Warfare and Biohazards

Security measures require inexpensive, highly selective and very sensitive small sensors that can be mass-produced and microfabricated. Such low cost sensors could be arranged as a sensor grid for large area coverage of sensitive infrastructure, like airports, public buildings, or traffic infrastructure. Threats can be of chemical, biological, radioactive or explosive nature. Microcantilever sensors are reported to offer very high sensitivities of explosives detection. Photomechanical chemical microsensors based on adsorption-induced and photoinduced stress changes due to the presence of diisopropyl methyl phosphonate (DIMP), which is a model compound for phosphorous-containing chemical warfare agents, and trinitrotoluene (TNT), an explosive are reported [15.97]. Further explosives frequently used include pentaerythritol tetranitrate (PETN) and hexahydro-1,3,5-triazine (RDX), often also with plastic fillers [15.98]. These compounds are very stable, if no detonator is present. Their explosive power, however, is very large, and moreover, the vapor pressures of PETN and RDX are very low, in the range of ppb and ppt. By functionalizing microcantilevers with self-assembled monolayers of 4-mercaptobeonzoic acid (4-MBA) PETN was detected at a level of 1400 ppt and RDX at a level of 290 ppt [15.99]. TNT was found to readily stick to Si surfaces, suggesting the use of microcantilevers for TNT detection, taking advantage of the respective adsorption/desorption kinetics [15.100, 101]. Detection of TNT via deflagration on a microcantilever is described by *Pinnaduwage* et al. [15.101]. They used piezoresistive microcantilevers where the cantilever deflection was measured optically via beam deflection.

TNT vapor from a generator placed 5 mm away from the microcantilever was observed to adsorb on its surface resulting in a decrease of resonance frequency. Application of an electrical pulse (10 V, 10 ms) to the piezoresistive cantilever resulted in deflagration of the TNT vapor and a bump in the cantilever bending signal. This bump was found to be related to the heat produced during deflagration. The amount of heat released is proportional to the area of the bump in the time versus bending signal diagram of the process. The deflagration was found to be complete, as the same resonance frequency as before the experiment was observed. The amount of TNT mass involved was determined as 50 pg. The technique was later extended to the detection of PETN and RDX, where much slower reaction kinetics was observed [15.99, 102]. Traces of 2,4-dinitrotoluene (DNT) in TNT can also be used for detection of TNT, because it is the major impurity in production grade TNT. Furthermore DNT is a decomposition product of TNT. The saturation concentration of DNT in air at 20 °C is 25 times higher than that of TNT. DNT was reported detectable at the 300 ppt level using polysiloxane polymer layers [15.103]. Microfabrication of electrostatically actuated resonant microcantilever beams in CMOS technology for detection of the nerve agent stimulant dimethylmethylphosphonate (DMMP) using polycarbosilane-coated beams [15.104] is an important step towards an integrated platform based on silicon microcantilevers, which besides compactness might also include telemetry [15.105]. Cu^{2+}/L-cysteine bilayer-coated microcantilever demonstrated high sensitivity and selectivity toward organo-phosphorus compounds in aqueous solution. The microcantilever was found to undergo bending upon exposure to nerve agent simulant DMMP at concentrations as low as 10^{-15} M due to the complexation of the phosphonyl group and the Cu^{2+}/L-cysteine bilayer on the microcantilever surface [15.106, 107].

15.7.2 Biochemical Environment

pH

Control of pH is often important in biochemical reactions. Therefore this section concerns measurement of pH using microcantilevers. The interfacial stress of self-assembled monolayers of mercaptohexadecanoic acid and hexadecanethiol depends on pH values and ionic strength [15.39]. SiO_2 and silicon nitride microcantilevers were also found to exhibit a deflection dependence with pH when coated with 4-aminobutyltriethoxysilane, 11-mercaptoundecanoic acid and Au/Al-coated over a pH range 2–12. Aminosilane-modified SiO_2/Au cantilevers performed robustly over pH range 2–8 (49 nm deflection/pH unit), while Si_3N_4/Au cantilevers performed well at pH 2–6 and 8–12 (30 nm deflection/pH unit) [15.108]. Microcantilevers with poly(methacrylic acid) (PMAA) and poly(ethylene glycol) dimethacrylate coating showed to be sensitive to pH changes [15.109]. Also hydrogel catings were found to be sensitive to pH [15.110]. The dependence of the micromechanical responses to different ionic strength and ion species present in the aqueous environment is discussed in [15.111], highlighting the critical role of counter- and co-ions on surface stress.

Glucose

Glucose sensing via microcantilevers is achieved by coating the cantilevers with the enzyme glucose oxidase on gold [15.112] or via polyethyleneimine (PEI) conjugation [15.113]. Glucose concentrations between 0.2 and 20 mM could be detected [15.114]. In another study a detection range between 2 and 50 mM is reported for glucose. No signal is observed for fructose, mannose and galactose [15.115].

Hydrogen Peroxide (H_2O_2)

Hydrogen peroxide is detected at the nM level using multilayer modified microcantilevers functionalized through a layer-by-layer nanoassembly technique via intercalation of the enzyme horseradish peroxidase. The magnitudes of bending were found to be proportional to the concentrations of hydrogen peroxide [15.116].

DNA, RNA

Specific DNA hybridization detection was observed via surface stress changes related to transduction of receptor-ligand binding into a direct nanomechanical response of microfabricated cantilevers without the need for external labeling or amplification. The differential deflection of the cantilevers was found to provide a true molecular recognition signal despite large responses of individual cantilevers. Hybridization of complementary oligonucleotides shows that a single base mismatch between two 12-mer oligonucleotides is clearly detectable [15.17]. The findings were confirmed or modeled by several groups [15.117, 118]. Hybridization in a complex nonspecific background was observed in a complement concentration range between 75 nM and 2 μM [15.40], following Langmuir model kinetics [15.119]. Enzymatic processes were directly performed on a microcantilever functionalized with DNA incorporating a Hind III restriction endonuclease site, followed by digestion with Hind III to produce DNA comprising a single-stranded end on the cantilever surface. Ligase was used to couple a second DNA molecule with a compatible end to the DNA on the cantilever [15.120]. Using gold nanoparticle labeled DNA, microcantilevers have been used to detect DNA strands with a specific sequence in dynamic mode, whereby a concentration of 23 pM could still be detected, as well as a single basepair mismatch [15.121]. Whereby adsorption of thiol functionalized single-stranded DNA is easily observed, hybridization cannot be observed if long hydrocarbon spacer molecules between single strand DNA and thiol anchor are used [15.122]. DNA hybridization is also observed using piezoresistive cantilevers [15.119, 123]. A different technique to read out the microcantilever deflections in an array is reported in [15.124]. There the optical beam deflection technique is combined with the scanning of a laser beam illuminating the cantilevers of an array sequentially. DNA hybridization is also reported using polymer SU-8 cantilevers [15.125]. Mukhopadhyay et al. report 20 nM hybridization sensitivity using piezoresistive cantilevers and DNA sequences with an overhang extension distal to the surface [15.126]. A larger array comprising 20 microcantilevers is described in [15.127]. Moreover, the authors present integration of the array with microfluidics. Surface stress changes in response to thermal dehybridization, or melting, is reported [15.128]. The dependence of salt concentration and hybridization efficiency is discussed in [15.129]. Two different DNA-binding proteins, the transcription factors SP1 and NF-kappa B are investigated [15.130]. Phase transition and stability issues of DNA are discussed in [15.131]. Differential gene expression of the gene 1-8U, a potential marker for cancer progression or viral infections, has been observed in a complex background. The measurements provide results within minutes at the picomolar level without target amplification, and are sensitive to base mismatches [15.132].

Proteins and Peptides

Microfabricated cantilevers were utilized to detect adsorption of low-density lipoproteins (LDL) and their oxidized form (oxLDL) on heparin, an to detect adsorption of bovine serum albumine and Immunoglobuline G (IgG) [15.133]. In [15.134] the activity, stability, lifetime and re-usability of monoclonal antibodies to myoglobin covalently immobilised onto microfabricated cantilever surfaces was investigated. Using piezoresistive microcantilevers the interaction of anti-bovine serum albumin (a-BSA) with bovine serum albumin (BSA) was studied [15.135]. Continuous label-free detection of two cardiac biomarker proteins (creatin kinase and myoglobin) is demonstrated using an array of microfabricated cantilevers functionalized with covalently anchored anti-creatin kinase and anti-myoglobin antibodies [15.41]. Label-free protein detection is reported using a microcantilever functionalized with DNA aptamers receptors for Taq DNA polymerase [15.136]. Label-free detection of C-reactive protein (CRP) using resonant frequency shift in piezoresistive cantilevers is described in [15.137], utilizing the specific binding characteristics of CRP antigen to its antibody, which is immobilized with Calixcrown SAMs on Au. Receptors on microcantilevers for serotonin, but insensitive to its biological precursor with similar structure tryptophan are described in [15.138]. Using single chain fragment antibodies instead of complete antibodies allowed to lower the limit of detection to concentrations of about 1 nM [15.139]. In [15.140] detection of prostate specific antigen (PSA) and C-reactive protein is reported. Detection of the human oestrogen receptor in free and oestradiol-bound conformation can be distinguished [15.141]. The Ca^{2+} binding protein calmodulin changes its conformation in presence of absence of Ca^{2+} resulting in a microcantilver deflection change [15.142]. No effect is observed upon exposure to K^+ and Mg^{2+}. Detection of activated cyclic adenosine monophosphate (cyclic AMP)-dependent protein kinase (PKA) is performed in dynamic mode employing a peptide derived from the heat-stable protein kinase inhibitor (PKI) [15.143]. Detection of streptavidin at 1–10 nM concentration is reported using biotin-coated cantilevers [15.144]. Using GST (glutathione-S-transferase) for detection of GST antibodies, a sensitivity of 40 nM was obtained [15.145]. A two-dimensional multiplexed real-time, label-free antibody–antigen binding assay by optically detecting nanoscale motions of two-dimensional arrays of microcantilever beams is presented in [15.146]. Prostate specific antigen (PSA) was detected at 1 ng/ml using antibodies covalently bound to one surface of the cantilevers. Conformational changes in membrane protein patches of bacteriorhodopsin proteoliposomes were observed with microcantilevers through prosthetic retinal removal (bleaching) [15.147]. Using an analog of the myc-tag decapeptide, binding of anti-myc-tag antibodies is reported [15.148].

Lipid Bilayers, Liposomes, Cells

Cantilever array sensors can sense the formation by vesicle fusion of supported phospholipid bilayers of 1,2-dioleoyl-sn-glycero-3-phosphocholine (DOPC) on their surface and can monitor changes in mechanical properties of lipid bilayers [15.149]. Liposomes are detected based on their interaction with protein C2A which recognizes the phosphatidylserine (PS) exposed on the surface of liposome [15.150]. Individual *Escherichia coli* (*E. coli*) O157:H7 cell–antibody binding events using microcantilevers operated in dynamic mode are reported [15.151]. The contractile force of self-organized cardiomyocytes was measured on biocompatible poly(dimethylsiloxane) cantilevers, representing a microscale cell-driven motor system [15.152]. Resonanting cantilevers were used to detect individual phospholipid vesicle adsorption in liquid. A resonance frequency shift corresponding to an added mass of 450 pg has been measured [15.153].

Spores, Bacteria and Viruses

Micromechanical cantilever arrays have been used for quantitative detection of vital fungal spores of *Aspergillus niger* and *Saccharomyces cerevisiae*. The specific adsorption and growth on concanavalin A, fibronectin or immunoglobulin G cantilever surfaces was investigated. Maximum spore immobilization, germination and mycelium growth was observed on the immunoglobulin G functionalized cantilever surfaces, as measured from shifts in resonance frequency within a few hours, being much faster than with standard Petri dish cultivation [15.154]. Short peptide ligands can be used to efficiently capture *Bacillus subtilis* (a simulant of Bacillus anthracis) spores in liquids. Fifth-mode resonant frequency measurements were performed before and after dipping microcantilever arrays into a static *B. subtilis* solution showing a substantial decrease in frequency for binding-peptide-coated microcantilevers as compared to that for control peptide cantilevers [15.155].

Medical

A bioassay of prostate-specific antigen (PSA) using microcantilevers has been presented in [15.156], covering a wide range of concentrations from 0.2 ng/ml to 60 µg/ml in a background of human serum albumin (HSA). Detection has been confirmed by another group using microcantilevers in resonant mode [15.157, 158]. The feasibility of detecting severe acute respiratory syndrome associated coronavirus (SARS-CoV) using microcantilever technology is studied in [15.159] by showing that the feline coronavirus (FIP) type I virus can be detected by a microcantilever modified by FIP type I anti-viral antiserum. A method for quantification of a prostate cancer biomarker in urine without sample preparation using monoclonal antibodies in described in [15.160].

15.7.3 Microcantilever Sensors to Measure Physical Properties

Besides chemical and biochemical sensing, microcantilevers can also detect changes in physical properties of surrounding media, such as gas or liquid, or of layers deposited on the cantilever itself.

Density and Viscosity

A piezoelectric unimorph cantilever as a liquid viscosity-and-density sensor was tested using water-glycerol solutions of different compositions, whereby the resonance frequency decreased while the width of the resonance peak increased with increasing glycerol content [15.161]. The viscosity of complex organic liquids with non-Newtonian behavior is studied in [15.162] in a wide range from 10 to 500 mm^2/s. Simultaneous determination of density and viscosity of water/ethanol mixtures based on resonance curves of microcantilevers is reported in [15.163]. A detailed theoretical study of viscoelastic effects on the frequency shift of microcantilever chemical sensors is given in [15.164]. Microcantilever deflection as a function of flow speed of viscous fluids is investigated in [15.165]. Viscosity of sugar solutions is tested using microcantilevers [15.166].

Gas and Flow Sensing

Gas sensing does not only involve chemical detection, but also pressure and flow sensing. *Brown* et al. [15.167] have studied the behavior of magnetically actuated oscillating microcantilevers at large deflections and have found hysteresis behavior at resonance. The amplitude at the actuation frequency changes depending on pressure due to damping. The authors have used cantilever in cantilever (CIC) structures, and have observed changes in deflection as gas pressure is varied. At atmospheric pressure, damping is large and the oscillation amplitude is relative small and hysteresis effects are absent. At lower pressure, abrupt changes in the oscillation amplitude occur with changes in the driving frequency. Since the change of amplitude and driving frequency, at which they occur are pressure dependent, these quantities can be used for accurate determination of gas pressure, demonstrated in the range between 10^{-3} and 10^2 Torr. *Brown* et al. [15.168] emphasize that microelectromechanical system pressure sensors will have a wide range of applications, especially in the automotive industry. Piezoresistive cantilever based deflection measurement has major advantages over diaphragms. The pressure range has been extended to 15–1450 Torr by means of design geometry adaptation. *Su* et al. [15.169] present highly sensitive ultrathin piezoresistive silicon microcantilevers for gas velocity sensing, whereby the deflection increases with airflow distribution in a steel pipe. The detection principle is based on normal pressure drag producing bending of the cantilever. The minimum flow speed measured was 7.0 cm/s, which is comparable to classical hot-wire anemometers. *Mertens* et al. [15.170] have investigated the effects of temperature and pressure on microcantilever resonance response in helium and nitrogen. Resonance response as a function of pressure showed three different regimes, which correspond to molecular flow, transition regimes and viscous flow, whereby the frequency variation of the cantilever is mainly due to changes in the mean free path of gas molecules. Effects observed allow measurement of pressures between 7.5×10^{-5} and 7500 Torr. *Mortet* et al. [15.171] present a pressure sensor based on a piezoelectric bimorph microcantilever with a measurement range between 75 and 6400 Torr. The resonance frequency shift is constant for pressures below 375 Torr. For higher pressures the sensitivity is typically a few ppm/mbar, but depends on the mode number. *Sievilä* et al. [15.172] present a cantilever paddle within a frame operating like a moving mirror to detect the displacements in the oscillating cantilever using a He-Ne laser in a Michelson interferometer configuration, whereby the cantilever acts as moving mirror element in one path of the interferometer. A fixed mirror serves a reference in the other arm of the interferometer.

Thermal Expansion
The thermal expansion of TaO_xN_y thin films deposited on a microcantilever was measured to examine the residual stress and the thermal expansion coefficient by observsing the changes in radius of curvature [15.173]. Thermal drift issues of resonaniting microcantilevers are discussed in detail in [15.174].

Infrared Imaging
Microcantilevers can also be used as uncooled, microcantilever-based infrared (IR) imaging devices by monitoring the bending of the microcantilever as a function of the IR radiation intensity incident on the cantilever surface. The infrared (thermal) image of the source is obtained by rastering a single microfabricated cantilever across the image formed at the focal plane of a concave mirror [15.175–177]. The method has later been refined such that photons are detected using the stress caused by photoelectrons emitted from a Pt film surface in contact with a semiconductor microstructure, which forms a Schottky barrier. The photoinduced bending of the Schottky barrier microstructure is due to electronic stress produced by photoelectrons diffusing into the microstructure [15.178]. The performance of IR imaging via microcantilevers has been enhanced by one-fold leg and two-fold legs beam structures with absorber plates [15.179–181].

15.8 Conclusions and Outlook

Cantilever-sensor array techniques have turned out to be a very powerful and highly sensitive tool to study physisorption and chemisorption processes, as well as to determine material-specific properties such as heat transfer during phase transitions. Experiments in liquids have provided new insights into such complex biochemical reactions as the hybridization of DNA or molecular recognition in antibody–antigen systems or proteomics.

Future developments must go towards technological applications, in particular to find new ways to characterize real-world samples such as clinical samples. The development of medical diagnosis tools requires an improvement of the sensitivity of a large number of genetic tests to be performed with small amounts of single donor-blood or body-fluid samples at low cost. From a scientific point of view, the challenge lies in optimizing cantilever sensors to improve their sensitivity to the ultimate limit: the detection of individual molecules.

Several fundamentally new concepts in microcantilever sensing are available in recent literature, which could help to achieve these goals: the issue of low quality factor of resonating microcantilevers in liquid has been elegantly solved by fabrication of a hollow cantilever that can be filled with biochemical liquids. Confining the fluid to the inside of a hollow cantilever also allows direct integration with conventional microfluidic systems, and significantly increases sensitivity by eliminating high damping and viscous drag [15.182]. Biochemical selectivity can be enhanced by using enantioselective receptors [15.183]. Other shapes for micromechanical sensors like microspirals could be advantageous for biochemical detection [15.184]. Miniaturization of microcantilevers into *true* nanometric dimensions, like by using single wall carbon nanotubes [15.185] or graphene sheets [15.186] will further increase sensitivity.

References

15.1 G. Binnig, C.F. Quate, C. Gerber: Atomic force microscope, Phys. Rev. Lett. **56**, 930–933 (1986)

15.2 R.J. Wilfinger, P.H. Bardell, D.S. Chhabra: The resonistor, a frequency sensitive device utilizing the mechanical resonance of a silicon substrate, IBM J. **12**, 113–118 (1968)

15.3 T.M.S. Heng: Trimming of microstrip circuits utilizing microcantilever air gaps, IEEE Trans. Microw. Theory Technol. **19**, 652–654 (1971)

15.4 K.E. Petersen: Micromechanical membrane switches on silicon, IBM J. Res. Dev. **23**, 376–385 (1979)

15.5 E.S. Kolesar: Electronic nerve agent detector, US Patent 4549427 (1983)

15.6 T. Itoh, T. Suga: Force sensing microcantilever using sputtered zinc-oxide thin-film, Appl. Phys. Lett. **64**, 37–39 (1994)

15.7 J.P. Cleveland, S. Manne, D. Bocek, P.K. Hansma: A nondestructive method for determining the spring constant of cantilevers for scanning force microscopy, Rev. Sci. Instrum. **64**, 403–405 (1993)

15.8 T. Thundat, R.J. Warmack, G.Y. Chen, D.P. Allison: Thermal, ambient-induced deflections of scanning force microscope cantilevers, Appl. Phys. Lett. **64**, 2894–2896 (1994)

15.9 J.K. Gimzewski, C. Gerber, E. Meyer, R.R. Schlittler: Observation of a chemical reaction using a micromechanical sensor, Chem. Phys. Lett. **217**, 589–594 (1994)

15.10 T. Thundat, G.Y. Chen, R.J. Warmack, D.P. Allison, E.A. Wachter: Vapor detection using resonating microcantilevers, Anal. Chem. **67**, 519–521 (1995)

15.11 H.P. Lang, R. Berger, C. Andreoli, J. Brugger, M. Despont, P. Vettiger, C. Gerber, J.K. Gimzewski, J.-P. Ramseyer, E. Meyer, H.-J. Güntherodt: Sequential position readout from arrays of micromechanical cantilever sensors, Appl. Phys. Lett. **72**, 383–385 (1998)

15.12 H. Ibach: Adsorbate-induced surface stress, J. Vac. Sci. Technol. A **12**, 2240–2245 (1994)

15.13 R. Berger, E. Delamarche, H.P. Lang, C. Gerber, J.K. Gimzewski, E. Meyer, H.-J. Güntherodt: Surface stress in the self-assembly of alkanethiols on gold, Science **276**, 2021–2024 (1997)

15.14 J.R. Barnes, R.J. Stephenson, M.E. Welland, C. Gerber, J.K. Gimzewski: Photothermal spectroscopy with femtojoule sensitivity based on micromechanics, Nature **372**, 79–81 (1994)

15.15 G.G. Stoney: The tension of metallic films deposited by electrolysis, Proc. R. Soc. **82**, 172–177 (1909)

15.16 F.J. von Preissig: Applicability of the classical curvature-stress relation for thin films on plate substrates, J. Appl. Phys. **66**, 4262–4268 (1989)

15.17 J. Fritz, M.K. Baller, H.P. Lang, H. Rothuizen, P. Vettiger, E. Meyer, H.-J. Güntherodt, C. Gerber, J.K. Gimzewski: Translating biomolecular recognition into nanomechanics, Science **288**, 316–318 (2000)

15.18 M. Alvarez, L.G. Carrascosa, M. Moreno, A. Calle, A. Zaballos, L.M. Lechuga, C.-A. Martinez, J. Tamayo: Nanomechanics of the formation of DNA self-assembled monolayers, hybridization on microcantilevers, Langmuir **20**, 9663–9668 (2004)

15.19 D. Sarid: *Scanning Force Microscopy with Applications to Electric, Magnetic, and Atomic Forces* (Oxford Univ. Press, New York 1991)

15.20 R. Berger, H.P. Lang, C. Gerber, J.K. Gimzewski, J.H. Fabian, L. Scandella, E. Meyer, H.-J. Güntherodt: Micromechanical thermogravimetry, Chem. Phys. Lett. **294**, 363–369 (1998)

15.21 J.E. Sader: Frequency response of cantilever beams immersed in viscous fluids with applications to the atomic force microscope, J. Appl. Phys. **84**, 64–76 (1998)

15.22 T. Bachels, F. Tiefenbacher, R. Schafer: Condensation of isolated metal clusters studied with a calorimeter, J. Chem. Phys. **110**, 10008–10015 (1999)

15.23 J.R. Barnes, R.J. Stephenson, C.N. Woodburn, S.J. O'Shea, M.E. Welland, T. Rayment, J.K. Gimzewski, C. Gerber: A femtojoule calorimeter using micromechanical sensors, Rev. Sci. Instrum. **65**, 3793–3798 (1994)

15.24 T.A. Brunt, T. Rayment, S.J. O'Shea, M.E. Welland: Measuring the surface stresses in an electrochemically deposited monolayer: Pb on Au(111), Langmuir **12**, 5942–5946 (1996)

15.25 R. Puers, D. Lapadatu: Electrostatic forces and their effects on capacitive mechanical sensors, Sens. Actuators A **56**, 203–210 (1996)

15.26 J. Fricke, C. Obermaier: Cantilever beam accelerometer based on surface micromachining technology, J. Micromech. Microeng. **3**, 190–192 (1993)

15.27 C. Lee, T. Itoh, T. Ohashi, R. Maeda, T. Suga: Development of a piezoelectric self-excitation, self-detection mechanism in PZT microcantilevers for dynamic scanning force microscopy in liquid, J. Vac. Sci. Technol. B **15**, 1559–1563 (1997)

15.28 T. Göddenhenrich, H. Lemke, U. Hartmann, C. Heiden: Force microscope with capacitive displacement detection, J. Vac. Sci. Technol. A **8**, 383–387 (1990)

15.29 J. Brugger, R.A. Buser, N.F. de Rooij: Micromachined atomic force microprobe with integrated capacitive read-out, J. Micromech. Microeng. **2**, 218–220 (1992)

15.30 C. Schönenberger, S.F. Alvarado: A differential interferometer for force microscopy, Rev. Sci. Instrum. **60**, 3131–3134 (1989)

15.31 M.J. Cunningham, S.T. Cheng, W.W. Clegg: A differential interferometer for scanning force microscopy, Meas. Sci. Technol. **5**, 1350–1354 (1994)

15.32 G. Meyer, N.M. Amer: Novel optical approach to atomic force microscopy, Appl. Phys. Lett. **53**, 2400–2402 (1988)

15.33 R. Berger, C. Gerber, J.K. Gimzewski, E. Meyer, H.-J. Güntherodt: Thermal analysis using a micromechanical calorimeter, Appl. Phys. Lett. **69**, 40–42 (1996)

15.34 L. Scandella, G. Binder, T. Mezzacasa, J. Gobrecht, R. Berger, H.P. Lang, C. Gerber, J.K. Gimzewski, J.H. Koegler, J.C. Jansen: Combination of single crystal zeolites and microfabrication: Two applications towards zeolite nanodevices, Microporous Mesoporous Mater. **21**, 403–409 (1998)

15.35 R. Berger, C. Gerber, H.P. Lang, J.K. Gimzewski: Micromechanics: a toolbox for femtoscale science: towards a laboratory on a tip, Microelectron. Eng. **35**, 373–379 (1997)

15.36 H.P. Lang, R. Berger, F.M. Battiston, J.-P. Ramseyer, E. Meyer, C. Andreoli, J. Brugger, P. Vettiger, M. Despont, T. Mezzacasa, L. Scandella, H.-J. Güntherodt, C. Gerber, J.K. Gimzewski: A chemical sensor based on a micromechanical cantilever array for the identification of gases and vapors, Appl. Phys. A **66**, 61–64 (1998)

15.37 M.K. Baller, H.P. Lang, J. Fritz, C. Gerber, J.K. Gimzewski, U. Drechsler, H. Rothuizen, M. Despont, P. Vettiger, F.M. Battiston, J.-P. Ramseyer, P. Fornaro, E. Meyer, H.-J. Güntherodt: A cantilever array based artificial nose, Ultramicroscopy **82**, 1–9 (2000)

15.38 S. Cesaro-Tadic, G. Dernick, D. Juncker, G. Buurman, H. Kropshofer, B. Michel, C. Fattinger, E. Delamarche: High-sensitivity miniaturized immunoassays for tumor necrosis factor α using microfluidic systems, Lab Chip **4**, 563–569 (2004)

15.39 J. Fritz, M.K. Baller, H.P. Lang, T. Strunz, E. Meyer, H.-J. Güntherodt, E. Delamarche, C. Gerber, J.K. Gimzewski: Stress at the solid-liquid interface of self-assembled monolayers on gold investigated with a nanomechanical sensor, Langmuir **16**, 9694–9696 (2000)

15.40 R. McKendry, J. Zhang, Y. Arntz, T. Strunz, M. Hegner, H.P. Lang, M.K. Baller, U. Certa, E. Meyer, H.-J. Güntherodt, C. Gerber: Multiple label-free biodetection and quantitative DNA-binding assays on a nanomechanical cantilever array, Proc. Natl. Acad. Sci. USA **99**, 9783–9787 (2002)

15.41 Y. Arntz, J.D. Seelig, H.P. Lang, J. Zhang, P. Hunziker, J.-P. Ramseyer, E. Meyer, M. Hegner, C. Gerber: Label-free protein assay based on a nanomechanical cantilever array, Nanotechnology **14**, 86–90 (2003)

15.42 A. Bietsch, M. Hegner, H.P. Lang, C. Gerber: Inkjet deposition of alkanethiolate monolayers and DNA oligonucleotides on gold: Evaluation of spot uniformity by wet etching, Langmuir **20**, 5119–5122 (2004)

15.43 A. Bietsch, J. Zhang, M. Hegner, H.P. Lang, C. Gerber: Rapid functionalization of cantilever array sensors by inkjet printing, Nanotechnology **15**, 873–880 (2004)

15.44 D. Lange, C. Hagleitner, A. Hierlemann, O. Brand, H. Baltes: Complementary metal oxide semiconductor cantilever arrays on a single chip: mass-sensitive detection of volatile organic compounds, Anal. Chem. **74**, 3084–3085 (2002)

15.45 C.A. Savran, T.P. Burg, J. Fritz, S.R. Manalis: Microfabricated mechanical biosensor with inherently differential readout, Appl. Phys. Lett. **83**, 1659–1661 (2003)

15.46 S. Okuyama, Y. Mitobe, K. Okuyama, K. Matsushita: Hydrogen gas sensing using a Pd-coated cantilever, Jpn. J. Appl. Phys. **39**, 3584–3590 (2000)

15.47 A. Fabre, E. Finot, J. Demoment, S. Contreras: Monitoring the chemical changes in Pd induced by hydrogen absorption using microcantilevers, Ultramicroscopy **97**, 425–432 (2003)

15.48 D.R. Baselt, B. Fruhberger, E. Klaassen, S. Cemalovic, C.L. Britton, S.V. Patel, T.E. Mlsna, D. McCorkle, B. Warmack: Design and performance of a microcantilever-based hydrogen sensor, Sens. Actuator. B Chem. **88**, 120–131 (2003)

15.49 Y.-I. Chou, H.-C. Chiang, C.-C. Wang: Study on Pd functionalization of microcantilever for hydrogen detection promotion, Sens. Actuator. B Chem. **129**, 72–78 (2008)

15.50 L. Scandella, G. Binder, T. Mezzacasa, J. Gobrecht, R. Berger, H.P. Lang, C. Gerber, J.K. Gimzewski, J.H. Koegler, J.C. Jansen: Combination of single crystal zeolites and microfabrication: Two applications towards zeolite nanodevices, Microporous Mesoporous Mater. **21**, 403–409 (1998)

15.51 R.L. Gunter, W.D. Delinger, T.L. Porter, R. Stewart, J. Reed: Hydration level monitoring using embedded piezoresistive microcantilever sensors, Med. Eng. Phys. **27**, 215–220 (2005)

15.52 X. Shi, Q. Chen, J. Fang, K. Varahramyan, H.F. Ji: Al_2O_3-coated microcantilevers for detection of moisture at ppm level, Sens. Actuator. B Chem. **129**, 241–245 (2008)

15.53 J. Zhou, P. Li, S. Zhang, Y.C. Long, F. Zhou, Y.P. Huang, P.Y. Yang, M.H. Bao: Zeolite-modified microcantilever gas sensor for indoor air quality control, Sens. Actuator. B Chem. **94**, 337–342 (2003)

15.54 A. Vidic, D. Then, C. Ziegler: A new cantilever system for gas and liquid sensing, Ultramicroscopy **97**, 407–416 (2003)

15.55 R. Berger, E. Delamarche, H.P. Lang, C. Gerber, J.K. Gimzewski, E. Meyer, H.J. Güntherodt: Surface stress in the self-assembly of alkanethiols on gold probed by a force microscopy technique, Appl. Phys. A **66**, S55–S59 (1998)

15.56 P.G. Datskos, I. Sauers: Detection of 2-mercaptoethanol using gold-coated micromachined cantilevers, Sens. Actuator. B Chem. **61**, 75–82 (1999)

15.57 A.N. Itakura, R. Berger, T. Narushima, M. Kitajima: Low-energy ion-induced tensile stress of self-assembled alkanethiol monolayers, Appl. Phys. Lett. **80**, 3712–3714 (2002)

15.58 M. Godin, P.J. Williams, V. Tabard-Cossa, O. Laroche, L.Y. Beaulieu, R.B. Lennox, P. Grutter: Surface stress, kinetics, and structure of alkanethiol self-assembled monolayers, Langmuir **20**, 7090–7096 (2004)

15.59 R. Desikan, I. Lee, T. Thundat: Effect of nanometer surface morphology on surface stress and adsorption kinetics of alkanethiol self-assembled monolayers, Ultramicroscopy **106**, 795–799 (2006)

15.60 S. Jeon, N. Jung, T. Thundat: Nanomechanics of a self-assembled monolayer on microcantilever sensors measured by a multiple-point deflection technique, Sens. Actuator. B Chem. **122**, 365–368 (2007)

15.61 K. Rijal, R. Mutharasan: Method for measuring the self-assembly of alkanethiols on gold at femtomolar concentrations, Langmuir **23**, 6856–6863 (2007)

15.62 S. Kohale, S.M. Molina, B.L. Weeks, R. Khare, L.J. Hope-Weeks: Monitoring the formation of self-assembled monolayers of alkanedithiols using a micromechanical cantilever sensor, Langmuir **23**, 1258–1263 (2007)

15.63 R. Desikan, S. Armel, H.M. Meyer, T. Thundat: Effect of chain length on nanomechanics of alkanethiol self-assembly, Nanotechnology **18**, 424028 (2007)

15.64 B. Rogers, L. Manning, M. Jones, T. Sulchek, K. Murray, B. Beneschott, J.D. Adams, Z. Hu, T. Thundat, H. Cavazos, S.C. Minne: Mercury vapor detection with a self-sensing, resonating piezoelectric cantilever, Rev. Sci. Instrum. **74**, 4899–4901 (2003)

15.65 H.F. Ji, Y.F. Zhang, V.V. Purushotham, S. Kondu, B. Ramachandran, T. Thundat, D.T. Haynie: 1,6-Hexanedithiol monolayer as a receptor for specific recognition of alkylmercury, Analyst **130**, 1577–1579 (2005)

15.66 A.R. Kadam, G.P. Nordin, M.A. George: Use of thermally induced higher order modes of a microcantilever for mercury vapor detection, J. Appl. Phys. **99**, 094905 (2006)

15.67 A.R. Kadam, G.P. Nordin, M.A. George: Comparison of microcantilever Hg sensing behavior with thermal higher order modes for as-deposited sputtered and thermally evaporated Au films, J. Vac. Sci. Technol. B **24**, 2271–2275 (2006)

15.68 H.F. Ji, E. Finot, R. Dabestani, T. Thundat, G.M. Brown, P.F. Britt: A novel self-assembled monolayer (SAM) coated microcantilever for low level caesium detection, Chem. Commun. **36**(6), 457–458 (2000)

15.69 T.A. Brunt, T. Rayment, S.J. O'Shea, M.E. Welland: Measuring the surface stresses in an electrochemically deposited metal monolayer: Pb on Au(111), Langmuir **12**, 5942–5946 (1996)

15.70 J. Mertens, E. Finot, M.H. Nadal, V. Eyraud, O. Heintz, E. Bourillot: Detection of gas trace of hydrofluoric acid using microcantilever, Sens. Actuator. B Chem. **99**, 58–65 (2004)

15.71 Y.J. Tang, J. Fang, X.H. Xu, H.F. Ji, G.M. Brown, T. Thundat: Detection of femtomolar concentrations of HF using an SiO_2 microcantilever, Anal. Chem. **76**, 2478–2481 (2004)

15.72 J. Mertens, E. Finot, O. Heintz, M.-H. Nadal, V. Eyraud, A. Cathelat, G. Legay, E. Bourillot, A. Dereux: Changes in surface stress, morphology and chemical composition of silica and silicon nitride surfaces during the etching by gaseous HF acid, Appl. Surf. Sci. **253**, 5101–5108 (2007)

15.73 T.L. Porter, T.L. Vail, M.P. Eastman, R. Stewart, J. Reed, R. Venedam, W. Delinger: A solid-state sensor platform for the detection of hydrogen cyanide gas, Sens. Actuator. B Chem. **123**, 313–317 (2007)

15.74 H.F. Ji, T. Thundat, R. Dabestani, G.M. Brown, P.F. Britt, P.V. Bonnesen: Ultrasensitive detection of CrO_4^{2-} using a microcantilever sensor, Anal. Chem. **73**, 1572–1576 (2001)

15.75 X.H. Xu, T.G. Thundat, G.M. Brown, H.F. Ji: Detection of Hg^{2+} using microcantilever sensors, Anal. Chem. **74**, 3611–3615 (2002)

15.76 S. Cherian, A. Mehta, T. Thundat: Investigating the mechanical effects of adsorption of Ca^{2+} ions on a silicon nitride microcantilever surface, Langmuir **18**, 6935–6939 (2002)

15.77 S. Cherian, R.K. Gupta, B.C. Mullin, T. Thundat: Detection of heavy metal ions using protein-functionalized microcantilever sensors, Biosens. Bioelectron. **19**, 411–416 (2003)

15.78 K. Liu, H.F. Ji: Detection of Pb^{2+} using a hydrogel swelling microcantilever sensor, Anal. Sci. **20**, 9–11 (2004)

15.79 P. Dutta, P.J. Chapman, P.G. Datskos, M.J. Sepaniak: Characterization of ligand-functionalized microcantilevers for metal ion sensing, Anal. Chem. **77**, 6601–6608 (2005)

15.80 F. Tian, V.I. Boiadjiev, L.A. Pinnaduwage, G.M. Brown, T. Thundat: Selective detection of

Cr(VI) using a microcantilever electrode coated with a self-assembled monolayer, J. Vac. Sci. Technol. A **23**, 1022–1028 (2005)

15.81 V.I. Boiadjiev, G.M. Brown, L.A. Pinnaduwage, G. Goretzki, P.V. Bonnesen, T. Thundat: Photochemical hydrosilylation of 11-undecenyltriethylammonium bromide with hydrogen-terminated Si surfaces for the development of robust microcantilever sensors for Cr(VI), Langmuir **21**, 1139–1142 (2005)

15.82 N.A. Carrington, L. Yong, Z.-L. Xue: Electrochemical deposition of sol-gel films for enhanced chromium(VI) determination in aqueous solutions, Anal. Chim. Acta **572**, 17–24 (2006)

15.83 S. Velanki, S. Kelly, T. Thundat, D.A. Blake, H.F. Ji: Detection of Cd(II) using antibody-modified microcantilever sensors, Ultramicroscopy **107**, 1123–1128 (2007)

15.84 H. Jensenius, J. Thaysen, A.A. Rasmussen, L.H. Veje, O. Hansen, A. Boisen: A microcantilever-based alcohol vapor sensor-application and response model, Appl. Phys. Lett. **76**, 2815–2817 (2000)

15.85 A. Hierlemann, D. Lange, C. Hagleitner, N. Kerness, A. Koll, O. Brand, H. Baltes: Application-specific sensor systems based on CMOS chemical microsensors, Sens. Actuator. B Chem. **70**, 2–11 (2000)

15.86 D. Lange, C. Hagleitner, A. Hierlemann, O. Brand, H. Baltes: Complementary metal oxide semiconductor cantilever arrays on a single chip: Mass-sensitive detection of volatile organic compounds, Anal. Chem. **74**, 3084–3095 (2002)

15.87 C. Vancura, M. Rüegg, Y. Li, C. Hagleitner, A. Hierlemann: Magnetically actuated complementary metal oxide semiconductor resonant cantilever gas sensor systems, Anal. Chem. **77**, 2690–2699 (2005)

15.88 L. Fadel, F. Lochon, I. Dufour, O. Français: Chemical sensing: Millimeter size resonant microcantilever performance, J. Micromech. Microeng. **14**, S23–S30 (2004)

15.89 Y.I. Wright, A.K. Kar, Y.W. Kim, C. Scholz, M.A. George: Study of microcapillary pipette-assisted method to prepare polyethylene glycol-coated microcantilever sensors, Sens. Actuator. B **107**, 242–251 (2005)

15.90 L.R. Senesac, P. Dutta, P.G. Datskos, M.J. Sepianiak: Analyte species and concentration identification using differentially functionalized microcantilever arrays and artificial neural networks, Anal. Chim. Acta **558**, 94–101 (2006)

15.91 F. Lochon, L. Fadel, I. Dufour, D. Rebière, J. Pistré: Silicon made resonant microcantilever: Dependence of the chemical sensing performances on the sensitive coating thickness, Mater. Sci. Eng. C **26**, 348–353 (2006)

15.92 S. Satyanarayana, D.T. McCormick, A. Majumdar: Parylene micro membrane capacitive sensor array for chemical and biological sensing, Sens. Actuator. B Chem. **115**, 494–502 (2006)

15.93 D. Then, A. Vidic, C. Ziegler: A highly sensitive self-oscillating cantilever array for the quantitative and qualitative analysis of organic vapor mixtures, Sens. Actuator. B Chem. **117**, 1–9 (2006)

15.94 P.J. Chapman, F. Vogt, P. Dutta, P.G. Datskos, G.L. Devault, M.J. Sepianiak: Facile hyphenation of gas chromatography and a microcantilever array sensor for enhanced selectivity, Anal. Chem. **79**, 364–370 (2007)

15.95 R. Archibald, P. Datskos, G. Devault, V. Lamberti, N. Lavrik, D. Noid, M. Sepianiak, P. Dutta: Independent component analysis of nanomechanical responses of cantilever arrays, Anal. Chim. Acta **584**, 101–105 (2007)

15.96 M. Alvarez, A. Calle, J. Tamayo, L.M. Lechuga, A. Abad, A. Montoya: Development of nanomechanical biosensors for detection of the pesticide DDT, Biosensor. Bioelectron. **18**, 649–653 (2003)

15.97 P.G. Datskos, M.J. Sepaniak, C.A. Tipple, N. Lavrik: Photomechanical chemical microsensors, Sens. Actuator. B Chem. **76**, 393–402 (2001)

15.98 L.A. Pinnaduwage, V. Boiadjiev, J.E. Hawk, T. Thundat: Sensitive detection of plastic explosives with self-assembled monolayer-coated microcantilevers, Appl. Phys. Lett. **83**, 1471–1473 (2003)

15.99 L.A. Pinnaduwage, A. Gehl, D.L. Hedden, G. Muralidharan, T. Thundat, R.T. Lareau, T. Sulchek, L. Manning, B. Rogers, M. Jones, J.D. Adams: A microsensor for trinitrotoluene vapour, Nature **425**, 474–474 (2003)

15.100 L.A. Pinnaduwage, D. Yi, F. Tian, T. Thundat, L.T. Lareau: Adsorption of trinitrotoluene on uncoated silicon microcantilever surfaces, Langmuir **20**, 2690–2694 (2004)

15.101 L.A. Pinnaduwage, A. Wig, D.L. Hedden, A. Gehl, D. Yi, T. Thundat, R.T. Lareau: Detection of trinitrotoluene via deflagration on a microcantilever, J. Appl. Phys. **95**, 5871–5875 (2004)

15.102 L.A. Pinnaduwage, T. Thundat, A. Gehl, S.D. Wilson, D.L. Hedden, R.T. Lareau: Desorption characteristics, of un-coated silicon microcantilever surfaces for explosive and common nonexplosive vapors, Ultramicroscopy **100**, 211–216 (2004)

15.103 L.A. Pinnaduwage, T. Thundat, J.E. Hawk, D.L. Hedden, R. Britt, E.J. Houser, S. Stepnowski, R.A. McGill, D. Bubb: Detection of 2,4-dinitrotoluene using microcantilever sensors, Sens. Actuator. B **99**, 223–229 (2004)

15.104 I. Voiculescu, M.E. Zaghloul, R.A. McGill, E.J. Houser, G.K. Fedder: Electrostatically actuated resonant microcantilever beam in CMOS technology for the detection of chemical weapons, IEEE Sensor. J. **5**, 641–647 (2005)

15.105 L.A. Pinnaduwage, H.F. Ji, T. Thundat: Moore's law in homeland defense: An integrated sensor platform based on silicon microcantilevers, IEEE Sensor. J. **5**, 774–785 (2005)

15.106 Y.M. Yang, H.F. Ji, T. Thundat: Nerve agents detection using a Cu^{2+}/L-cysteine bilayer-coated microcantilever, J. Am. Chem. Soc. **125**, 1124–1125 (2003)

15.107 H.F. Ji, X.D. Yan, J. Zhang, T. Thundat: Molecular recognition of biowarfare agents using micromechanical sensors, Expert Rev. Mol. Diagn. **4**, 859–866 (2004)

15.108 H.F. Ji, K.M. Hansen, Z. Hu, T. Thundat: Detection of pH variation using modified microcantilever sensors, Sens. Actuator. B Chem. **72**, 233–238 (2001)

15.109 R. Bashir, J.Z. Hilt, O. Elibol, A. Gupta, N.A. Peppas: Micromechanical cantilever as an ultrasensitive pH microsensor, Appl. Phys. Lett. **81**, 3091–3093 (2002)

15.110 Y.F. Zhang, H.F. Ji, D. Snow, R. Sterling, G.M. Brown: A pH sensor based on a microcantilever coated with intelligent hydrogel, Instrum. Sci. Technol. **32**, 361–369 (2004)

15.111 M. Watari, J. Galbraith, H.P. Lang, M. Sousa, M. Hegner, C. Gerber, M.A. Horton, R.A. McKendry: Investigating the molecular mechanisms of in-plane mechanochemistry on cantilever arrays, J. Am. Chem. Soc. **129**, 601–609 (2007)

15.112 A. Subramanian, P.I. Oden, S.J. Kennel, K.B. Jacobson, R.J. Warmack, T. Thundat, M.J. Doktycz: Glucose biosensing using an enzyme-coated microcantilever, Appl. Phys. Lett. **81**, 385–387 (2002)

15.113 X.D. Yan, H.F. Ji, Y. Lvov: Modification of microcantilevers using layer-by-layer nanoassembly film for glucose measurement, Chem. Phys. Lett. **396**, 34–37 (2004)

15.114 J.H. Pei, F. Tian, T. Thundat: Glucose biosensor based on the microcantilever, Anal. Chem. **76**, 292–297 (2004)

15.115 X.D. Yan, X.H.K. Xu, H.F. Ji: Glucose oxidase multilayer modified microcantilevers for glucose measurement, Anal. Chem. **77**, 6197–6204 (2005)

15.116 X.D. Yan, X.L. Shi, K. Hill, H.F. Ji: Microcantilevers modified by horseradish peroxidase intercalated nano-assembly for hydrogen peroxide detection, Anal. Sci. **22**, 205–208 (2006)

15.117 K.M. Hansen, H.F. Ji, G.H. Wu, R. Datar, R. Cote, A. Majumdar, T. Thundat: Cantilever-based optical deflection assay for discrimination of DNA single-nucleotide mismatches, Anal. Chem. **73**, 1567–1571 (2001)

15.118 M.F. Hagan, A. Majumdar, A.K. Chakraborty: Nanomechanical forces generated by surface grafted DNA, J. Phys. Chem. B **106**, 10163–10173 (2002)

15.119 R. Marie, H. Jensenius, J. Thaysen, C.B. Christensen, A. Boisen: Adsorption kinetics and mechanical properties of thiol-modified DNA-oligos on gold investigated by microcantilever sensors, Ultramicroscopy **91**, 29–36 (2002)

15.120 K.A. Stevenson, A. Mehta, P. Sachenko, K.M. Hansen, T. Thundat: Nanomechanical effect of enzymatic manipulation of DNA on microcantilever surfaces, Langmuir **18**, 8732–8736 (2002)

15.121 M. Su, S.U. Li, V.P. Dravid: Microcantilever resonance-based DNA detection with nanoparticle probes, Appl. Phys. Lett. **82**, 3562–3564 (2003)

15.122 M. Alvarez, L.G. Carrascosa, M. Moreno, A. Calle, A. Zaballos, L.M. Lechuga, C. Martinez, J. Tamayo: Nanomechanics of the formation of DNA self-assembled monolayers and hybridization on microcantilevers, Langmuir **20**, 9663–9668 (2004)

15.123 R.L. Gunter, R. Zhine, W.G. Delinger, K. Manygoats, A. Kooser, T.L. Porter: Investigation of DNA sensing using piezoresistive microcantilever probes, IEEE Sensor. J. **4**, 430–433 (2004)

15.124 M. Alvarez, J. Tamayo: Optical sequential readout of microcantilever arrays for biological detection, Sens. Actuator. B Chem. **106**, 687–690 (2005)

15.125 M. Calleja, M. Nordstrom, M. Alvarez, J. Tamayo, L.M. Lechuga, A. Boisen: Highly sensitive polymer-based cantilever-sensors for DNA detection, Ultramicroscopy **105**, 215–222 (2005)

15.126 R. Mukhopadhyay, M. Lorentzen, J. Kjems, F. Besenbacher: Nanomechanical sensing of DNA sequences using piezoresistive cantilevers, Langmuir **21**, 8400–8408 (2005)

15.127 L.M. Lechuga, J. Tamayo, M. Alvarez, L.G. Carrascosa, A. Yufera, R. Doldan, E. Peralias, A. Rueda, J.A. Plaza, K. Zinoviev, C. Dominguez, A. Zaballos, M. Moreno, C. Martinez, D. Wenn, N. Harris, C. Bringer, V. Bardinal, T. Camps, C. Vergnenegre, C. Fontaine, V. Diaz, A. Bernad: A highly sensitive microsystem based on nanomechanical biosensors for genomics applications, Sens. Actuator. B Chem. **118**, 2–10 (2006)

15.128 S.L. Biswal, D. Raorane, A. Chaiken, H. Birecki, A. Majumdar: Nanomechanical detection of DNA melting on microcantilever surfaces, Anal. Chem. **78**, 7104–7109 (2006)

15.129 J.C. Stachowiak, M. Yue, K. Castelino, A. Chakraborty, A. Majumdar: Chemomechanics of surface stresses induced by DNA hybridization, Langmuir **22**, 263–268 (2006)

15.130 F. Huber, M. Hegner, C. Gerber, H.J. Guntherodt, H.P. Lang: Label free analysis of transcription factors using microcantilever arrays, Biosens. Bioelectron. **21**, 1599–1605 (2006)

15.131 S.L. Biswal, D. Raorane, A. Chaiken, A. Majumdar: Using a microcantilever array for detecting phase transitions and stability of DNA, Clin. Lab. Med. **27**, 163–163 (2007)

15.132 J. Zhang, H.P. Lang, F. Huber, A. Bietsch, W. Grange, U. Certa, R. McKendry, H.-J. Güntherodt, M. Hegner, C. Gerber: Rapid and label-free nanomechanical detection of biomarker transcripts in human RNA, Nat. Nanotechnol. **1**, 214–220 (2006)

15.133 A.M. Moulin, S.J. O'Shea, M.E. Welland: Microcantilever-based biosensors, Ultramicroscopy **82**, 23–31 (2000)

15.134 C. Grogan, R. Raiteri, G.M. O'Connor, T.J. Glynn, V. Cunningham, M. Kane, M. Charlton, D. Leech: Characterisation of an antibody coated microcantilever as a potential immuno-based biosensor, Biosens. Bioelectron. **17**, 201–207 (2002)

15.135 A. Kooser, K. Manygoats, M.P. Eastman, T.L. Porter: Investigation of the antigen antibody reaction between anti-bovine serum albumin (a-BSA) and bovine serum albumin (BSA) using piezoresistive microcantilever based sensors, Biosens. Bioelectron. **19**, 503–508 (2003)

15.136 C.A. Savran, S.M. Knudsen, A.D. Ellington, S.R. Manalis: Micromechanical detection of proteins using aptamer-based receptor molecules, Anal. Chem. **76**, 3194–3198 (2004)

15.137 J.H. Lee, K.H. Yoon, K.S. Hwang, J. Park, S. Ahn, T.S. Kim: Label free novel electrical detection using micromachined PZT monolithic thin film cantilever for the detection of C-reactive protein, Biosens. Bioelectron. **20**, 269–275 (2004)

15.138 Y.F. Zhang, S.P. Venkatachalan, H. Xu, X.H. Xu, P. Joshi, H.F. Ji, M. Schulte: Micromechanical measurement of membrane receptor binding for label-free drug discovery, Biosens. Bioelectron. **19**, 1473–1478 (2004)

15.139 N. Backmann, C. Zahnd, F. Huber, A. Bietsch, A. Pluckthun, H.P. Lang, H.J. Güntherodt, M. Hegner, C. Gerber: A label-free immunosensor array using single-chain antibody fragments, Proc. Natl. Acad. Sci. USA **102**, 14587–14592 (2005)

15.140 K.W. Wee, G.Y. Kang, J. Park, J.Y. Kang, D.S. Yoon, J.H. Park, T.S. Kim: Novel electrical detection of label-free disease marker proteins using piezoresistive self-sensing micro-cantilevers, Biosens. Bioelectron. **20**, 1932–1938 (2005)

15.141 R. Mukhopadhyay, V.V. Sumbayev, M. Lorentzen, J. Kjems, P.A. Andreasen, F. Besenbacher: Cantilever sensor for nanomechanical detection of specific protein conformations, Nano Lett. **5**, 2385–2388 (2005)

15.142 X. Yan, K. Hill, H. Gao, H.F. Ji: Surface stress changes induced by the conformational change of proteins, Langmuir **22**, 11241–11244 (2006)

15.143 H.S. Kwon, K.C. Han, K.S. Hwang, J.H. Lee, T.S. Kim, D.S. Yoon, E.G. Yang: Development of a peptide inhibitor-based cantilever sensor assay for cyclic adenosine monophosphate-dependent protein kinase, Anal. Chim. Acta, **585**, 344–349 (2007)

15.144 W. Shu, E.D. Laue, A.A. Seshia: Investigation of biotin-streptavidin binding interactions using microcantilever sensors, Biosens. Bioelectron. **22**, 2003–2009 (2007)

15.145 V. Dauksaite, M. Lorentzen, F. Besenbacher, J. Kjems: Antibody-based protein detection using piezoresistive cantilever arrays, Nanotechnology **18**, 125503 (2007)

15.146 M. Yue, J.C. Stachowiak, H. Lin, R. Datar, R. Cote, A. Majumdar: Label-free protein recognition two-dimensional array using nanomechanical sensors, Nano Lett. **8**, 520–524 (2008)

15.147 T. Braun, N. Backmann, M. Vogtli, A. Bietsch, A. Engel, H.P. Lang, C. Gerber, M. Hegner: Conformational change of bacteriorhodopsin quantitatively monitored by microcantilever sensors, Biophys. J. **90**, 2970–2977 (2006)

15.148 B.H. Kim, O. Mader, U. Weimar, R. Brock, D.P. Kern: Detection of antibody peptide interaction using microcantilevers as surface stress sensors, J. Vac. Sci. Technol. B **21**, 1472–1475 (2003)

15.149 I. Pera, J. Fritz: Sensing lipid bilayer formation and expansion with a microfabricated cantilever array, Langmuir **23**, 1543–1547 (2007)

15.150 S.J. Hyun, H.S. Kim, Y.J. Kim, H.I. Jung: Mechanical detection of liposomes using piezoresistive cantilever, Sens. Actuator. B Chem. **117**, 415–419 (2006)

15.151 B. Ilic, D. Czaplewski, M. Zalalutdinov, H.G. Craighead, P. Neuzil, C. Campagnolo, C. Batt: Single cell detection with micromechanical oscillators, J. Vac. Sci. Technol. B **19**, 2825–2828 (2001)

15.152 J. Park, J. Ryu, S.K. Choi, E. Seo, J.M. Cha, S. Ryu, J. Kim, B. Kim, S.H. Lee: Real-time measurement of the contractile forces of self-organized cardiomyocytes on hybrid biopolymer microcantilevers, Anal. Chem. **77**, 6571–6580 (2005)

15.153 S. Ghatnekar-Nilsson, J. Lindahl, A. Dahlin, T. Stjernholm, S. Jeppesen, F. Hook, L. Montelius: Phospholipid vesicle adsorption measured in situ with resonating cantilevers in a liquid cell, Nanotechnology **16**, 1512–1516 (2005)

15.154 N. Nugaeva, K.Y. Gfeller, N. Backmann, H.P. Lang, M. Duggelin, M. Hegner: Micromechanical cantilever array sensors for selective fungal immobilization and fast growth detection, Biosens. Bioelectron. **21**, 849–856 (2005)

15.155 B. Dhayal, W.A. Henne, D.D. Doorneweerd, R.G. Reifenberger, P.S. Low: Detection of Bacillus subtilis spores using peptide-functionalized cantilever arrays, J. Am. Chem. Soc. **128**, 3716–3721 (2006)

15.156 G. Wu, R.H. Datar, K.M. Hansen, T. Thundat, R.J. Cote, A. Majumdar: Bioassay of prostate-specific antigen (PSA) using microcantilevers, Nat. Biotechnol. **19**, 856–860 (2001)

15.157 K.S. Hwang, J.H. Lee, J. Park, D.S. Yoon, J.H. Park, T.S. Kim: In-situ quantitative analysis of a prostate-specific antigen (PSA) using a nanomechanical PZT cantilever, Lab. Chip. **4**, 547–552 (2004)

15.158 J.H. Lee, K.S. Hwang, J. Park, K.H. Yoon, D.S. Yoon, T.S. Kim: Immunoassay of prostate-specific antigen (PSA) using resonant frequency shift of piezo-

15.158 electric nanomechanical microcantilever, Biosens. Bioelectron. **20**, 2157–2162 (2005)
15.159 S. Velanki, H.-F. Ji: Detection of feline coronavirus using microcantilever sensors, Meas. Sci. Technol. **17**, 2964–2968 (2006)
15.160 D. Maraldo, F.U. Garcia, R. Mutharasan: Method for quantification of a prostate cancer biomarker in urine without sample preparation, Anal. Chem. **79**, 7683–7690 (2007)
15.161 W.Y. Shih, X.P. Li, H.M. Gu, W.H. Shih, I.A. Aksay: Simultaneous liquid viscosity and density determination with piezoelectric unimorph cantilevers, J. Appl. Phys. **89**, 1497–1505 (2001)
15.162 A. Agoston, F. Keplinger, B. Jakoby: Evaluation of a vibrating micromachined cantilever sensor for measuring the viscosity of complex organic liquids, Sens. Actuator. A Phys. **123/124**, 82–86 (2005)
15.163 N. McLoughlin, S.L. Lee, G. Hahner: Simultaneous determination of density and viscosity of liquids based on resonance curves of uncalibrated microcantilevers, Appl. Phys. Lett. **89**, 184106 (2006)
15.164 U. Sampath, S.M. Heinrich, F. Josse, F. Lochon, I. Dufour, D. Rebière: Study of viscoelastic effect on the frequency shift of microcantilever chemical sensors, IEEE Trans. Ultrason. Ferroelectr. **53**, 2166–2173 (2006)
15.165 A. Jana, A. Raman, B. Dhayal, S.L. Tripp, R.G. Reifenberger: Microcantilever mechanics in flowing viscous fluids, Appl. Phys. Lett. **90**, 114110 (2007)
15.166 M. Hennemeyer, S. Burghardt, R. Stark: Cantilever micro-rheometer for the characterization of sugar solutions, Sensors **8**, 10–22 (2008)
15.167 K.B. Brown, Y. Ma, W. Allegretto, R.P.W. Lawson, F.E. Vermeulen, A.M. Robinson: Microstructural pressure sensor based on an enhanced resonant mode hysteresis effect, J. Vac. Sci. Technol. B **19**, 1628–1632 (2001)
15.168 K.B. Brown, W. Allegretto, F.E. Vermeulen, A.M. Robinson: Simple resonating microstructures for gas pressure measurement, J. Micromech. Microeng. **12**, 204–210 (2002)
15.169 Y. Su, A.G.R. Evans, A. Brunnschweiler, G. Ensell: Characterization of a highly sensitive ultra-thin piezoresistive silicon cantilever probe and its application in gas flow velocity sensing, J. Micromech. Microeng. **12**, 780–785 (2002)
15.170 J. Mertens, E. Finot, T. Thundat, A. Fabre, M.H. Nadal, V. Eyraud, E. Bourillot: Effects of temperature and pressure on microcantilever resonance response, Ultramicroscopy **97**, 119–126 (2003)
15.171 V. Mortet, R. Petersen, K. Haenen, M. D'Olieslaeger: Wide range pressure sensor based on a piezoelectric bimorph microcantilever, Appl. Phys. Lett. **88**, 133511 (2006)
15.172 P. Sievilä, V.P. Rytkönen, O. Hahtela, N. Chekurov, J. Kauppinen, I. Tittonen: Fabrication and characterization of an ultrasensitive acousto-optical cantilever, J. Micromech. Microeng. **17**, 852–859 (2007)
15.173 C.A. Jong, T.S. Chin, W.L. Fang: Residual stress and thermal expansion behavior of TaO_xN_y films by the micro-cantilever method, Thin Solid Films **401**, 291–297 (2001)
15.174 F. Shen, P. Lu, S.J. O'Shea, K.H. Lee, T.Y. Ng: Thermal effects on coated resonant microcantilevers, Sens. Actuator. A Phys. **95**, 17–23 (2001)
15.175 P.G. Datskos, P.I. Oden, T. Thundat, E.A. Wachter, R.J. Warmack, S.R. Hunter: Remote infrared radiation detection using piezoresistive microcantilevers, Appl. Phys. Lett. **69**, 2986–2988 (1996)
15.176 P.I. Oden, P.G. Datskos, T. Thundat, R.J. Warmack: Uncooled thermal imaging using a piezoresistive microcantilever, Appl. Phys. Lett. **69**, 3277–3279 (1996)
15.177 E.A. Wachter, T. Thundat, P.I. Oden, R.J. Warmack, P.G. Datskos, S.L. Sharp: Remote optical detection using microcantilevers, Rev. Sci. Instrum. **67**, 3434–3439 (1996)
15.178 P.G. Datskos, S. Rajic, I. Datskou: Detection of infrared photons using the electronic stress in metal-semiconductor cantilever interfaces, Ultramicroscopy **82**, 49–56 (2000)
15.179 C.B. Li, B.B. Jiao, S.L. Shi, D.P. Chen, T.C. Ye, Q.C. Zhang, Z.Y. Guo, F.L. Dong, Z.Y. Miao: A novel uncooled substrate-free optical-readable infrared detector: design, fabrication and performance, Meas. Sci. Technol. **17**, 1981–1986 (2006)
15.180 X. Yu, Y. Yi, S. Ma, M. Liu, X. Liu, L. Dong, Y. Zhao: Design and fabrication of a high sensitivity focal plane array for uncooled IR imaging, J. Micromech. Microeng. **18**, 057001 (2008)
15.181 B. Jiao, C. Li, D. Chen, T. Ye, Y. Ou, L. Dong, Q. Zhang: An optical readout method based uncooled infrared imaging system, Int. J. Infrared Millim. Waves **29**, 261–271 (2008)
15.182 T.P. Burg, S.R. Manalis: Suspended microchannel resonators for biomolecular detection, Appl. Phys. Lett. **83**, 2698–2700 (2003)
15.183 P. Dutta, C. Tipple, N. Lavrik, P. Datskos: Enantioselective sensors based on antibody-mediated nanomechanics, Anal. Chem. **75**, 2342–2348 (2003)
15.184 H.F. Ji, Y.Q. Lu, H.W. Du, X.H. Xu, T. Thundat: Spiral springs and microspiral springs for chemical and biological sensing, Appl. Phys. Lett. **88**, 063504 (2006)
15.185 G. Singh, P. Rice, R.L. Mahajan: Fabrication and mechanical characterization of a force sensor based on an individual carbon nanotube, Nanotechnology **18**, 475501 (2007)
15.186 A. Sakhaee-Pour, M.T. Ahmadian, A. Vafai: Applications of single-layered graphene sheets as mass sensors and atomistic dust detectors, Solid State Commun. **145**, 168–172R (2008)

16. Biological Molecules in Therapeutic Nanodevices

Stephen C. Lee, Bharat Bhushan

In this chapter, we discuss the incorporation of molecules into nanodevices as functional device components. Our primary focus is on biological molecules, although we also discuss the use of organic molecules as functional components of supramolecular nanodevices. Our primary device interest is in devices used in human therapy and diagnosis, though when it is informative, we discuss other nontherapeutic nanodevices containing biomolecular components. We discuss design challenges associated with devices built from prefabricated components (biological macromolecules) but that are not as frequently associated with fully synthetic nanodevices. Some design challenges (abstraction of device object properties, inputs, and outputs) can be addressed using existing systems engineering approaches and tools (including unified modeling language), whereas others (selection of optimal biological macromolecules from the billions available) have not been fully addressed. We discuss various assembly strategies applicable to biological macromolecules and organic molecules (self-assembly, chemoselective conjugation) and their advantages and disadvantages. We provide an example of a functional mesoscale device, a planar field-effect transistor (FET) protein sensor, that depends on nanoscale components for its function. We also use the sensor platform to illustrate how protein and other molecular engineering approaches can address nanoscale technological problems, and argue that protein engineering is a legitimate nanotechnology in this application. In developing the functional FET sensor, both direct adsorption of protein analyte receptors as well as linkage of receptors to the sensing surface through a polymer layer were tested. However, in the realized FET sensor, interfaces consist of a polymer layer linked to the semiconductor surface and to an analyte receptor (a protein). Nanotribology and other surface-science investigations of the interfaces revealed phenomena not previously documented for nanoscale protein interfaces (lubrication by directly adsorbed proteins, increases in friction force associated with polymer-mediated increases in sample compliance). Furthermore, the studies revealed wear of polymer and receptor proteins from semiconductor surfaces by an atomic force microscopy (AFM) tip which was not a concerted process, but rather depth of wear increased with increasing load on the cantilever. These studies also revealed that the polymer–protein interfaces were disturbed by nanonewton forces, suggesting that interfaces of immunoFET protein sensors translated to in vivo use must likely be protected from, or hardened to endure, abrasion from tissue. The results demonstrate that nanoscience (in this case, nanotribology) is needed to design and characterize functional planar immunoFET sensors, even though the sensors themselves are mesoscale devices. The results further suggest that modifications made to the sensor interfaces to address these nanoscale challenges may be best accomplished by protein and interfacial engineering approaches.

16.1	Definitions and Scope		454
	16.1.1	Design Issues	455
	16.1.2	Identification of Biomolecular Components	456
	16.1.3	Design Paradigms	457
	16.1.4	Utility and Scope of Therapeutic Nanodevices	461
16.2	Assembly Approaches		461
	16.2.1	Low-Throughput Construction Methods	461
	16.2.2	Supramolecular Chemistry and Self-Assembly	463
	16.2.3	Chemoselective Conjugation	469

- 16.2.4 Unnatural Amino Acids to Support Chemoselective Conjugation of Biologically Produced Proteins . 471
- 16.3 **Sensing Devices** 471
  - 16.3.1 Planar FET Protein Sensors 472
  - 16.3.2 Biotechnology Approaches to the "Fundamental Limitations" of Planar ImmunoFETs 473
  - 16.3.3 Nanotribology of Protein–Sensing Interfaces on Micromachined Surfaces 475
- 16.4 **Concluding Remarks: Barriers to Practice** 478
  - 16.4.1 You Do Not Know What You Do Not Know: the Consequences of Certainty 479
  - 16.4.2 Are Proteins and Molecules Legitimately Part of Nanotechnology? 479
- **References** 480

16.1 Definitions and Scope

Nanotechnology is a field in rapid flux and development, as this volume shows, and definition of its metes and bounds, as well as identification of subdisciplines embraced by it, can be difficult and controversial. The term *nanotechnology* means many things to many people, and aspects of multiple disciplines, from physics to information technology to biotechnology, legitimately fall into the intersection of the Venn diagram of disciplines that defines nanotechnology. The breadth of the field allows almost any interested party to contribute to it, but the same ambiguity can render the field diffuse and amorphous. If nanotechnology embraces everything, what then is it? Consideration of the scope of the field may be useful.

To frame the discussion, we will define nanotechnology as the discipline that aims to satisfy desired objectives using materials and devices whose valuable properties are based on a specific nanometer-scale element of their structures. As opposed to nanoscience, nanotechnology is application oriented, so nanoscience is important to nanotechnology primarily to the extent that it is relevant to device design, function or application.

The meaning of *therapeutic* is self-explanatory and refers here to intervention in *human* disease processes (although many of the approaches discussed are equally applicable to veterinary medicine). We confine our discussion mostly to therapeutics used in vivo, because such applications clearly benefit from the low invasiveness that ultrasmall, but multipotent, nanotherapeutics potentially offer. It is debatable whether imaging, diagnostic or sensing devices can be considered therapeutic in this context, though as we will see, sensing/diagnostic functionalities are often inextricable elements of therapeutic nanodevices, and it is difficult to consider smart nanotherapeutics without discussion of their sensing capabilities. Therapeutics incorporating diagnostic capabilities (via their capability for sensing or imaging contrast delivery) are now recognized as their own class of drug entities, referred to as *theranostics* (see below).

Our definition of nanotechnology is both broader and narrower than more common definitions. First, our definition embraces macroscale structures whose useful properties derive from their nanoscale aspects. Second, we have a *device-centric* bias: we are interested in devices that perform multistep work processes. Third, consistent with our device-centric bias, the term *specific* (as in, *specific nanometer-scale elements*) is intended to exclude materials whose utility derives solely from properties inherent to being finely divided (high surface-to-volume ratios, for instance), or other material, chemical, and physical properties unless those properties contribute to specific device function. We made this exclusion based on our assessment that therapeutic nanodevices are more intriguing than nanomaterials per se (see below), although we will engage these attributes where they are germane to specific device or therapeutic applications. Fourth, our definition implies that limited nanotechnology has been available since the 1970s in the form of biotechnology. Based on their nanoscale structures, individual biological macromolecules (such as proteins) often exhibit the coordinated, modular multifunctionality that is characteristic of purpose-built devices (Fig. 16.1).

Biological macromolecules rely on the deployment of specific chemical functionalities with specific relative distributions in space with nanometer (and greater) resolution for their function, so the inclusion of molecular engineering aspects of biotechnology under the nanotechnology rubric is legitimate, despite the discomfort this may cause to traditionally trained engineers.

As we will see, intervention in human disease often requires inclusion of biomolecules in therapeutic devices, as frequently no functional synthetic analogue of active proteins and nucleic acids is available. As we will discuss, specific nanoscale device problems also can be addressed with biotechnology and protein engineering approaches, so the legitimacy of inclusion of biotechnology in nanotechnology is now beyond debate.

An analogous argument can be made that organic chemistry is an early form of nanotechnology. Compared with organic small molecules, protein functional capabilities and properties are generally more complex and more dependent on their conformation in three-dimensional space at nanometer and subnanometer scales. The nanotechnology sobriquet, therefore, *may* be more appropriate to biotechnology than to organic chemistry. However, supramolecular chemistry and therapeutic supramolecular devices depend on specific design features of organic molecules and assemblies thereof, so organic chemistry might be viewed as an even earlier version of nanotechnology, based on an argument very similar to that we use for biotechnology.

As described above, this chapter focuses primarily on nanoscale therapeutic devices as opposed to therapeutic nanomaterials. Devices are integrated functional structures and not admixtures of materials, compounds or substances. Devices exhibit desirable emergent properties inherent to their design, properties that emerge as the result of the spatial and/or temporal organization, coordination, and regulation of action of individual components. The organization of components in devices allows them to perform multistep, cogent work processes that cannot be mimicked by simple admixtures of individual components. In fact, if device functions can be mimicked well by simple mixtures of components, the labor involved in configuring and constructing a nanoscale device is not warranted. Our device definition thus excludes nanomaterials used as drug formulation excipients (pharmacologically inert materials included in formulations that improve pharmacophore uptake, biodistribution, pharmacokinetic, handling, storage or other properties), but embraces those same materials as integral components of drug-delivery or other clinical devices.

16.1.1 Design Issues

The biotechnology industry historically has focused on production of individual soluble protein and nucleic acid molecules for pharmaceutical use, with only limited attention paid to functional supramolecular

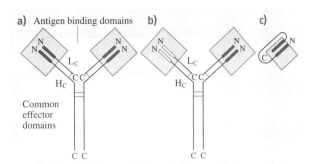

Fig. 16.1a–c Antibodies resemble purpose-built devices with distinct functional domains [16.1]. Native immunoglobulin class G (IgG) antibodies are composed of four polypeptide chains: two heavy chains (Hc) and two light chains (Lc), joined by interchain disulfide linkages (*lines* between Hc and Lc moieties). Amino and carboxy termini of individual polypeptide chains are indicated by N and C. Antigen-binding domains are responsible for specific antigen recognition, vary from antibody to antibody, and are indicated by the *thicker lines*. Common effector functions (Fc receptor binding, complement fixation, etc.) are delimited to domains of the antibodies that are constant from molecule to molecule. (a) A native IgG antibody is monospecific but bivalent in its antigen binding capacity. (b) An engineered, bispecific, bivalent antibody capable of recognizing two distinct antigens. (c) An engineered antibody fragment (single-chain Fv or SCFv) that is monospecific and monovalent can recognize only one antigenic determinant and is engineered to lack common effector functions. This construct is translated as a single, continuous polypeptide chain (hence the name SCFv) because a peptide linker (indicated by the *connecting line* in the figure) is incorporated to connect the carboxy-end of the Hc fragment and the amino-end of the Lc fragment

structures [16.2–8]. This bias toward free molecules flies in the face of the obvious importance of integrated supramolecular structures in biology and, to the casual observer, may seem an odd gap in attention and emphasis on the part of biotechnologists. The bias toward single-molecule, protein therapeutics follows from the fact that biotechnology developed as an industrial activity, governed by market considerations. Of the potential therapeutics that might be realized from biotechnology, single-protein therapeutics are among the easiest to realize from both technical and regulatory perspectives, and so warrant extensive industrial attention. This is changing, however, and more complex entities (actual supramolecular therapeutic devices) have and will appear with increasing frequency in the 21st century.

New materials derived from micro-/nanotechnology provide the opportunity to complement the traditional limits of biotechnology by providing scaffolds that can support higher-level organization of multiple biomolecules to perform work activities that could not be performed by mixtures of free, soluble molecules. Such supramolecular structures have been called nanobiotechnological devices [16.9], nanobiological devices [16.2–8], and semisynthetic nanodevices, and they figure prominently in therapeutic nanotechnology. Some such devices have sensing and diagnostic capabilities, and therefore belong to the already discussed class of theranostic agents (see [16.9] for a recent example of a multifunctional theranostic nanoparticle, and [16.10] for a recent review of such multifunctional nanoparticulate magnetic resonance imaging contrast agents in cardiovascular disease).

16.1.2 Identification of Biomolecular Components

Design of nanodevices is similar to design of other engineered structures, providing that the special properties of the materials (relating to their nanoscale aspects such as quantum, electrical, mechanical, biological properties, etc.), as well as their pharmacological properties, are considered. Therapeutics interact with patients on multiple levels, ranging from organismal to molecular, but it is reasonable to expect that most nanotherapeutics will interface with patients at the nanoscale [16.2–16]. Typically, this means interaction between therapeutics and patient biological macromolecules, supramolecular structures, organelles, cells or tissues, which in turn often dictates the incorporation of biological macromolecules (and other biostructures) into nanodevices [16.2–10]. Incorporating biological structures into (nanobiological) devices presents special challenges that do not occur in other aspects of engineering.

Unlike fully synthetic devices, semibiological nanodevices must incorporate prefabricated biological components (proteins, nucleic acids or derivatives thereof), and therefore intact nanodevices are seldom made entirely de novo. As a corollary, knowledge of properties of nanobiological device components is usually incomplete, as the molecules were not made by human design, so their properties are not known a priori and must be discovered. Therefore the range of activities inherent to any nanobiological device design may be much less obvious and less well defined than for fully synthetic devices. Further complicating the issue, the activities of biological molecules are often multifaceted (many genes and proteins exhibit pleiotropic activities), and the full range of functionality of individual biological molecules in interactions with other biological systems (as in nanotherapeutics) is often not known. This makes the design and prototyping of biological nanodevices an empirically intensive, iterative process [16.2–16].

Paralleling the paucity of information typically available about individual proteins, the number of distinct natural proteins in the biologic world is unknown but exponentially high (certainly in excess of 10^{13} distinct molecules [16.17]). When one considers engineered proteins, particularly those made by high-throughput mutagenesis methods, the number of existing protein sequences rises additional orders of magnitude. Most of these molecules remain to be discovered, so their individual properties (that might be critical to device designers, such as functional pH and temperature ranges, ionic requirements, cofactor requirements, radiation tolerance, resistance to degradation, etc.) are mostly unknown. Among the relative handful of proteins that are known, properties are typically incompletely known, and those properties that are known often are not those of greatest interest in selecting a protein as a nanodevice component. In fact, existing protein databases focus primarily on pharmacological properties or evolutionary relationships between proteins. We have proposed the building of databases useful specifically for nanobiotechnology, though this has yet to occur [16.5, 17, 18]. Existing protein databases, based as they are on phylogenetic, protein sequence, protein structure or protein primary function information, are not satisfactory for nanodevice design, if such design is to be realized as a discipline in and of itself, and if the immense power of biologic nanotechnology is to be realized to any significant extent. As it is, inventors of nanobiotechnological devices often rely on their personal knowledge of biology and biochemistry to select biological components for devices. It is obvious then, that the devices designed by even the most sophisticated biologists are very likely to be suboptimal. As a corollary, this means that nanobiological device designers must have significant biological expertise. This constitutes a huge barrier to entry for nonbiologists who could otherwise contribute importantly to the field. Thus, the lack of an appropriate protein database and accompanying search tools implies an immense opportunity cost for the field of biologic nanotechnology.

Biological macromolecules have properties, particularly those relating to their stability, that can limit their use in device contexts. In general, proteins, nucleic acids, lipids, and other biomolecules are more labile to physical/biochemical insult than are many synthetic materials. With the possible exceptions of topical agents or oral delivery and endosomal uptake of nanotherapeutics (both involving exposure to low pH), device lability in the face of *physical* insult is generally a major consideration only in ex vivo settings (relating to device storage, sterilization, etc.), because physical conditions that would destroy the device would be bad for the patient as well. However, living organisms remodel themselves constantly in response to stress, development, pathology, and external stimuli. For instance, epithelial tissues and blood components are constantly eliminated and regenerated, and bone and vasculature are continuously remodeled. The metabolic facilities responsible (circulating and tissue-bound proteases and other enzymes, various clearance organs, the immune system, etc.) can potentially process biological components of nanobiological therapeutic devices as well as endogenous materials, leading to partial or complete degradation of nanotherapeutic structure, function, or both. Furthermore, immune and wound responses protect the host against pathogenic organism incursions, by mechanisms that involve sequestering and degrading pathogens. Nanobiological therapeutics are subject to the actions of these host defense systems as well as normal remodeling processes. Various strategies to stabilize biomolecules and structures in heterologous in vivo environments are applicable to nanobiological therapeutics. Conversely, instability of active biocomponents can offer a valuable and simple way to delimit the activity of nanotherapeutics containing biomolecules.

16.1.3 Design Paradigms

Several early attempts to codify the canonical properties of ideal nanobiological devices, and therapeutic nanodevices in particular, have been made [16.1, 5, 11–15, 19, 20]. These attempts to codify design constitute a limited set of design guidelines that are summarized in Table 16.1. Naturally occurring, functional biological components generally exist in the context of higher-order systems that support the organisms of which they are a part. In general, nanobiological devices contain biological components that retain their function in new (device) contexts. In other words, one must incorporate into the device enough of a functional biological unit (nucleic acid, protein, oligomeric protein complex, organelle, cell, etc.) to allow that unit to perform the function for which it was selected. If one wishes, for example, to appropriate the specific antigen-recognition property of an antibody for a device function (say, in targeting, as discussed later), it is not necessary to incorporate the entire 150 000 atomic mass unit (AMU) antibody, the bulk of which is devoted to functions other than antigen recognition (Fig. 16.1) [16.21], but it *is* critical to incorporate the approximately 20 000 AMU of the antibody essential for specific antibody–antigen binding (the variable domain, Fig. 16.1).

Device function is the result of the summed and various activities of biological and synthetic device

Table 16.1 Some ideal characteristics of nanodevices [16.1, 5, 11–15, 19, 20]

(a) Characteristics of all nanobiological devices
Biological molecules must retain function.
Device function is the result of the summed activities of device components.
The relative organization of device components drives device function.
Device functions can be unprecedented in the biological world.

(b) Desirable characteristics of therapeutic platforms
Therapeutics should be minimally invasive.
Therapeutics should have the capacity to target sites of disease.
Therapeutics should be able to sense disease states in order to: • Report conditions at the disease site to clinicians   • Administer metered therapeutic interventions.
Therapeutic functions should be segregated into standardized modules.
Modules should be interchangeable to tune therapeutic function.

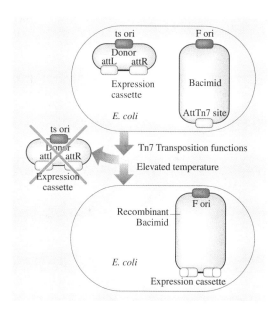

Fig. 16.2 The bacmid molecular cloning system as it is represented in the molecular biology literature. Bacmid is a molecular device designed to allow efficient production of recombinant insect viruses (baculovirus) in *Escherichia coli* [16.22–24]. Baculovirus is replicated in *E. coli* by the F plasmid origin of replication ("F ori"), and as such, is called a bacmid. The bacmid also includes an engineered transposable DNA element 7 (Tn7) attachment site isolated from the chromosome of an enteric bacteria (AttTn7). AttTn7 can receive Tn7 elements transposed from other cellular locations. A donor plasmid (donor) is replicated by a temperature-sensitive plasmid pSC101 origin of replication ("ts ori"). The donor also incorporates an expression cassette containing both the gene of interest for ultimate expression in insect cells and a selectable genetic marker operable in *E. coli*. The expression cassette is flanked by DNA sequences (attL and attR) that are recognized by the Tn7 transposition machinery. Tn7 transposition machinery resides elsewhere in the same *E. coli* cell. When donor plasmid is introduced into *E. coli* containing bacmid, Tn7 transposition machinery causes the physical relocation of expression cassettes from donor plasmid to bacmid. Unreacted donor plasmid is conveniently removed by elevating the incubation temperature, causing the ts pSC101 replicon to cease to function and, in turn, causing the donor to be lost. If selection for the genetic markers within the expression cassette is applied at this point, the only *E. coli* that survive are those containing recombinant bacmid (i. e., those that have received the gene for insect cell expression by transposition from the donor). Recombinant bacmid are conveniently isolated from *E. coli* and introduced into insect cell culture, where expression of the gene of interest occurs (after [16.25]) ◂

components. The nanobiological device designer can exert control over the relative organization of biological device components, which allows biomolecular components of devices abstracted from their native context to contribute to overall device functions that are entirely different from those in which they participated in their original, organismal contexts.

All of these features are illustrated in the bacmid, or Bac-to-Bac, system, a commercially available molecular cloning device ([16.22–24] and Fig. 16.2). Bacmid configures prokaryotic and eukaryotic genetic elements from multiple sources into a device for producing recombinant eukaryotic viral genomes in bacteria, a function that is unprecedented in nature. The system is feasible because of the modularity of the genetic elements involved and because of the strict control of the relative arrangement of genetic elements allowed by recombinant deoxyribonucleic acid (DNA) technology. Analogous cloning devices based on comparable arrangements of bacterial and eukaryotic regulatory and structural genes are reviewed in [16.18]. Other nucleic acid devices using genetic control elements from phylogenetically various sources are being developed to preprogram the micro- and nanoscale architectural properties and physiological behavior of living things [16.25, 26].

Bacmid provides an example of a nontherapeutic nanobiological device and illustrates some specific design approaches for building functional devices with biocomponents. Bacmid complies fully with those design recommendations of Table 16.1 that are not therapeutic device specific. However, Table 16.1 provides guidelines only. Systems engineering approaches commonly used in software and computer engineering provide a more rigorous framework to consider nanodevice design [16.17, 18]. Systems tools such as unified modeling language (UML) allow more precise depiction of nanobiological devices than do most text descriptions. A simplified UML use-case of bacmid is presented in Fig. 16.3.

Most importantly, UML forces designers to abstract knowledge of component (object) functions and properties and to express them in terms of object inputs and outputs. This makes device designs modular by explic-

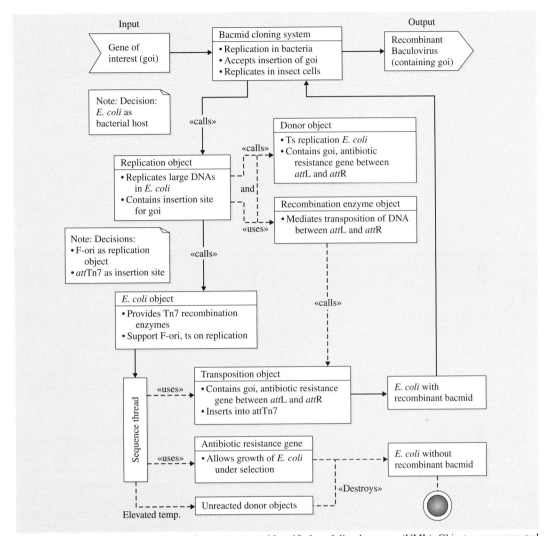

Fig. 16.3 The bacmid system represented as a use-case with unified modeling language (UML). Objects are represented by *boxes*, split horizontally into compartments. The *topmost compartment* contains the object name, the *middle compartment* contains the object attributes, and the *bottom compartment* contains the object operations (if different from the object attributes). Notes contain expository information such as design decisions. Sequence threads indicate serial events in process flow. Process flow is indicated by *solid arrows*. *Dashed arrows* indicate communication between objects and, for the purposes of illustrating bacmid, are labeled as *uses* (the object utilizes a second object), *calls* (the object elicits an action from a second object), and *destroys* (a step that causes the loss or destruction of one or more objects). End of process is indicated. The abstraction of the function of individual objects driven by UML may make other analogous devices (similar to bacmid, but substituting one or more isofunctional objects) obvious [16.17, 18]. In protein-containing devices, systematic identification of isofunctional objects could be facilitated using protein databases modified as suggested in [16.17] (after [16.18])

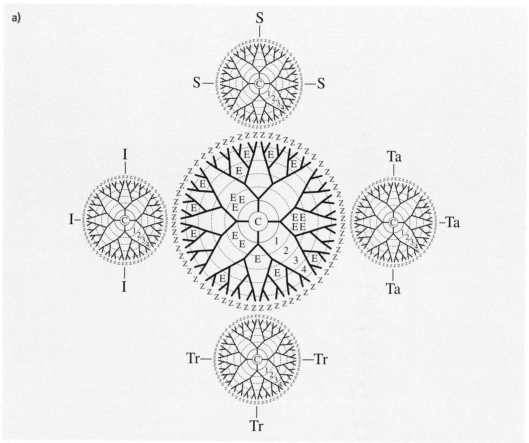

Fig. 16.4a,b A hypothetical, modular nanotherapeutic agent patterned after the dendrimer-based cluster agent for oncology of *Baker* [16.27–29]. Self-assembling, multidendrimer structures as shown are sometimes referred to as tecto(dendrimers). Each polyamidoamine (PAMAM) dendrimer subunit is grown from an initiator core (C), and the tunable surface groups of the dendrimers are represented by Z. In this device, each dendrimer subunit has a specific, dedicated function in the device: the central dendrimer encapsulates small-molecule therapeutics (E), whereas other functional components are segregated to other dendrimer components. These include biochemical targeting/tethering functions (Ta), therapeutic triggering functions to allow activation of prodrug portions of the device by an external operator (Tr), metal or other constituents for imaging (I), and sensing functions (S) to mediate intrinsically controlled activation or release of therapeutic. This design constitutes a therapeutic platform [16.14, 16] because of its modular design. The depicted device is only one possible configuration of an almost infinite number of analogous therapeutics that can be tuned to fit particular therapeutic needs by interchanging functional modules (after [16.16]). (**b**) A more sophisticated assembly strategy for dendrimer therapeutics that utilizes self-assembly of biomolecules [16.30–32]. Individual dendritic polymers (spheres representing generation 5 and generation 7 PAMAM dendrimers) are conjugated with single-stranded (ss) oligonucleotides (light and dark gray lines). When oligonucleotides of two dendritic polymers are complementary, individual oligos hybridize, forming double-stranded (ds)DNA complexes, linking the two dendritic polymers. An interesting feature of this assembly system is the rigidity of short dsDNA (< 50 base pair) segments. This allows assembly of objects at precise, tunable nanoscale distances (after [16.30]) ▲ ▶

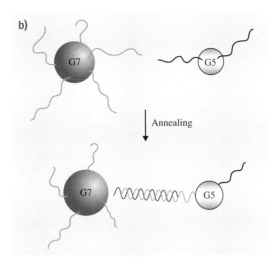

In nanotherapeutic applications, devices should be noninvasive and target therapeutic payloads to sites of disease to maximize therapeutic benefit while minimizing undesired side-effects. This, of course, implies the existence of therapeutic effector functions in these nanodevices, to give devices the ability to remediate a physiologically undesirable condition. Beyond that, several desirable attributes relate to sensing of biomolecules, cells or physical conditions (sensing disease itself, identification of residual disease, and potentially responding to intrinsic or externally supplied triggers for payload release). Other properties relate to communication between device subunits (for instance, between sensor and effector domains of the device) or between the device and an external operator. With appropriate design, device functions can be modular, as illustrated by the early, hypothetical dendrimer-based therapeutic shown in Fig. 16.4a [16.11, 16, 27–29].

itly identifying object properties and limitations: any object producing the same outputs and accepting the same inputs can be substituted for the original component in the functional nanodevice. By this means, UML notation transforms nanodevice designs from one-off individual designs into general designs for a class of functionally similar devices [16.18]. In general, biologist inventors have intuitively used these systems design approaches to design devices, as we discuss [16.18]. However, UML-enforced explicit callout of component properties that are necessary and sufficient for device function makes design of isofunctional devices easier. An argument might be made that, in some cases, UML expression of a device design, an example of which is shown in Fig. 16.3, may render related devices (that are implied in the UML depiction) obvious. If so, broad use of UML in nanobiotechnology might have significant intellectual property implications.

16.1.4 Utility and Scope of Therapeutic Nanodevices

Therapeutic nanotechnology will be useful when the underlying biology of the disease states involved is amenable to intervention at the nanoscale. While several disease states and physiological conditions (cancer, vaccination, cardiovascular disease, etc.) are particularly accessible to nanoscale interventions, some nanotechnological approaches may be applicable more broadly, in indications we currently cannot predict. Much as was the case with the introduction of recombinant protein therapeutics, nanotherapeutics may present regulatory and pharmacoeconomic challenges related to their novelty and their cost of goods (COGs). However, there is little doubt that nanobiological devices providing clear patient benefit, and whose production, regulatory approval, and distribution are amenable to feasible business models, will enter clinical practice.

16.2 Assembly Approaches

Assembly of components into devices is amenable to multiple approaches. In the case of devices comprising a single molecule or processed from a single crystal (some microfabricated structures, single polymers, or grafted polymeric structures) assembly may not be an issue. Integration of multiple, separately microfabricated components may sometimes be necessary and may sometimes drive the need for assembly, even for silicon devices. Furthermore, many therapeutic nanodevices contain multiple, chemically diverse components that must be assembled precisely to support their harmonious contribution to device function.

16.2.1 Low-Throughput Construction Methods

Low-throughput device construction methods are more applicable to construction of prototypes for research purposes than to production of commercial therapeutics. For example, direct-write technologies can obtain high (nanometer-scale) resolution; electron-beam (e-beam) lithography is a technique requiring no mask and can yield resolutions on the order of tens of nanometers, depending on the resist materials used [16.33]. Resolution in e-beam lithography is ultimately limited by electron scattering in the resist and electron optics, and like most direct-write approaches, e-beam lithography is limited in its throughput. Parallel approaches involving simultaneous writing with up to 1000 shaped e-beams are under development [16.33] and may mitigate limitations in manufacturing rate.

Atomic force microscopy (AFM) approaches utilize an ultrafine cantilever tip (typically with tip diameters of 50 nm or less) in contact with, or tapping, a surface or stage. The technique can be used to image molecules, analyze molecular biochemical properties (such as ligand–receptor affinity [16.34]), or manipulate materials at the nanoscale. In the latter mode, force microscopy has been used to manipulate atoms to build individual nanostructures since the mid-1980s, though the manufacturing throughput of *manual* placement of atoms and nanoscale components by force microscopy is limited, even with highly multiplexed arrays of cantilevers.

Dip-pen nanolithography (DPN) is a force microscopy methodology that can achieve high-resolution features (features of 100 nm or less) in a single step. In DPN, the AFM tip is coated with molecules to be deployed on a surface, and the molecules are transferred from the AFM tip to the surface as the coated tip contacts it. DPN also can be used to functionalize surfaces with two or more constituents and is well suited for deployment of functional biomolecules on synthetic surfaces with nanoscale precision [16.35, 36]. DPN suffers the limitations of synthetic throughput typical of AFM construction strategies.

Much as multibeam strategies might improve throughput in e-beam lithography [16.34], multiple tandem probes may significantly increase assembly throughput for construction methods that depend on force microscopy, but probably not sufficiently to allow manufacture of bulk quantities of nanostructures, as will likely be needed for consumer nanotherapeutic devices. As standard of care evolves increasingly toward tailored courses of therapy [16.37], individual therapeutics will become increasingly multicapable and powerful. Potentially, fewer copies of a nanotherapeutic may be required per patient, and each patient's nanotherapeutic may be tailored to him/her. It is not inconceivable that this might make relatively low-throughput synthesis/assembly methods practical, though this remains to be seen. If tailored therapeutics become the standard, the pharmaceutical industry will be irrevocably changed, with the concept of the *blockbuster* that can produce multiple billions of dollars in revenue per year through sales of a single agent to a broad population made no longer relevant. Business models in the pharmaceutical industry may be changed beyond recognition, and in the fullness of time, pharmacies could begin to resemble the formularies of old, with the capacity to make a specific preparation for an individual patient on site, but now using sophisticated bio- and nanotechnologies.

Individualized therapies for one patient are likely to differ from those for other patients by their incorporation of engineered molecules that are immunologically, toxicologically or physiologically tolerable by each individual host. Device functional properties will be tuned to integrate with individual host physiologies. One facile way to achieve patient-specific theranostics might be to design a modular device whose properties are tuned by substitution of biochemically isofunctional components, but whose other properties (immunological, toxicological, etc.) are suited to individual patients. This sounds like the modular nanodevice construction strategy we advocate [16.17, 18]. Should this come to pass, pharmaceutical companies will have to find a way to claim as intellectual property immense collections of similar function, but individually tailored, nanotherapeutic devices. Patenting each device individually is economically and logistically impossible, and a patent on a single specific nanotherapeutic might be easily *invented around* by competitors. One possible solution may be to structure patent claims broadly, around UML representations of nanotherapeutics that explicitly call out the critical characteristics of objects to identify alternate, isofunctional objects that might be substituted for the objects of the device in its original design [16.18]. This may capture a group of analogous, related theranostics, rather than an individual nanotherapeutic, or it may make some or all of the analogous devices obvious in the eyes of the patent examiner. This would likely preclude the proliferation of devices, each captured by its own patent, but template off a single central invention, as occurred with analogues of bacmid [16.18].

For the moment, though, ideal manufacturing approaches for nanotherapeutic devices resemble either industrial polymer chemistry, occurring in bulk, in convenient buffer systems, or in massively parallel industrial microfabrication approaches. In any case, therapy for a single patient may involve multiple billions of individual nanotherapeutic units, so each individual nanotherapeutic structure must require only minimal input from a human technician. Self-assembly, when feasible, allows device construction without ongoing human intervention.

16.2.2 Supramolecular Chemistry and Self-Assembly

Exploitation of the self-assembly properties of dendrimers and its exploitation to build a modular therapeutic has already been alluded to in Fig. 16.4a. Self-assembly has long been recognized as a potentially critical labor-saving approach to the construction of nanostructures [16.38], and many organic and inorganic materials have self-assembly properties that can be exploited to build structures with controlled configurations. Self-assembly processes are usually driven by thermodynamic forces and generally result in structures that are not covalently linked. Intra-/intermolecular forces driving assembly can be electrostatic or hydrophobic interactions, hydrogen bonds, and van der Waals interactions between and within subunits of the self-assembling structures or the assembly environment. Thus, final configurations are limited by the ability to *tune* the properties of the subunits and control the assembly environment to generate particular structures.

Highly hydrophobic carbon nanotubes spontaneously assemble into higher-order [16.39] structures (nanoropes and multiwall carbon nanotubes) as the result of hydrophobic interactions between individual tubes [16.39]. C_{60} fullerenes and single-wall carbon nanotubes (SWCNT) also spontaneously assemble (Fig. 16.5) into higher-order nanostructures called *peapods* [16.40], in which fullerene molecules are encapsulated in nanotubes. The fullerenes of peapods modulate the local electronic properties of the SWCNT in which they are encapsulated and may allow tuning of carbon nanotube electrical properties, perhaps as in one-dimensional (1-D) carbon nanotube field-effect transistors (FETs) [16.41]. Several self-assembled carbon structures and carbon-structure-containing devices are depicted in Fig. 16.5.

In drug delivery, the most familiar self-assembled nanostructures are micelles [16.42–44]. These structures are formed from the association of block copolymer subunits (Fig. 16.6), each individual subunit containing hydrophobic and hydrophilic domains. Micelles spontaneously form when the concentration of their subunits exceeds the critical micelle concentration (CMC) in a solvent in which one of the polymeric domains is insoluble (Fig. 16.6). The CMC is determined by the insoluble polymeric domain, and can be adjusted by control of the chemistry and length of the immiscible domain, as well as by control of solvent conditions. Micelles formed at low concentrations from low-CMC polymers are stable at high dilution. Micelles formed from polymer monomers with high CMCs can dissociate upon dilution, a phenomenon that might be exploited to control release of therapeutic cargos. If desired, micelles can be stabilized by covalent cross-linking to generate shell-stabilized structures [16.42–44]. Size dispersity and other properties of micelles can be manipulated by control of solvent conditions, incorporation of excipients (to modulate polymer packing properties), temperature, and agitation. From the standpoint of size, reasonably monodisperse preparations (polydispersities of 1–5%) of nanoscale micellar structures can be prepared.

The immense versatility of industrial polymer chemistry allows micellar structures to be tuned chemically to suit the task at hand. They can be modified for targeting (by appending ligands that recognize particular targeting sites to their surfaces) or to support higher-order assembly of micelles. They can be made to imbibe therapeutic or other molecules for delivery and caused to dissociate or disgorge themselves of payloads at desired times or bodily sites under the influence of local physical or chemical conditions. The tunability of these and other properties at the level of monomeric polymer subunits (as well as the level of assembled higher-order structures) makes micelles potentially powerful nanoscale vehicles for the delivery of drugs or imaging contrast agents.

In its most sophisticated manifestation, synthetic polymer self-assembly strategies can allow the generation of three-dimensional structures of highly defined nanoscale morphology. The production of controlled self-assembled structures can be affected by synthesis and assemble specific organic chemical compounds of controlled structure and chirality. Nanoscale molecular assemblies of this nature are said to be the result of supramolecular chemistries. Supramolecular chemistries exploit designer knowledge of molecular geometries, intramolecular interactions (hydrophobic, metal chelation, hydrogen bonding, and dipole interactions), and intramolecular packing properties to drive formation of nanoscale structures of controlled shape [16.45, 46]. As the referenced reviews [16.45, 46] show, supramolecular chemistry can produce space-

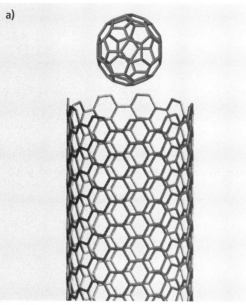
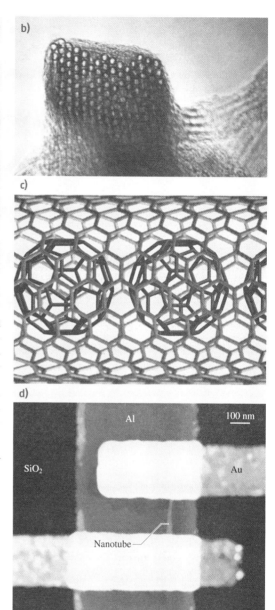

Fig. 16.5 (a) Shown to scale are two highly defined carbon nanostructures: a C_60 fullerene and (10, 10) single-wall carbon nanotube (SWCNT). **(b)** A self-assembled nanorope composed of carbon nanotubes that assemble by virtue of hydrophobic interactions [16.39–41]. **(c)** Schematic depiction of another self-assembled carbon nanostructure (a peapod) consisting of the fullerenes and SWCNT of (a), with the fullerenes encapsulated in the SWCNT [16.40]. Fullerene encapsulation in the peapod modulates the local electronic properties of the SWCNT. **(d)** A nanotube field-effect transistor (FET) consisting of gold source and drain electrodes on an aluminum stage with a carbon nanotube serving as the FET channel [16.41] (after [16.39–41])

filling nanostructures of stunning regularity, beauty, and elegance. However, what is most exciting in the context of therapeutic nanodevices (in our opinion) is the tunable biological activity of some supramolecular assemblies.

Incorporation of pharmacophores into supramolecular structures can render the structures biologically active. *Tysseling-Mattiace* and colleagues recently reported a supramolecular assembly (IKVAV peptide amphiphile, or IKVAV PA) with biologic activity in vitro and in vivo [16.47, 48]. IKVAV PA is based on amphiphilic polymer monomers, as are micelles; but unlike micelles, IKVAV PA is assembled from polymeric monomers whose chemistries are tuned to allow them to preferentially assemble into linear nanoscale filaments rather than spheres (Fig. 16.7). IKVAV PA nanofibers further coalesce into a gel under physiological salt and pH conditions. Thus, aqueous solutions of IKVAV PA spontaneously form a gel when injected in vivo.

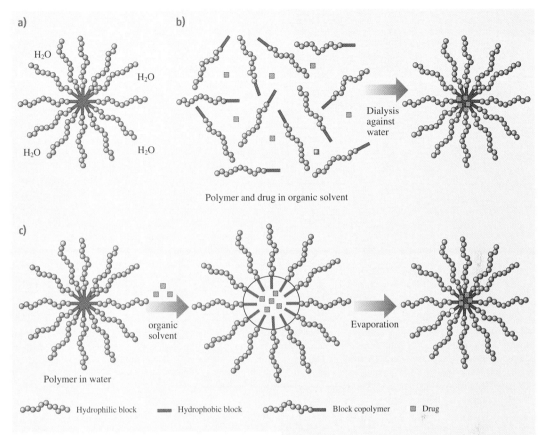

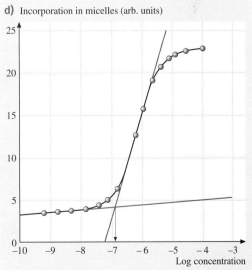

Fig. 16.6a–d Micellar drug-delivery vehicles and their self-assembly from block copolymers [16.42–44]. (**a**) Morphology of a micelle in aqueous buffer. Hydrophobic and hydrophilic polymer blocks, copolymers containing the hydrophilic polymer blocks, copolymers containing the blocks, micelles generated from the block copolymers, and (hydrophobic) drugs for encapsulation in the micelles are indicated. (**b**) Micelle self-assembly and charging with drug occurring simultaneously when the drug–polymer formulation is transitioned from organic to aqueous solvent by dialysis. (**c**) Preformed micelles can be passively imbibed with drugs in organic solvent. Organic solvent is then removed by evaporation, resulting in compression of the (now) drug-bearing hydrophobic core of the micelle. (**d**) An illustration of concentration-driven micelle formation. At and above the critical micelle concentration (CMC), block copolymer monomers assemble into micelles rather than exist as free block copolymer molecules. The arrow indicates the CMC for this system (after [16.42–44])

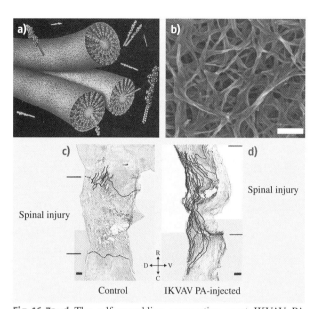

Fig. 16.7a–d The self-assembling neuroactive agent IKVAV PA (IKVAV peptide amphiphile) promotes regeneration of axons across sites of mechanically induced spinal injury in mice. (**a**) Schematic representation showing individual PA molecules assembled into a bundle of nanofibers interwoven to produce IKVAV PA. (**b**) Scanning electron micrograph image showing the network of nanofibers in vitro; scale bar in (**b**) indicates 200 nm. (**c,d**) Representative tracings of descending motor axon fibers within a distance of 500 μm rostral of the lesion in vehicle-injected (**c**) and IKVAV PA-injected (**d**) mice. The *dotted lines* demarcate the borders of the area of spinal cord injury. *Colored lines* indicate descending motor neurons impinging on the lesion. *Scale bars* in (**c,d**) indicate 100 μm. R – rostral; C – caudal; D – dorsal; V – ventral. IKVAV PA also promotes lesser, but still significant, regeneration of sensory neurons (not shown [16.48]). IKVAV PA-induced improvements in regeneration of axons are associated with significant behavioral improvements in mice. The self-assembly properties of IKVAV PA is essential to its function: IKVAV peptide alone does not promote regeneration [16.47, 48] (after [16.48])

tissue is a key inhibitor of neurite outgrowth to repair the break.

In a murine model of induced spinal cord injury, IKVAV PA was injected into the site of spinal breaks. Relative to controls, injected animals exhibited reduced neurite apoptosis, reduced glial cell differentiation and scar tissue formation, and increased transit of the site of injury by descending (motor) neurons and by ascending (sensory) neurons (Fig. 16.7). These histological improvements in treated animals were accompanied by behavioral improvements (significant recovery of injury-induced motor and sensory deficits) that persisted longitudinally after the IKVAV PA was known to have been cleared from the site of injury. These improvements in motor and sensory function were unprecedented in the spinal injury model, and will be highly significant if translated to human therapy.

Furthermore, equimolar amounts of IKVAV peptide injected into the sites of injury produced no significant neurological improvements relative to control animals. That is, the structure provided by the amphiphilic polymer assembly is necessary to IKVAV PA therapeutic function. Consequently, IKVAV PA fits our definition of a nanodevice, and consists of an assembly module (the amphiphilic polymer monomers) and a bioactive module (IKVAV).

Design of self-assembling polymeric units that form specific nanoscale structures requires organic chemical synthetic capabilities and modeling skills that are not widely distributed in the physical and biological nanotechnology communities, but they do constitute genuine nanotechnologies, in as much as supramolecular chemistry facilitates multicomponent functional structures whose functions depend on structurally defined nanoscale components. The potential of supramolecular structures in therapy is immense.

Supramolecular chemistry of synthetic peptides has provided pharmacologically active antimicrobial structures (Fig. 16.8) [16.49, 50]. In the anti-infective architecture, individual peptide components are flat, circular molecules. The planar character of the toroidal subunits is a consequence of the alternating chirality of alternating DL amino acids (AAs) in the primary sequence of the peptide rings, currently only possible for synthetic peptides. Ribosomes recognize and incorporate into nascent polypeptides only L amino acids, and so, as a result of AA chirality and bond strain, peptides made by ribosomes cannot be flat, closed toroids like those of the peptidyl anti-infective agents. Alternation of D and L AAs is not possible in proteins made by ribosomal synthesis, though as we discuss below, this

The designers incorporated the peptide isoleucine–lysine–valine–alanine–valine (IKVAV) into the polar head group of each amphiphilic polymer monomer. IKVAV is a neuroactive peptide derived from the protein laminin, and is presented at immensely high valency on the surface of assembled IKVAV PA nanofilaments. In the central nervous system, IKVAV peptide is known to inhibit differentiation to glial cells and promote neuronal outgrowth. Glial cells have a primary role in laying down scar tissue after spinal cord injury, and scar

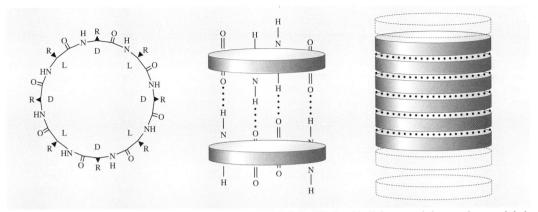

Fig. 16.8 A self-assembling peptide antibiotic nanostructure [16.49, 50]. Peptide linkages and the α-carbons and their pendant R-groups are indicated. The synthetic peptide rings are planar as a result of the alternating chirality (D or L) of their amino acid (AA) constituents. R-groups of AAs radiate out from the center of the toroid structure. Individual toroids self-assemble (stack) as a result of hydrogen-bonding interaction between amine and carboxy groups of the peptide backbones of adjacent toroids. The surface chemistry of multitoroid stacks is tuned at the level of the AA sequence and, therefore, R-group content of the synthetic peptide rings. The chemical properties of the stacked toroid surfaces allow them to intercalate into the membranes of pathogenic organisms, with lethal consequences. The specific membrane preferences for intercalation of the compound are tuned by control of the R-group content of the toroids (after [16.49,50])

may change. Much as in α-helical domains of ribosomally synthesized proteins, however, the AA R-groups (which are of varying hydrophobic or hydrophilic chemical specificities) are arranged in the plane of the closed D, L rings, extending out from the center of the rings. Hydrogen bonds between individual rings govern self-assembly of the toroids into rod-like stacks, while the R-groups dominate interactions between multiple stacks of toroids and other macromolecules and structures. The planar toroidal subunits can be administered as monomers and self-assemble into multitoroid rods (Fig. 16.8) at the desired site of action (in biological membranes). The peptide toroid R-groups are chemically tuned so that the rod structures into which they spontaneously assemble intercalate preferentially in specific lipid bilayers (i.e., in pathogen versus host membranes). Moreover, the assembled rods may undergo an additional level of self-assembly into multirod structures, spanning pathogen membranes [16.49, 50]. Intercalation of the self-assembled rods into pathogen membranes reduces the integrity of pathogen membranes selectively, potentially rendering agents toxic to pathogens but not to hosts.

Other self-assembling, nanoscale antibiotics with morphologies more like that of micellar structures [16.42–44] have been described (N8N antimicrobial nanoemulsion [16.51]). They, as well as the self-assembling D, L peptide toroids [16.49, 50], represent critically needed, novel antibacterial agents. Resistance to traditional, microbially derived antibiotics often is tied to detoxifying functions associated with secondary metabolite synthesis; these detoxifying functions are essential for the viability of many antibiotic-producing organisms [16.52]. The genes encoding such detoxifying functions are rapidly disseminated to other microorganisms, accounting for the rapid evolution of drug-resistant organisms that has bedeviled antimicrobial chemotherapy for the last half-century. Synthetic nanoscale antibiotics, such as peptide toroids [16.49, 50] and N8N antimicrobial nanoemulsion [16.51], act by mechanisms entirely distinct from those of traditional secondary metabolite antibiotics, and presumably no native detoxifying genes exist for these supramolecular antibiotics *because* they are structures not produced by evolution. Therefore, novel nanoscale antimicrobials may not be subject to the unfortunately rapid rise in resistant organisms associated with most secondary metabolite antibiotics, though this remains to be seen. As bacterial infection continues to reemerge as a major cause of morbidity and mortality in the developed world, as a consequence of increasing antibiotic-resistant pathogens, novel, nanoscale antibiotics will become increasingly important.

Biological macromolecules undergo self-assembly at multiple levels, and like all instances of such assembly, biological self-assembly processes are driven by thermodynamic forces. Some biomolecules undergo intramolecular self-assembly (as in protein folding from linear peptide sequences, Fig. 16.9). Higher-order structures are, in turn, built by self-assembly of smaller self-assembled subunits (for instance, structures assembled by hybridization of multiple oligonucleotides, enzyme complexes, fluid mosaic membranes, ribosomes, organelles, cells, tissues, etc.). Assemblies of higher-order structures, such as oligomeric proteins, are driven by the same forces that drive intramolecular protein folding.

Naturally occurring proteins are nonrandom copolymers of 20 chemically distinct amino acid (AA) subunits. The precise order of AAs (i. e., via interactions between AA side-chains) drives the linear polypeptide chains to form specific secondary structures (the α-helices and β-sheet structures seen in Fig. 16.9). The secondary structures have their own preferences for association, which in turn leads to the formation of the tertiary and quaternary structures that constitute the folded protein structures. In its entirety, this process produces consistent structures that derive their biological functions from strict control of the deployment of chemical specificities (the AA side-chains) in three-dimensional space.

Biomolecules can be used to drive assembly of nanostructures, either as free molecules or when conjugated to heterologous nanomaterials (Fig. 16.9). For instance, three-dimensional nanostructures can be made by DNA hybridization [16.53–56]. Such DNA nano-

structures can exhibit tightly controlled topographies but have limited integrity in terms of geometry [16.56], due to the flexibility of double-stranded (ds)DNA of 200 or more base pairs in length.

Often the domains of biomolecules responsible for assembly and recognition are small, continuous, and discrete enough that they can be abstracted from their native context as assembly modules and appended to

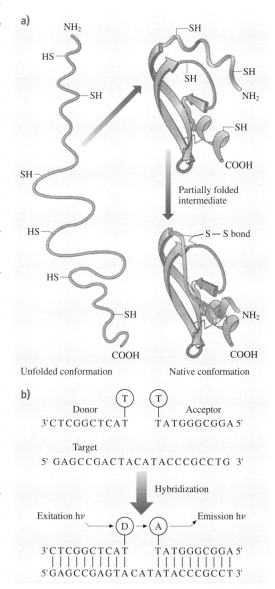

Fig. 16.9a,b Self-assembly of biological macromolecules. (a) Linear peptide chains (with amino- and carboxy-ends, as well as sulfhydryl groups of cysteine residues indicated) undergo a multistep folding process that involves the formation of secondary structures (α-helices, indicated by heavy helical regions, and β-sheet regions, indicated by the heavy arrows) that themselves associate into a tertiary structure. Final conformation is stabilized by the formation of intrachain disulfide linkages involving cysteine thiol groups. (b) A fluorescence transfer device that depends on self-assembly of biomolecules. The device is composed of donor (D) and acceptor (A) molecules brought into close proximity (within a few angstroms) by the base-pair hybridization of complementary oligonucleotides. When the structure is assembled, acceptor and donor are energetically coupled, and fluorescence transfer can occur [16.53, 54] (after [16.54]) ▶

other nanomaterials for direct formation of controlled nanoscale architectures. Antibodies and other specific biological affinity reagents can be used to assemble hybrid nanostructures (Fig. 16.9). Oligonucleotides are particularly intriguing in this application.

Unlike longer DNA segments, shorter dsDNA segments (of 50 base pairs or less) are rigid. The rigidity of short ds oligonucleotides has been cleverly exploited to build multidendrimer complexes (Fig. 16.4b) [16.30–32]. Dendrimers in these complexes are held at very precise distances from one another, defined by the length of the dsDNA segments. Nanometer-scale interdendrimer distances measured by AFM and other methods are within 10% of distances predicted from the dsDNA length [16.30, 31]. The length of the ds oligonucleotide is defined by the DNA sequence, and is easily tuned. This assembly strategy is currently being used to build multidendrimer anticancer devices, analogous to those envisioned in Fig. 16.4a, albeit with greater synthetic control and stability. The implications of the strategy goes well beyond dendrimer complexes, and might be used to impose precise distance relationships on nanoscale components or between nanoscale components and surfaces in devices.

16.2.3 Chemoselective Conjugation

Several chemoselective bioconjugate approaches have arisen from the field of protein semisynthesis [16.20, 57, 58]. These protein synthetic chemistries allow site-specific conjugation of polypeptides to heterologous materials in bulk, as the result of conjugation between exclusive, mutually reactive electrophile–nucleophile pairs, one on the polypeptide, the other on its conjugation partner. Chemoselective conjugation strategies have been applied to the synthesis of multiple nanobiological devices [16.1, 1–8, 15, 57–60].

Proteins are profoundly dependent on their three-dimensional shapes for their activities: chemical derivatization at critical AA sites can profoundly and negatively impact protein bioactivity. Because conjugation can be directed to preselected sites via chemoselective approaches, and since the sites of conjugation in the protein can be chosen because the proteins tolerate adducts at those positions, proteins coupled to nanomaterials by such chemoselective methodologies often retain their biological activity. In contrast, protein bioactivity in conjugates often is lost or profoundly impaired when proteins are coupled to nanomaterials using promiscuous chemistries. For instance, promiscuous chemistries [1-ethyl-3-(3-dimethylaminopropyl) carbodiimide (EDC) conjugation [16.61]] used to conjugate cytokines to nanoparticles tend to inactivate human interleukin (hIL)-3 and other cytokines. The same protein–particle bioconjugates retain bioactivity if judiciously chosen chemoselective conjugation strategies are used [16.1, 59, 60].

The potential utility of chemoselective conjugation for incorporation of active biological structures into semisynthetic nanodevices cannot be overestimated. In fact, some protein engineering methods that can change protein topography, and therefore the spatial relationships between proteins and the device nanocomponents to which they are conjugated, can be used to control protein-device spatial arrangements only in circumstances where the association between the protein and device is nonrandom and oriented, as can occur via chemoselective conjugation.

Protein sequences are said to be circularly permuted (CP) when their parental amino- and carboxy-ends are ligated together, and new amino- and carboxy-ends are introduced elsewhere in the protein sequence. There are as many potential CP variants of a given protein as there are AAs in the protein sequence. In each distinct variant, the CP sequence initiates (and ends) at a different AA of the parental sequence.

The fact that many proteins tolerate circular permutation of their primary AA sequences came as a surprise to molecular biologists. Nonetheless, multiple examples of biofunctional CP proteins have been documented [16.62–71]. We recently developed a method that displays all possible CP variants of a protein on the surfaces of phage particles. Phage-presented CP proteins can be conveniently screened for function by their affinity for a target (for a known ligand of the parent protein, for instance). We call this process scanning circular permutation [16.72, 73], and it allows functional screening of every possible CP variant of a protein of interest.

One of the remarkable findings of the scanning CP work was the finding that a large fraction (as many as $\frac{1}{3}$, and possibly more) of the distinct CP variants of a test protein were bioactive [16.72]. Many of the new amino- and carboxy-ends of functional CP variants fell within known secondary structural domains of the parent protein, and yet CP variants interrupting those secondary structural domains exhibited full bioactivity. This all suggests that the linear order of protein AA and secondary structural domains as they occur in natural proteins may not be requisite for protein structure and function, and further that multiple biofunctional CP variants of many proteins may be possible [16.62–72].

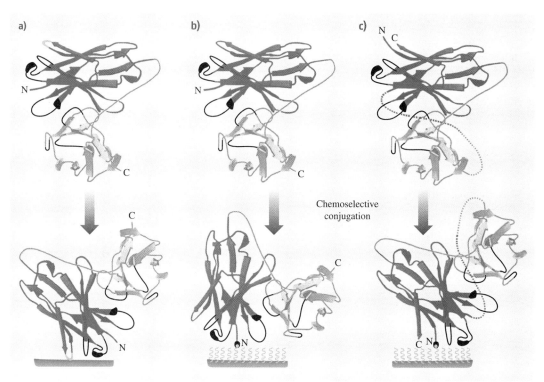

Fig. 16.10a–c Examples of protein engineering manipulations that could potentially alter the orientations of proteins (SCFvs) deployed on micromachined surfaces. Ribbon diagrams of SCFvs (VH–VL configuration) to be deposited on a SiO_2 surface. VH is shown in *dark brown*, VL in *light brown*. Complementarity-determining regions (CDRs) of VH and VL are shown in *black*. CDRs direct specific antigen recognition. Micromachined metal oxide surface is represented by a brown bar. A polymeric layer on the surface is represented by *wavy lines*. N- and C-ends of SCFvs are indicated. Chemoselective ligation between N-ends of SCFvs and polymer is indicated. (**a**) Affinity peptide–SCFv: surface-specific affinity peptide selected from a display library (for instance, SiO_2- [16.74, 75] or Al_2O_3-binding [16.76] peptide) is inserted into the SCFv antibody fragment (*gray line*). The affinity peptide binds to the surface, effectively orienting the SCFv and determining the proximity of the CDRs to the SiO_2 surface. (**b**) Parent SCFv: chemoselective conjugation of a modified SCFv (with an N-terminal aldehyde; [16.1, 59, 60, 72, 73] to the polymer layer. Note the position of the VH CDRs. (**c**) A circularly permuted variant of a SCFv: chemoselective conjugation of a circularly permuted [16.72, 73], but otherwise comparable, SCFv. In (**b**) and (**c**), note that chemoselective conjugation produces a consistent orientation of SCFvs and that, relative to the parent SCFv, CP alters the proximity of the CDRs to the surface (after [16.73])

Scanning CP [16.72] provides a high-throughput way to identify biofunctional CP variants and to build collections of CP versions of a single parent sequence.

When proteins are conjugated to a polymer interface by a single, specific amino acid (as by chemoselective conjugation [16.20, 57, 58]), their orientations are more or less regular (limited by the conformational freedoms of the protein and interfacial polymers themselves). Typically, chemoselective conjugation is limited solely to protein N-termini, producing consistent orientation relative to the surface. Circular permutation alters protein topology to modulate distances between protein functional domains (such as antigen-combining sites of SCFvs or single chain fragment variables) and N-termini (where conjugation uniquely occurs, and so proximal to the surface). Together, chemoselective conjugation and circular permutation modulate protein orientation on surfaces to allow tuning of distance to surfaces and protein functional domains (like antigen combining sites) [16.72, 73]. This effect is illustrated for a hypothetical antibody fragment (so-called single-chain Fv or SCFv) in Fig. 16.10. The potential utility of changing these distances is discussed in the context of a nanobiotechnological sensor system below. In this sensor, certain key nanoscale distances are the primary determinants of overall sensor function, and we will show how circular permutagenesis and other molecular biology approaches can be used to address problems occurring at the nanoscale.

16.2.4 Unnatural Amino Acids to Support Chemoselective Conjugation of Biologically Produced Proteins

Early protein chemoselective conjugation methods depended on either protein or peptide chemical synthesis to introduce a nucleophile or electrophile into the protein sequence for chemoselective conjugation, or on some postsynthesis processing of a native AA to a chemically reactive group [16.20, 57, 58]. Typically, the reactive group has been at an end of a polypeptide (most typically, at the N-end). Recently, more elegant and versatile methods to introduce reactive groups to polypeptides via ribosomal synthesis have developed. These methods utilize biologically produced proteins made in expression host systems engineered to introduce unnatural AAs (UAAs) at specific sites in proteins. UAAs can have a variety of functional side-chains, including chromophores, fluors, photoreactive groups or chemically reactive groups. To the extent that reactive functionalities of UAAs can participate in chemoselective conjugations with other chemical groups, this approach can be used to prepare proteins for chemoselective conjugation to other surfaces or nanoscale moieties. Importantly, since the UAAs are introduced by ribosomal protein synthesis, the position of the UAA, and therefore the site of chemoselective conjugation, is not limited to the end of the protein sequence. Chemoselective conjugation can potentially be made to occur at any AA of the polypeptide sequence. Such UAA-incorporating systems have been developed in prokaryotic, lower eukaryotic [16.77, 78], and mammalian expression systems [16.79], potentially allowing proteins with UAAs to be synthesized in host systems of varying expression properties or with host-specific posttranslational modifications.

These systems for ribosomal introduction of UAAs at specific sites in biologically produced proteins have several general features in common [16.77–79]. They exploit suppression of nonsense codons (of which there are three in *E. coli*, and which, in wild-type strains, stop protein translation) by a mutant transfer ribonucleic acid (RNA) (a so-called suppressing tRNA) whose anticodon loop is homologous to the nonsense codon that will drive UAA insertion into nascent protein. Not only must this mutant tRNA recognize the nonsense codon, but it must also have the capacity to be charged with the desired UAA by a corresponding aminoacyl-tRNA synthase. Generating aminoacyl-tRNA synthases that can recognize both the suppressing tRNA and the desired UAA requires extensive genetic engineering effort. To ensure specificity of UAA incorporation, the aminoacyl-tRNA synthase cannot recognize any native AA, and cannot direct charging of any of the host's other tRNAs.

As current (2008) metabolic engineering capabilities go, UAA-incorporating systems must rank as the most sophisticated described to date. Nonetheless, their output is engineered protein with reactive groups at predetermined AA positions within them. These reactive groups can be conjugated chemoselectively to other nanoscale components, to effect changes in the spatial relationship of particular protein functional domains and the components to which proteins are conjugated. A collection of proteins differing from one another only by the position of the UAA that participates in chemoselective conjugation could have a utility to nanotechnologists that closely parallels that of a collection of circularly permuted proteins. As Fig. 16.10 shows, circular permuted proteins chemoselectively conjugated to surfaces can be used to tune distance between underlying surfaces and protein functional domains. Incorporation of UAAs and chemoselective conjugation therefore might also be used for analogous purposes in nanobiological devices.

16.3 Sensing Devices

The need for *smart* therapy is a key theme of therapeutic nanotechnology, and of pharmacology as a whole. We discussed above the emerging class of theranostic agents that include sensing or diagnostic functions.

Drugs with narrow therapeutic windows should ideally be delivered only to their desired site of action and be pharmacologically active only when their activity is needed. These strategies can limit undesired secondary effects of therapy, some of which can be debilitating or life threatening. One possible approach to this issue is the incorporation of sensing capability (specifically, the capacity to recognize appropriate contexts for therapeutic activity) into nanotherapeutic devices. Sensing capability may allow self-regulation of a therapeutic device, reporting to an external clinician/device operator (as in imaging applications, see below), or both.

Biosensors are typically considered to be multifunctional, multicomponent devices [16.80]. Usually a biosensor system is composed of a signal transducer, a sensor interface, a biological detection (bioaffinity) agent, and an associated assay methodology, with each system component governed by its inherent operational considerations. Signal transducers are moieties that are sensitive to a physicochemical change in their environment and that undergo some detectable change in chemistry, structure or state as the result of analyte (the thing to be sensed) recognition. Analytes for nanotherapeutic application could be biomolecules, such as proteins, small molecules (organic or inorganic), and ions (salts or hydrogen ions), or physical conditions (such as redox state or temperature). Interfaces are the sensor components that interact directly with the analyte. For sensor use in nanotherapeutic devices, immobilized or otherwise captured biological molecules (proteins, nucleic acids) often constitute the active parts of biosensor interfaces. Whatever the chemical nature of the interface, it determines the selectivity, sensitivity, and stability of the sensing system and also is a dominant determinant of sensor operational limits. Assay methodology determines the need (or lack thereof) for analyte tracers, the number of analytical reagents, and the complexity and rapidity of the sensing process.

16.3.1 Planar FET Protein Sensors

To illustrate the potential of biotechnology to solve nanoscale problems, we will focus on protein-sensing immunoFETs (a field-effect transistor sensor which binds analyte via an antibody molecule or fragment thereof on its sensing surface). Field-effect transistors (FETs) consist of a current source and current drain separated from each other by a semiconductor channel through which current flows. A *gate* to which electrical bias is applied is positioned above the channel to modulate channel conductance. Depending on gate bias, current flow through the FET channel is increased or decreased (Fig. 16.11) [16.81–85]. Electrical properties of FET sensing channels can be modulated by any proximal electric field/charge acting as a gate, allowing configuration of sensors [16.81–91], including, in principal, protein sensors. Receptor molecules (antibody fragments, peptides, aptamers, etc.) that recognize protein analyte molecules are deployed on the sensing channel, and analytes bind to the interface via the receptors. Charges of protein analytes induce a dipole between the surface and the underlying depletion region of the semiconductor, eliciting a gating effect which affects current through the FET. Differently charged analytes interact differentially with charge carriers in the FET, producing analyte-charge-specific responses. FET sensors can be constructed from nanowires (so-called 1-D FETs). One-dimensional structures *may* have much enhanced sensitivity due to their high surface-area-to-volume ratios [16.82–85]. An example of a 1-D FET was shown in Fig. 16.5, and 1-D FETs much like those pictured are widely accepted as nanotechnological devices, though meso- to macroscale planar FET sensors often are not. The reason for considering 1-D FETs as *real* nanotechnology is that 1-D devices contain a nanoscale component (the semiconducting nanowire). They also additionally contain nanoscale receptor molecules. Thus, although physical nanotechnologists often chafe (very unreasonably) at this argument, planar FETs are nanodevices for the

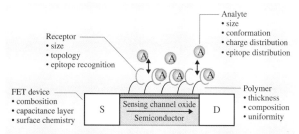

Fig. 16.11 A protein-sensing FET and some of the parameters influencing its function. Current source and drain (S, D) are shown, as is the semiconductor (*gray box*). Capacitance layer/sensing channel oxide is shown in *brown*. Current flowing from the source to the drain in the semiconductor (I_{SD}) is indicated by the *black arrow*, and increases or decreases in response to charges brought to the sensing channel surface as a result of analyte binding. Receptors (SCFvs, streptavidin, etc.) are shown as *gray C-shaped lines* and bind analyte (*gray balls, labeled A*) specifically but reversibly and are conjugated to the sensing channel by an interfacial polymer. Parameters that might be chosen judiciously to optimize the sensor are shown. Note that interfacial optimization occurs primarily on the nanoscale, with specific changes to molecular sizes and orientations

same reason as 1-D nanowire FETs are. Planar FETs, though they themselves may be meso- to macroscale, depend on obligately nanoscale components (biomolecular receptors for analytes) for their function. Here we focus on *planar* FETs as a platform to demonstrate how biotechnology can solve a problem occurring at the nanoscale.

The potential utility of FET protein sensing in physiological environments is obvious, but has only recently been realized [16.90, 91]. FET sensor sensitivity to analyte charge is limited by multiple factors [16.81–91], though the key problem cited is typically regarded to be shielding by buffer ions in the high-salt (150 mM Na^+) physiological environment (Fig. 16.12) [16.81–84, 86]. This led to an assessment that planar immunoFETs, owing to the size of antibodies (10–12 nm) and the maximal distance over which analyte charges were thought to be detectable by the FET (sometimes called the Debye length, < 2 nm), were fundamentally infeasible. The classical analysis of immunoFETs reasonably considers ion shielding, but is so fatally flawed in its consideration of receptor protein structures on FET surfaces that it is irrelevant. To biochemists, the fallacies of the classical model are obvious, and biochemists encountering the classical model immediately dismiss it as meaningless, though much of the sensing community has clung tenaciously to the moribund model. We have discussed the fatal flaws of the classical analysis elsewhere [16.90, 91] and will repeat the discussion only briefly:

1. Use of intact antibody as receptors as in Fig. 16.12 is unnecessary: *much* smaller (10×), antigen-binding antibody fragments are available.
2. Antibodies do not adsorb exclusively at C3 domain (as shown in Fig. 16.12): antibodies bind surfaces in a distribution of orientations.
3. Antibodies contain a flexible region at the crotch of the Y called the *hinge* that allows antigen-combining sites vast freedom in their relative positions. The implicit assumption illustrated in Fig. 16.12 is that antibodies are rigid bodies, which is wrong.
4. The effect of antigen (analyte) three-dimensional (3-D) shape and charge distribution on charge proximity to the sensing channel is not considered in classical analysis.
5. The diversity available in native and phage-presented antibody repertoires was not exploited. The impact on sensor properties of analyte receptors (antibodies or fragments thereof) of differing valency, conformations, antigen (analyte) affinity or analyte epitope recognition properties were not considered.
6. Improvement of antibody receptor properties by protein engineering is not considered in the classical analysis.
7. Data directly shows that planar FET protein sensors can detect a biologically important analyte at concentrations which occur in real biological systems under physiological salt conditions [16.90, 91].

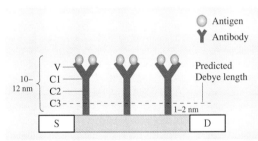

Fig. 16.12 The biochemical basis of the classical analysis interpreted as indicating the infeasibility of planar immunoFETs operating in physiological buffer. Artist's conception of antibodies on FET sensing channel. Antibody domains (constant 1, 2, 3 as C1, C2, C3 and variable as V) are indicated, as is the predicted Debye length in physiological buffer (150 mM salt, about 1–2 nm). FET source and drain (S and D) are shown. Shown is the classical (but wrong, see text) depiction of antibodies adsorbed onto a FET sensing channel solely by their C3 domains. This flawed representation is central to the classical objection to the feasibility of immunoFETs (after [16.81])

Suffice it to say that, if planar immunoFETs are not feasible, the reasoning of the classical assessment has little to do with why they might be infeasible. However, we know from our work that planar FET protein sensors operating at physiological salt concentrations are feasible [16.90, 91], though we show that the distance between the analyte charge and the sensing channels is a key parameter for sensor sensitivity. We will focus on biotechnology approaches to engineer the nanometer distance between bound analyte charges and FET sensing surfaces as an existence proof of the utility of biotechnology to nanotechnology.

16.3.2 Biotechnology Approaches to the "Fundamental Limitations" of Planar ImmunoFETs

First, in the classical model [16.81–84, 86], chemoselective conjugation of receptors (antibodies) to sensing interfaces is not considered. Chemoselective conjugation would have the desirable effect of orienting the antibody layer consistently, from one antibody molecule to the next. Even though the classical representation (as mimicked in Fig. 16.12) shows antibodies adsorbed onto sensing surfaces in an oriented fashion, it is well known to immunologists that antibodies adsorb to surfaces in a stochastic distribution of orientations, binding to the surfaces via a random distribution of antibody surface regions. That is to say that the antibody layer as contemplated in the classical assessment is misrepresented, and the lack of orientation of the antibodies at real interfaces would interfere with many of the available biotechnology strategies to control the distance between the analyte charge and the sensing surface. However, the entire classical assessment of immunoFETs [16.81–84, 86] is based on the incorrect representation of the extent of order in the protein orientation shown in Fig. 16.12. This critical flaw alone invalidates the classical assessment, but there is much more wrong with the assessment than just this point.

Secondly, use of intact antibodies (IgGs) as immunoFET receptors is unnecessary: specific epitope recognition function can be isolated on fragments several-fold smaller than intact antibodies [16.21, 92–99]. Single-chain Fv antibody fragments (single-chain fragment variables, SCFvs) are less than half the size of intact antibodies. Single-domain (camelid) antibody fragments, called variable heavy–heavy (VHH), are smaller still: about 10% of the mass of intact antibodies [16.97–99]. There are multiple methods to convert existing antibody genes (isolated from mammalian B-cells) to genes that will direct biologic production of either SCFvs or VHHs recognizing the same antigen (analyte), as well as methods to isolate de novo SCFvs or VHHs that are directed to antigens of interest that are well known to those skilled in the art [16.97–99]. The simple expedient of using an antibody fragment as an immunoFET receptor reduces the distance between the sensing surface and the bound analyte charge to nearly the calculated distance over which buffer counterion shielding is expected to occur (Fig. 16.12 [16.81–84, 86]). Again, failure to consider this fact is in itself sufficient to invalidate the classical assessment. Antibody fragments are convenient substrates for chemoselective conjugation, and we exploit the protein orientation chemoselective approaches in our SCFv layers to engineer analyte–semiconductor distances.

Thirdly, no consideration is given in the classical model to engineering the antibody molecules or fragments thereof to optimize their function as immunoFET receptors. This was a major oversight, even at the time when the classical model was promulgated in the early 1990s. The tools available for antibody engineering are vast and sophisticated. We have discussed a few of them above (receptor circular permutagenesis [16.72, 73] and introduction of UAAs [16.77–79], both useful in the context of chemoselective conjugation of the antibody or fragments thereof to the sensing interface). As Fig. 16.10 illustrates, the effect of circular permutagenesis of a SCFv is to change the orientation of its antigen-combining sites to the site of chemoselective conjugation. Effectively, the orientation (relative to the immunoFET sensing channel) and therefore the distance charges of analytes bound to antibody fragments is different when analytes are bound to different CP antibody variants. Since charges of protein analytes are not very mobile, the critical distance between analyte charges and sensing surfaces is different with different circularly permuted variants of a single antibody fragment. In the context of the immunoFET, it should be possible to identify those circularly permuted antibody fragments that produce the greatest sensor sensitivity, presumably because they minimize the distance between bound analyte charges and FET sensing surfaces. Similarly, the optimal AA position for a UAA in an antibody fragment could be determined for an antibody fragment used as a FET receptor from the standpoint of immunoFET sensitivity. The approach could tune the distance between underlying surfaces and analyte charges, much as circular permutagenesis can.

There are still many other biotechnology approaches that might be used to optimize receptors for immunoFETs that have been ignored to date. We and multiple others have identified peptides that can specifically bind microfabricated surfaces [16.74–76]. Such peptides might be inserted into SCFv or VHH sequences and used to adhere the SCFv or VHH to the micromachined surface, as we suggest [16.73, 74]. Presumably, variable positions of peptide insertion could lead to differential orientations of antibody fragments on the immunoFET surface, with the potential benefits in terms of FET sensitivity already described for circular permutagenesis or UAA insertion described above. This has the potentially great advantage of adhering immunoFET receptors directly to the sensing surface, but our experience with direct adsorption of receptor proteins on micromachined surfaces have proven direct adsorption to be unsatisfactory in our hands (see below).

The foregoing is a very incomplete list of potential biotechnology approaches to addressing a limitation correctly identified (buffer counterion shielding of analyte charges) but incorrectly analyzed [16.81–84, 86]. Many more approaches are possible, although we will not attempt to enumerate them here. The point is that planar immunoFETs have a nanoscale limitation that can be addressed by biotechnology. Hence, in this application, biotechnology is a bona fide nanotechnology.

16.3.3 Nanotribology of Protein-Sensing Interfaces on Micromachined Surfaces

In FET protein sensors, the critical importance of the sensing interface is clear, implying the need to characterize the interface, and nanotribology and surface characterization have led to the discovery of additional interfacial design parameters and properties that must be considered in immunoFET design. Parameters of interest were thickness of interfaces (related to, but distinct from, the critical distance between bound analyte charges and semiconductor surfaces), surface roughness (smoothness), and interfacial robustness (resistance to wear). Resistance to wear is a critical issue if sensors are to be subjected to abrasion, as by tissue in an in vivo environment.

These studies were performed on silicon surfaces with a thermally grown oxide layer (silica surfaces [16.100, 104–106]) and on a sputter-coated aluminum layer with a native oxide (aluminum surfaces [16.100]). Silicon surfaces simulate the sensing surfaces of a metal–oxide–semiconductor field-effect transistor (MOSFET), whereas aluminum simulates the sensing surface of an AlGaN (aluminum–gallium nitride) heterojunction FET. Native Al_2O_3 oxide which develops on sputter-coated aluminum surfaces is also the most prevalent oxide on the surfaces of AlGaN/GaN heterojunction FETs. All studies were performed with the model receptor protein streptavidin, and not with an antibody fragment.

Lee et al. and *Bhushan* et al. studied the step-by-step morphological changes resulting on silicon surfaces during deposition of a model protein (streptavidin) [16.104, 105]. These studies revealed, among other things, that streptavidin adsorbed to silicon interacted with the surface very weakly. Most of the adsorbed streptavidin could be removed from the surface by rinsing with an aqueous buffer [16.105]. Streptavidin adherence by adsorption to micropatterned asperities on the silicon surface and to the edges of silicon fragments was modestly stronger than to the flat silicon surface, although adhesion was still weak [16.104, 105]. Adhering streptavidin to the silicon surface via a biotinylated silane polymer (3-aminopropyltriethoxysilane, APTES Fig. 16.13) significantly enhanced resistance to wear [16.100, 105, 106].

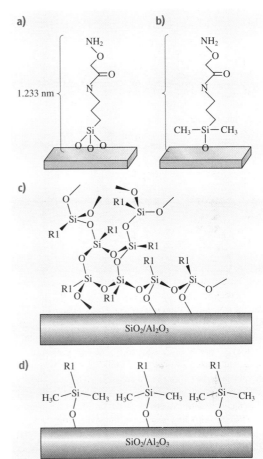

Fig. 16.13a–d Tri- and monoalkoxy silane constituents of the sensor interface on micromachined substrates. In (a) and (b), oxyamine-terminated polymers for chemoselective conjugation are shown. (a) shows the trialkoxy interfacial polymer and (b) its monoalkoxy analogue. As determined by summing bond lengths, the length of polymer in nm is shown. Both ellipsometry (Fig. 16.14) and AFM analysis show that interfaces made with the monoalkoxy version of polymers are thinner and less rough, presumably due to reduced intramolecular silane polymerization [16.100–103]. (c) shows a hypothetical film made from trialkoxy polymer of (a), and (d) shows a hypothetical film film made with the monoalkoxy polymer of (b). R1 is $(CH_2CH_2NHCOCH_2ONH_2)$ (after [16.100])

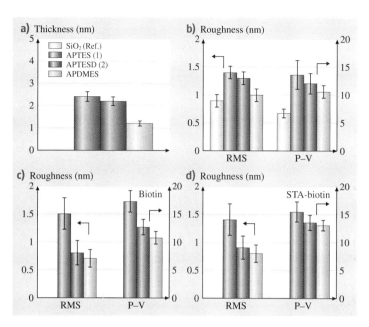

Fig. 16.14a–d Representation of nanotribological properties of films on silicon/SiO$_2$ surfaces. APTES (1) films were deposited by the method used in our published HFET (heterojunction field effect transistor) protein sensors [16.91, 92], and APTES (2) and APDMES films were deposited by the improved method described in [16.101]. (**a**) Silane film thickness (ellipsometry), (**b**) surface roughness [root mean square (RMS) and peak-to-valley (P–V) values] of silane films, (**c**) surface roughness (RMS and P–V values) of the same three silane films following biotinylation, and (**d**) surface roughness (RMS and P–V values) of the same three silane films following streptavidin binding to the biotin (after [16.100])

Ellipsometry of APTES-coated silicon surfaces showed the thickness of the polymer layer to be several-fold the expected monolayer thickness [16.100]. The unexpected thickness of the film likely represents multilayering by APTES, a phenomenon that had been previously reported [16.101–103]. Despite the fact that the sensor with a multilayered APTES layer was sufficiently sensitive to detect analyte [16.90, 91], thinner interfaces would theoretically increase sensor sensitivity to analyte. Multilayering is related to the trivalency of APTES siloxane residues and is influenced by the deposition method [16.101–103]. Consequently, APTES polymer films were also constructed using a more optimized deposition protocol [16.101]. The more optimal method produced thinner APTES films, as expected (Fig. 16.14). However, peak-to-valley values for APTES films deposited by either protocol were substantially greater than the summed bond lengths of APTES [16.100, 104–106]. The monosiloxane reagent, aminopropyldimethylethoxysilane (APDMES, Fig. 16.13), which has a much diminished capacity to produce multilayers, was also used to build polymer films on silicon. Strikingly, APDMES produced the thinnest films, with thicknesses comparable to the summed bond lengths of the APDMES polymer (Fig. 16.14).

The thicker silane films (based on APTES) were also substantially rougher than the APDMES film.

APTES films applied by our original procedure were roughest, followed by APTES films deposited by the more optimal procedure, and APDMES films were the smoothest and most regular. This roughness ranking was conserved after subsequent steps in interface construction (biotinylation of and streptavidin binding to the biotinylated silane polymer, Fig. 16.14 [16.100, 106]). The results observed on silicon surfaces were replicated on aluminum surfaces, with APDMES providing a smoother, thinner interfacial polymer layer than did APTES [16.100]. Furthermore, the rank order of roughness following biotinylation was also conserved for APTES films deposited by our two protocols and APDMES films on aluminum [16.100].

Bhushan et al. [16.100] studied friction and wear of streptavidin deposited by physical adsorption and via a biotinylated APTES polymer layer on the silicon surface. The coefficient of friction for streptavidin-adsorbed silicon surfaces is less than for uncoated silicon [16.100], because the streptavidin acts as a lubricant film. Consistent with this interpretation, coefficients of friction were dependent on the concentration of streptavidin (that is, on the surface density of adsorbed streptavidin molecules, which increases when higher concentrations of streptavidin were used in coating the surfaces), and decreased at higher streptavidin concentrations. At higher densities of adsorbed streptavidin, the surface is more uniform and smoother, and the silicon substrate is more fully covered with protein than at lower concentration. This implies that the streptavidin forms a nearly continuous lubricant film at higher concentration. When streptavidin is linked to the surface through a biotinylated silane polymer, the coefficient of friction increases. This follows from the increased com-

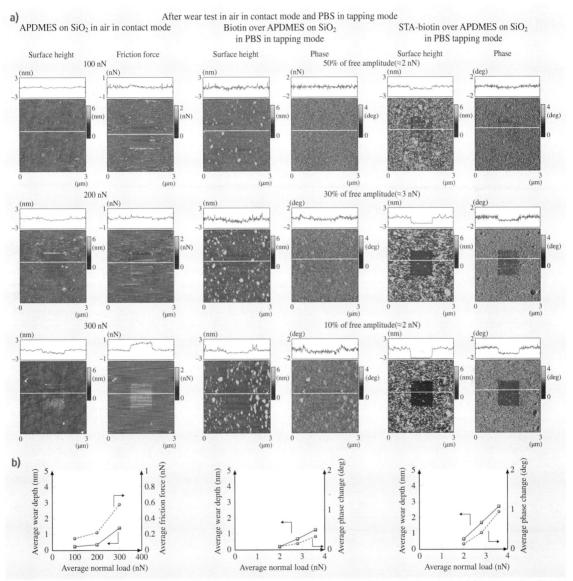

Fig. 16.15a,b Differential wear of interfacial polymer–protein films by AFM operated in tapping mode. (**a**) AFM surface height and friction force/phase angle images and cross-sectional profiles obtained after wear testing in air in contact mode/phosphate buffered saline (PBS) in tapping mode at a range of normal loads and free amplitudes, and (**b**) plot of average wear depth and average friction force/phase angle as a function of average normal load, for APDMES on silicon/SiO_2, biotin over APDMES on silicon/SiO_2, and on STA–biotin on APDMES and silicon/SiO_2 (after [16.100])

pliance of the underlying silane polymer. When normal load is applied to the surface, the surface is compressed, owing to the increased mechanical compliance afforded by interfacial biotinylated polymer length, flexibility or layering, all resulting in a larger contact area between the AFM tip and interfacial molecules. Since the size of streptavidin is much larger than that of the biotinylated silane polymer, the tightly packed streptavidin molecules cause very little lateral deflection of the tip. In streptavidin surfaces deposited by linkage to a polymer bound to the silicon surface, high contact area and low lateral deflection cause the friction force to increase relative to the same applied normal load on streptavidin surfaces deposited by adsorption [16.100, 106]. Together, these effects produce a *cushioning effect* that runs counter to the lubricating effect observed when protein (streptavidin) is directly adsorbed to the surfaces, increasing the coefficients of friction for streptavidin deposited by a polymer interface. The observations of streptavidin's lubricating properties and polymer–protein interfacial cushioning properties have been replicated on aluminum surfaces [16.100]. We think these are likely general properties of many protein interfaces on micromachined surfaces. We expect that many proteins (such as SCFvs adsorbed onto various metal oxide surfaces) will produce analogous lubricating properties, and that cushioning-mediated friction increases will occur for many polymer–protein interfaces, regardless of the composition of the underlying smooth, micromachined surfaces.

Interestingly, streptavidin bound to silicon surfaces via biotinylated silane polymers had differential resistance to wear, with APTES interfaces (which exhibit the most multilayering) being less robust than interfaces based on APDMES (with limited/no multilayering). The most highly multilayered polymer interfaces had, by definition, larger extents of intermonomer cross-linking, and were less resistant to wear by cantilevers [16.100]. This was interpreted by analogy to graphene layers in graphite, in that bonds between monomers of the polymer layer were more numerous than bonds between the polymer layer and the silicon substrate. Hence, when the polymer layer tore under stress from the cantilever, a larger section of the cross-linked polymer sheet tore free. It was expected that,

if the polymer layer were a true, un-cross-linked, self-assembled monolayer, individual polymer monomers would have no way to transfer force applied on them to their neighboring monomers, and so would be individually plucked off the interface by the tip.

These studies also [16.100, 106] occasioned further insights into the mechanism of wear of polymer–protein interfaces on micromachined surfaces. The polymer–protein interfaces (biotinylated silane polymer–streptavidin) on silicon were sufficiently delicate that their wear could not be investigated by contact-mode AFM at any load tested. Contact mode always immediately stripped the surface, even at the lowest loads achievable. Consequently, wear experiments were performed at varying loads in tapping mode, wherein the free amplitude (and hence the load) of the cantilever was systematically varied (Fig. 16.15). As one might expect, as load increased, wear increased. The finding implies that wear of APDMES–streptavidin interfaces on silicon is not a concerted process, but proceeds sequentially more deeply as the applied force increases. That said, all tested interfaces were very delicate, and required only nanonewton forces to remove them from the surfaces.

It can reasonably be expected that unprotected immunoFET sensors in vivo will experience at least nanonewton abrasive forces. Such forces are likely to be sufficient to strip receptor layers from the immunoFET, and thereby cause sensor failure. This implies that the sensor polymer–receptor interfaces must either be hardened somehow to resist encountered abrasive insult, or that the sensor interfaces must be sheltered from direct, abrasive insult by tissue. This problem is not satisfactorily solved, but is currently under investigation. However, the point remains that the primacy of the importance of the immunoFET interface drove surface-science and nanotribological study of the interface. As a consequence, we documented properties new to nanoscale tribology of protein interfaces (lubricating and cushioning effects). Nanotribological characterization of models of immunoFET surfaces also revealed another nanoscale parameter that must be addressed before the immunoFETs can be deployed in in vivo clinical or research applications: the robustness to mechanical insult (or lack thereof) of FET receptor layers.

16.4 Concluding Remarks: Barriers to Practice

The effort to produce nanotherapeutic devices is highly interdisciplinary, with a sweep of knowledge that is difficult for any one investigator to master fully. The effort required constitutes a major barrier to entry to the field.

Previous versions of this chapter ended with a discussion of the complexity of biology (being much greater than encountered in traditional engineering disciplines), the fashion in which biologists prioritize information, and cultural differences between engineers and biological scientists [16.107]. The chapter discussed some of the ways in which biologists and engineers hold and disseminate information (approaches so vastly different as to interfere with design of nanobiological devices). It also discussed the (then largely unmet) need for significant biological education for biomedical engineers, at least those who will be involved in producing devices that interact with living systems in vivo, or contain components derived from living systems. The consensus regarding the need for biologic training for biomedical engineers appears to be shifting in favor of more biology, in no small part as a result of the increasing prevalence of biologic components in biomedical nano- and microdevices and attempts to transfer those technologies to the clinic. However, there are still conundrums related to how biomedical and other engineers see themselves and see biology that result in lost opportunities.

16.4.1 You Do Not Know What You Do Not Know: the Consequences of Certainty

As discussed above, planar immunoFETs and similar bioFETs operating at physiological salt concentrations were regarded as infeasible from 1991 up until recently [16.81–84, 86]. However, there are now examples of such planar FET protein sensors that can detect protein analytes in physiological buffers at analyte concentrations known to occur in vivo [16.90, 91]. While this might seem to be a breakthrough in sensing technology, this development was delayed for years longer than it needed to be by the intransigent adherence of biosensor specialists to a flawed assessment of feasibility [16.81–84, 86].

The irony was that, with the right FET platform, planar FET sensing of proteins in physiological buffers has always been possible. The sensing community could have discovered this fact 20 years ago, and potentially could have moved to making functional immunoFET sensors for clinical use at that time, but for stubborn adherence to a flawed model of planar immunoFET interface structure. Until as little as a year ago, funding agencies and journals responded to applications or papers discussing planar FETs for protein sensing at physiological salt concentrations as if the documents reflected the proposer's or author's naivete. The direct response was often that the issue of planar FET sensors had been dealt with since 1991, and did not warrant further consideration. This response might have made sense had the classical assessment been theoretically sound, or if it were unsound in some way that required research to discover. However, the assessment was unsound from its first enunciation, largely because it did not take into account knowledge about protein and antibody structure that was available long before 1991. In the original proposal of the assessment, this may have been a result of the (then) poorly appreciated interdisciplinary nature of biosensor technology, or naivete regarding antibody structures and properties.

The classical assessment was promulgated by electrical engineers who had little knowledge of antibody structure, so they made mistakes in aspects of the assessment dependent on understanding antibodies. In its general form, this problem is a pitfall to which all interdisciplinary scientists are vulnerable. By definition, one cannot know what one does not know, and to some extent ignorance must be forgiven, as it is the default human condition. However, it should not have taken a working planar bioFET to overturn the classical assessment. The classical assessment was *dead on arrival*. Immunologists knew on cursory inspection that the classical assessment was nearly without relevance to real protein sensing. Unfortunately, engineers did not. In many cases, they failed to reexamine the assumptions of the classical assessment, in favor of reiterating its conclusion. If anything was done improperly, it was in failing to consider reasoned discussion of the flaws in the assessment (as above). Doing so may have cost the sensor field 20 years.

This is mentioned without the intent to embarrass persons who proposed or adhered to the classical assessment, but rather to point out that dogmatic certainty has consequences. Error is an ongoing hazard in any interdisciplinary field. As biomedical nanotechnologists, we mention this story with trepidation for our own pronouncements.

16.4.2 Are Proteins and Molecules Legitimately Part of Nanotechnology?

There is a school of thought that holds that nanotechnology deals solely with quantum and other phenomena that manifest only when matter is subdivided into nanoscale particles or structures. This reasoning would exclude supramolecular chemistry depending on organic molecules and devices built from protein or nucleic acid functional components from nanotechnology. If this reasoning is accepted as valid, nanotechnology so defined may have little intersection with human health.

As the reader might assume, such reasoning does not appeal to us. We hope we have presented a compelling case that molecules are essential components of nanotherapeutics. Organic or biologic macromolecules (proteins, nucleic acids, etc.) are necessary for many interactions between living things and therapeutic devices. In part, this is because these molecules are dimensionally and chemically appropriate to interact with the biocomponent of the patient (i.e., the argument of *speaking* to the patient's body in a language it understands). Also, synthetic nanostructures with the precise chemical and topological complexity of biological macromolecules cannot yet be produced or tuned with the same efficacy as biotechnology produces and tunes the properties of proteins (i.e., the argument that biomolecules are the only show in town). Moreover, with the examples of the FET protein sensor and the IKVAV PA amphiphile, we have shown that biotechnology and organic chemistry (respectively) can be used to address problems occurring at the nanoscale. In this sense, they fall reasonably under the rubric of nanotechnology.

The molecular nanocomponents of the immunoFET and IKVAV PA are tuned to suit (at least partially) using systematic knowledge of nanocomponent properties. It is our opinion that engineering is any activity that uses knowledge of component properties to control the behavior of materials and devices containing those components to achieve a desired end. In the case of the biological molecules with which we work, knowledge of properties is frequently fragmentary, limiting our ability to design individual protein molecules de novo, and driving us to use molecular genetic screening techniques to identify variant proteins with desired properties. Nonetheless, we see ourselves as engineers because we use knowledge of our materials (proteins) to its limits to optimize functional devices [immunoFETs, magnetic resonance imaging (MRI) contrast agents]. In the future, protein structure is likely to become more amenable to quantitative modeling and design, and the inclusion of protein engineering as a genuine engineering discipline will become less controversial.

At any rate, fortunately for us, and for biomedical nanotechnologists, the restrictive view of the definition of nanotechnology and engineering is less widely held than in the past. These fields will likely become increasingly welcoming to biologists as time passes.

References

16.1 S.C. Lee, R. Parthasarathy, T. Duffin, K. Botwin, T. Beck, G. Lange, J. Zobel, D. Jansson, D. Kunneman, E. Rowold, C.F. Voliva: Antibodies to PAMAM dendrimers: Reagents for immune detection assembly and patterning of dendrimers. In: *Dendrimers and Other Dendritic Polymers*, ed. by D. Tomalia, J. Frechet (Wiley, London 2001) pp. 559–566

16.2 S.C. Lee: Biotechnology for nanotechnology, Trends Biotechnol. **16**, 239–240 (1998)

16.3 S.C. Lee: Engineering the protein components of nanobiological devices. In: *Biological Molecules in Nanotechnology: The Convergence of Biotechnology, Polymer Chemistry and Materials Science*, ed. by S.C. Lee, L. Savage (IBC, Southborough 1998) pp. 67–74

16.4 S.C. Lee: How a molecular biologist can wind up organizing nanotechnology meetings. In: *Biological Molecules in Nanotechnology: The Convergence of Biotechnology, Polymer Chemistry and Materials Science*, ed. by S.C. Lee, L. Savage (IBC, Southborough 1998)

16.5 S.C. Lee: The nanobiological strategy for construction of nanodevices. In: *Biological Molecules in Nanotechnology: The Convergence of Biotechnology, Polymer Chemistry and Materials Science*, ed. by S.C. Lee, L. Savage (IBC, Southborough 1998) pp. 3–14

16.6 S.C. Lee: A biological nanodevice for drug delivery, National Science and Technology Council. IWGN Workshop Report: Nanotechnology Research Directions. International Technology Research Institute, World Technology Division (Kluwer, Baltimore 1999) pp. 91–92

16.7 S.C. Lee, R. Parthasarathy, K. Botwin: Proteinpolymer conjugates: Synthesis of simple nanobiotechnological devices, Polym. Prepr. **40**, 449–450 (1999)

16.8 L. Jelinski: Biologically related aspects of nanoparticles, nanostructured materials and nanodevices. In: *Nanostructure Science and Technology*, ed. by R.W. Siegel, E. Hu, M.C. Roco (Kluwer, Dordrecht 1999) pp. 113–130

16.9 B.R. Smith, J. Heverhagen, M. Knopp, P. Schmalbrock, J. Shapiro, M. Shiomi, N. Moldovan, M. Ferrari, S.C. Lee: Magnetic resonance imaging of atherosclerosis in vivo using biochemically targeted ultrasmall superparamagnetic iron oxide

16.10 A.J. Nijdam, T.R. Nicholson III, J. Shapiro, B.R. Smith, J.T. Heverhagen, P. Schmalbrock, M.V. Knopp, A. Kebbel, D. Wang, S.C. Lee: Biochemically targeted nanoparticulate contrast agents for magnetic resonance imaging diagnosis of cardiovascular disease, Curr. Nanosci. **5**, 88–102 (2009)

16.11 J.R. Baker Jr.: Therapeutic nanodevices. In: *Biological Molecules in Nanotechnology: The Convergence of Biotechnology, Polymer Chemistry and Materials Science*, ed. by S.C. Lee, L. Savage (IBC, Southborough 1998) pp. 173–183

16.12 R. Duncan: Drug targeting: Where are we now and where are we heading?, J. Drug Target. **5**, 1–4 (1997)

16.13 R. Duncan, S. Gac-Breton, R. Keane, Y.N. Sat, R. Satchi, F. Searle: Polymer-drug conjugates, PDEPT and PELT: Basic principles for design and transfer from the laboratory to clinic, J. Cont. Release **74**, 135–146 (2001)

16.14 D.S. Goldin, C.A. Dahl, K.L. Olsen, L.H. Ostrach, R.D. Klausner: Biomedicine. The NASA-NCI collaboration on biomolecular sensors, Science **292**, 443–444 (2001)

16.15 S.C. Lee: Dendrimers in nanobiological devices. In: *Dendrimers and Other Dendritic Polymers*, ed. by D. Tomalia, J. Frechet (Wiley, London 2001) pp. 548–557

16.16 J.R. Baker Jr., A. Quintana, L. Piehler, M. Banazak-Holl, D. Tomalia, E. Racka: The synthesis and testing of anti-cancer therapeutic nanodevices, Biomed. Microdevices **3**, 61–69 (2001)

16.17 K.D. Bhalerao, E. Eteshola, M. Keener, S.C. Lee: Nanodevice design through the functional abstraction of biological macromolecules, Appl. Phys. Lett. **87**, 143902–143904 (2005)

16.18 S.C. Lee, K. Bhalerao, M. Ferrari: Object oriented design tools for supramolecular devices and biomedical nanotechnology, Ann. New York Acad. Sci. **1013**, 110–123 (2004)

16.19 J.M. Harris, N.E. Martin, M. Modi: Pegylation: A novel process for modifying pharmacokinetics, Clin. Pharmacokin. **40**, 539–551 (2001)

16.20 S.B.H. Kent: Building proteins through chemistry: Total chemical synthesis of protein molecules by chemical ligation of unprotected protein segments. In: *Biological Molecules in Nanotechnology: The Convergence of Biotechnology, Polymer Chemistry and Materials Science*, ed. by S.C. Lee, L. Savage (IBC, Southborough 1998) pp. 75–92

16.21 C.A. Janeway, P. Travers, M. Walport, J.D. Capra: *Immunobiology* (Elsevier, London 1999)

16.22 S.C. Lee, M.S. Leusch, V.A. Luckow, P. Olins: Method of producing recombinant viruses in bacteria, US Patent 5348886 (1993)

16.23 M.S. Leusch, S.C. Lee, P.O. Olins: A novel hostvector system for direct selection of recombinant baculoviruses (bacmids) in E. coli, Gene **160**, 191–194 (1995)

16.24 V.A. Luckow, S.C. Lee, G.F. Barry, P.O. Olins: Efficient generation of infectious recombinant baculoviruses by site-specific, transposon-mediated insertion of foreign DNA into a baculovirus genome propagated in E. coli, J. Virol. **67**, 4566–4579 (1993)

16.25 T. Gardner, C.R. Cantor, J.J. Collins: Construction of a genetic toggle switch in E. coli, Nature **403**, 339–342 (2000)

16.26 J. Hasty, F. Isaacs, M. Dolnik, D. McMillen, J.J. Collins: Designer gene networks: Towards fundamental cellular control, Chaos **11**, 107–220 (2001)

16.27 S. Uppuluri, D.R. Swanson, L.T. Piehler, J. Li, G. Hagnauer, D.A. Tomalia: Core shell tecto(dendrimers). I. Synthesis and characterization of saturated shell models, Adv. Mater. **12**, 796–800 (2000)

16.28 A.K. Patri, I.J. Majoros, J.R. Baker Jr.: Dendritic polymer macromolecular carriers for drug delivery, Curr. Opin. Chem. Biol. **6**, 466–471 (2002)

16.29 A. Quintana, E. Raczka, L. Piehler, I. Lee, A. Myc, I. Majoros, A.K. Patri, T. Thomas, J. Mule, J.R. Baker Jr.: Design and function of a dendrimer-based therapeutic nanodevice targeted to tumor cells through the folate receptor, Pharma. Res. **19**, 1310–1316 (2002)

16.30 Y. Choi, T. Thomas, A. Kotlyar, M. Islam, J. Baker Jr.: Synthesis and functional evaluation of DNA-assembled polyamidoamine dendrimer clusters for cancer cell-specific targeting, Chem. Biol. **12**, 35–43 (2005)

16.31 Y. Choi, A. Mecke, B.G. Orr, M.M. Banaszak Holl, J.R. Baker Jr.: DNA-directed synthesis of generation 7 and 5 PAMAM dendrimer nanoclusters, Nano Lett. **4**, 391–397 (2004)

16.32 D.G. Mullen, A.M. Desai, J.N. Waddell, X.-M. Cheng, C.V. Kelly, D.Q. McNerny, I.J. Majoros, J.R. Baker Jr., L.M. Sander, B.G. Orr, M.M. Banaszak Holl: The implications of stochastic synthesis for the conjugation of functional groups to nanoparticles, Bioconjug. Chem. **19**, 1748–1752 (2008)

16.33 T.R. Groves, D. Pickard, B. Rafferty, N. Crosland, D. Adam, G. Schubert: Maskless electron beam lithography: Propects, progress and challenges, Microelectron. Eng. **61**, 285–293 (2002)

16.34 M. Guthold, R. Superfine, R. Taylor: The rules are changing: Force measurements on single molecules and how they relate to bulk reaction kinetics and energies, Biomed. Microdevices **3**, 9–18 (2001)

16.35 L.M. Demers, D.S. Ginger, S.-J. Park, Z. Li, S.-W. Chung, C.A. Mirkin: Direct patterning of modified oligonucleotides on metals and insulatos by dip-pen nanolithography, Science **296**, 1836–1838 (2002)

16.36 K.-B. Lee, S.-J. Park, C.A. Mirkin, J.C. Smith, M. Mrksich: Protein nanoarrays generated by dip-pen nanolithography, Science **295**, 1702–1705 (2002)

16.37 M. Ferrari, J. Liu: The engineered course of treatment, Mech. Eng. **123**, 44–47 (2001)

16.38 K.E. Drexler: *Engines of Creation: The Coming Era of Nanotechnology* (Anchor Books, New York 1986)

16.39 J. Cumings, A. Zetti: Low-friction nanoscale linear bearing realized frommultiwall carbon nanotubes, Science **289**, 602–604 (2000)

16.40 D.J. Hornbaker, S.-J. Kahng, S. Mirsa, B.W. Smith, A.T. Johnson, E.J. Mele, D.E. Luzzi, A. Yazdoni: Mapping the one-dimensional electronic states of nanotube peapod structures, Science **295**, 828–831 (2002)

16.41 C. Dekker: Carbon nanotubes as molecular quantum wires, Phys. Today **28**, 22–28 (1999)

16.42 M.-C. Jones, J.-C. Leroux: Polymeric micellesa new generation of colloidal drug carriers, Eur. J. Pharma. Biopharma. **48**, 101–111 (1999)

16.43 I. Uchegbu: Parenteral drug delivery: 1, Pharma. J. **263**, 309–318 (1999)

16.44 I. Uchegbu: Parenteral drug delivery: 2, Pharma. J. **263**, 355–359 (1999)

16.45 J.D. Hartgerink, E.R. Zubarev, S.I. Stupp: Supramolecular one-dimensional objects, Curr. Opin. Solid State Mater. Sci. **5**, 355–361 (2001)

16.46 L.C. Palmer, Y.S. Velichko, M. Olvera De La Cruz, S.I. Stupp: Supramolecular self-assembly codes for functional structures, Philos. Trans. R. Soc. A **365**, 1417–1433 (2007)

16.47 G.A. Silva, C. Catherine, K.L. Niece, E. Beniash, D.A. Harrington, J.A. Kessler, S.I. Stupp: Selective differentiation of neural progenitor cells by high-epitope density nanofibers, Science **303**, 1352–1355 (2004)

16.48 V.M. Tysseling-Mattiace, V. Sahni, K.L. Niece, D. Birch, C. Czeisler, M. Fehlings, S.I. Stupp, J.A. Kessler: Self-assembling nanofibers inhibit glial scar gormation and promote axon elongation after spinal cord injury, J. Neurosci. **28**, 3814–3823 (2008)

16.49 S. Fernandez-Lopez, H.-S. Kim, E.C. Choi, M. Delgado, J.R. Granja, A. Khasanov, K. Kraehenbuehl, G. Long, D.A. Weinberger, K.M. Wilcoxen, M. Ghardiri: Antibacterial agents based on the cyclic D,L-alpha-peptide architecture, Nature **412**, 452–455 (2001)

16.50 A. Saghatelian, Y. Yokobayashi, K. Soltani, M.R. Ghadiri: A chiroselective peptide replicator, Nature **409**, 777–778 (2001)

16.51 T. Hamouda, A. Myc, B. Donovan, A.Y. Shih, J.D. Reuter, J.R. Baker Jr.: A novel surfactant nanoemulsion with a unique non-irritant topical antimicrobial activity against bacteria, enveloped viruses and fungi, Microbiol. Res. **156**, 1–7 (2001)

16.52 J. Davies: Aminoglycoside-aminocyclitol antibiotics and their modifying enzymes. In: *Antibiotics in Laboratory Medicine*, ed. by V. Lorian (Williams and Wilkins, Baltimore 1984) pp. 474–489

16.53 M.J. Heller: Utilization of synthetic DNA for molecular electronic and photonic-based device applications. In: *Biological Molecules in Nanotechnology: The Convergence of Biotechnology, Polymer Chemistry and Materials Science*, ed. by S.C. Lee, L. Savage (IBC, Southborough 1998) pp. 59–66

16.54 Z. Ma, S. Taylor: Nucleic acid triggered catalytic drug release, Proc. Natl. Acad. Sci. USA **97**, 11159–11163 (2000)

16.55 R.C. Merkle: Biotechnology as a route to nanotechnology, Trends Biotechnol. **17**, 271–274 (1999)

16.56 N.C. Seeman, J. Chen, Z. Zhang, B. Lu, H. Qiu, T.-J. Fu, Y. Wang, X. Li, J. Qi, F. Liu, L.A. Wenzler, S. Du, J.E. Mueller, H. Wang, C. Mao, W. Sun, Z. Shen, M.H. Wong, R. Sha: A bottom-up approach to nanotechnology using DNA. In: *Biological Molecules in Nanotechnology: The Convergence of Biotechnology, Polymer Chemistry and Materials Science*, ed. by S.C. Lee, L. Savage (IBC, Southborough 1998) pp. 45–58

16.57 G. Lemieux, C. Bertozzi: Chemoselective ligation reactions with proteins, oligosaccharides and cells, Trends Biotechnol. **16**, 506–512 (1998)

16.58 R. Offord, K. Rose: Multicomponent synthetic constructs. In: *Biological Molecules in Nanotechnology: The Convergence of Biotechnology, Polymer Chemistry and Materials Science*, ed. by S.C. Lee, L. Savage (IBC, Southborough 1998) pp. 93–105

16.59 S.C. Lee, R. Parthasarathy, K. Botwin, D. Kunneman, E. Rowold, G. Lange, J. Zobel, T. Beck, T. Miller, W. Hood, J. Monahan, R. Jansson, J.P. McKearn, C.F. Voliva: Biochemical and immunological properties of cytokines conjugated to dendritic polymers, Biomed. Microdevices Biomems Biomed. Nanotechnol. **6**, 191–201 (2004)

16.60 S.C. Lee, R. Parthasarathy, T. Duffin, K. Botwin, T. Beck, G. Lange, J. Zobel, D. Kunneman, E. Rowold, C.F. Voliva: Recognition properties of antibodies to PAMAM dendrimers and their use in immune detection of dendrimers, Biomed. Microdevices **3**, 51–57 (2001)

16.61 G.T. Hermanson: *Bioconjugate Chemistry* (Academic, San Diego 1996)

16.62 S. Topell, R. Glockshuber: Circular permutation of the green fluorescent protein, Meth. Mol. Biol. **183**, 31–48 (2002)

16.63 A. Rojas, S. Garcia-Vallve, J. Palau, A. Romeu: Circular permutations in proteins, Biologia **54**, 255–277 (1999)

16.64 T.U. Schwartz, R. Walczak, G. Blobel: Circular permutation as a tool to reduce surface entropy triggers crystallization of the signal recognition particle receptor beta subunit, Protein Sci. **13**, 2814–2818 (2004)

16.65 U. Heinemann, M. Hahn: Circular permutation of polypeptide chains: Implications for protein folding and stability, Prog. Biophys. Mol. Biol. **64**, 121–143 (1996)

16.66 A. Buchwalder, H. Szadkoski, K. Kirschner: A fully active variant of dihydrofolate reductase with a circularly permuted sequence, Biochem. **31**, 1621–1630 (1992)

16.67 L.S. Mullins, K. Wesseling, J.M. Kuo, J.B. Garrett, F.M. Raushel: Transposition of protein sequences: Circular permutation of ribonuclease T1, J. Am. Chem. Soc. **116**, 5529–5533 (1994)

16.68 M. Hahn, K. Piotukh, R. Borriss, U. Heinemann: Native-like in vivo folding of a circularly permuted jellyroll protein shown by crystal structure analysis, Proc. Natl. Acad. Sci. USA **91**, 10417–10421 (1994)

16.69 Y.R. Yang, H.K. Schachznan: Aspartate transcarbamoylase containing circularly permuted catalytic polypeptide chains, Proc. Natl. Acad. Sci. USA **90**, 11980–11984 (1993)

16.70 X. Lin, G. Koelsch, J.A. Loy, J. Tang: Rearranging the domains of pepsinogen, Protein Sci. **4**, 159–166 (1995)

16.71 M.L. Vignais, C. Corbier, G. Mulliert, C. Branlant, G. Branlant: Circular permutation within the coenzyme binding domain of the tetrameric glyceraldehyde-3-phosphate dehydrogenase from *Bacillus stearothermophilus*, Protein Sci. **4**, 994–1000 (1995)

16.72 E. Eteshola, C.D. Van Valkenburgh, S. Merlin, E. Rowold, J. Adams, R. Ibdah, L.E. Pegg, A. Donelly, E. Rowold, J. Klover, S.C. Lee: Screening of a library of circularly permuted proteins on phage to manipulate protein topography, J. Nanoeng. Nanosyst. **219**, 45–55 (2006)

16.73 E. Eteshola, M.T. Keener, M.A. Elias, J. Shapiro, L.J. Brillson, B. Bhushan, S.C. Lee: Engineering functional protein interfaces for immunologically modified field effect transistors (ImmunoFETs) by molecular genetic means, J. R. Soc. Interface **5**, 123–127 (2008)

16.74 E. Eteshola, L. Brillson, S.C. Lee: Selection and characteristics of peptides that bind thermally grown silicon dioxide films, Biomol. Eng. **22**, 202–204 (2005)

16.75 R.R. Naik, L.L. Brott, S.J. Clarson, M.O. Stone: Silica-precipitating peptides isolated from a combinatorial phage display peptide library, J. Nanosci. Nanotechnol. **2**, 95–100 (2002)

16.76 E.M. Krauland, B.R. Peelle, K.D. Wittrup, A.M. Belcher: Peptide tags for enhanced cellular and protein adhesion to single-crystalline sapphire, Biotechnol. Bioeng. **97**, 1009–1020 (2007)

16.77 L. Wang, J. Xie, P.G. Schultz: Expanding the genetic code, Annu. Rev. Biophys. Biomol. Struct. **35**, 225–249 (2006)

16.78 J. Xie, P.G. Schultz: A chemical toolkit for proteins – an expanded genetic code, Nat. Rev. Mol. Cell Biol. **7**, 775–782 (2006)

16.79 N. Hino, A. Hayashi, K. Sakamoto, S. Yokoyama: Site-specific incorporation of non-natural amino acids into proteins in mammalian cells with an expanded genetic code, Nat. Protoc. **1**, 2957–2962 (2006)

16.80 K. Rogers: Principles of affinity-based biosensors, Mol. Biotechnol. **14**, 109–129 (2000)

16.81 M.J. Schoning, A. Poghossian: Recent advances in biologically sensitive field-effect transistors (BioFETS), Analyst **127**, 1137–1151 (2002)

16.82 P. Bergveld, J. Hendrikes, W. Olthuis: Theory and application of the material work function for chemical sensors based on the field effect principle, Meas. Sci. Technol. **9**, 1801–1808 (1998)

16.83 W. Olthius, P. Bergveld, J. Kruise: The exploitation of ISFETs to determine acid-base behavior of proteins, Electrochim. Acta **43**, 3483–3488 (1997)

16.84 R.B. Schasfoort, R.P. Kooyman, P. Bergveld, J. Greve: A new approach to ImmunoFET operation, Biosens. Bioelectron. **5**, 103–124 (1990)

16.85 Y. Cui, Q. Wei, H. Park, C.M. Lieber: Nanowire nanosensors for highly sensitive and selective detection of biological and chemical species, Science **293**, 1289 (2001)

16.86 P. Bergveld: A critical evaluation of direct electrical protein detection methods, Biosens. Bioelectron. **6**, 55–72 (1991)

16.87 J.I. Hahm, C.M. Lieber: Direct ultrasensitive electrical detection of DNA and DNA sequence variations using nanowire nanosensors, Nano Lett. **4**, 51–54 (2004)

16.88 A. Star, J.C.P. Gabriel, K. Bradley, G. Gruner: Electronic detection of specific protein binding using nanotube FET devices, Nano Lett. **3**, 459–463 (2003)

16.89 G. Zheng, F. Patolsky, Y. Cui, W.U. Wang, C.M. Lieber: Multiplexed electrical detection of cancer markers with nanowire sensor arrays, Nat. Biotechnol. **23**, 1294–1301 (2005)

16.90 J. Shapiro, S. Gupta, E. Eteshola, M. Elias, X. Wen, W. Lu, L.J. Brillson, S.C. Lee: Challenges in optimization of nanobiotechnological devices illustrated by partial optimization of a protein biosensor, Proc. 2nd Int. Congr. Nanobiotechnol. Nanomed., NanoBio 2007 (Int. Association Nanotechnology, San Jose 2007), (CD)

16.91 S. Gupta, M. Elias, X. Wen, J. Shapiro, L. Brillson, W. Lu, S.C. Lee: Detection of clinically relevant levels of biological analyte under physiologic buffer using planar field effect transistors, Biosens. Bioelectron. **24**, 505–511 (2008)

16.92 K. Decanniere, A. Desmyter, M. Lauwereys, M.A. Ghahroudi, S. Muyldermans, L. Wyns: A single-domain antibody fragment in complex with rnase a: non-canonical loop structures and

16.93 K. Decanniere, T.R. Transue, A. Desmyter, D. Maes, S. Muyldermans, L. Wyns: Degenerate interfaces in antigen-antibody complexes, J. Mol. Biol. **313**, 473–478 (2001)

16.94 A. Desmyter, K. Decanniere, S. Muyldermans, L. Wyns: Antigen specificity and high affinity binding provided by one single loop of a camel single-domain antibody, J. Biol. Chem. **276**, 26285–26290 (2001)

16.95 A. Muruganandam, J. Tanha, S. Narang, D. Stanimirovic: Selection of phage-displayed llama single-domain antibodies that transmigrate across human blood-brain barrier endothelium, FASEB Journal **16**, 240–242 (2002)

16.96 S. Muyldermans: Single domain camel antibodies: Current status, Mol. Biotechnol. **74**, 277–302 (2001)

16.97 S. Muyldermans, M. Lauwereys: Unique single-domain antigen binding fragments derived from naturally occurring camel heavy-chain antibodies, J. Mol. Recogn. **12**, 131–140 (1999)

16.98 L. Riechmann, S. Muyldermans: Single domain antibodies: comparison of camel VH and camelised human VH domains, J. Immunol. Meth. **231**, 25–38 (1999)

16.99 M.S. Hayden, L.K. Gilliland, J.A. Ledbetter: Antibody engineering, Curr. Opin. Immunol. **9**, 201–212 (1997)

16.100 B. Bhushan, K.J. Kwak, S. Gupta, S.C. Lee: Nanoscale adhesion, friction and wear studies of biomolecules on polymer-coated silica and alumina based surfaces, J. R. Soc. Interface **6**, 719–733 (2009)

16.101 Y. Han, D. Mayer, A. Offenhausser, S. Ingebrandt: Surface activation of thin silicon oxides by wet cleaning and silanization, Thin Solid Films **510**, 175–180 (2006)

16.102 K. Kallury, P.M. MacDonald, M. Thompson: Effect of surface water and base catalysis on the silanization of silica by (aminopropyl)alkoxysilanes studied by x-ray photoelectron spectroscopy and 13C cross-polarization/magic angle spinning nuclear magnetic resonance, Langmuir **10**, 492–499 (1994)

16.103 J.H. Moon, J.W. Shin, S.Y. Kim, J.W. Park: Formation of uniform aminosilane thin layers: An imine formation to measure relative surface density of the amine group, Langmuir **12**, 4621–4624 (1996)

16.104 B. Bhushan, D.R. Tokachichu, M.T. Keener, S.C. Lee: Morphology and adhesion of biomolecules on silicon based surfaces, Acta Biomater. **1**, 327–341 (2005)

16.105 S.C. Lee, M.T. Keener, D.R. Tokachichu, B. Bhushan, P.D. Barnes, B.J. Cipriany, M. Gao, L.J. Brillson: Protein binding on thermally grown silicon dioxide, J. Vac. Sci. Technol. B **23**, 1856–1865 (2005)

16.106 B. Bhushan, D.R. Tokachichu, M.T. Keener, S.C. Lee: Nanoscale adhesion, friction and wear studies on silicon based surfaces, Acta Biomater. **2**, 39–49 (2006)

16.107 S.C. Lee, M. Reugsegger, P.D. Barnes, B.R. Smith, M. Ferrari: Therapeutic nanodevices. In: *Springer Handbook of Nanotechnology*, ed. by B. Bhushan (Springer, Berlin, Heidelberg 2007) pp. 461–504

17. G-Protein Coupled Receptors: Progress in Surface Display and Biosensor Technology

Wayne R. Leifert, Tamara H. Cooper, Kelly Bailey

Signal transduction by G-protein coupled receptors (GPCRs) underpins a multitude of physiological processes. Ligand recognition by these receptors leads to activation of a generic molecular switch involving heterotrimeric G-proteins and guanine nucleotides. With growing interest and commercial investment in GPCRs in areas such as drug targets, orphan receptors, high-throughput screening of drugs, biosensors etc., greater attention will focus on assay development to allow for miniaturization, ultrahigh throughput, and eventually, microarray/biochip assay formats that will require nanotechnology-based approaches. Stable, robust, cell-free signaling assemblies comprising receptor and appropriate molecular switching components will form the basis of future GPCR/G-protein platforms which should be adaptable for such applications as microarrays and biosensors. This chapter focuses on cell-free GPCR assay nanotechnologies and describes some molecular biological approaches for the construction of more sophisticated, surface-immobilized, homogeneous, functional GPCR sensors. The latter points should greatly extend the range of applications to

17.1	The GPCR:G-Protein Activation Cycle 488
17.2	Preparation of GPCRs and G-Proteins 489
	17.2.1 Expression Systems for Recombinant GPCRs/G-Proteins. 490
17.3	Protein Engineering in GPCR Signaling ... 490
	17.3.1 Fluorescent Proteins 490
	17.3.2 Engineering of Promiscuous Gα Proteins 491
	17.3.3 Protein Engineering for Surface Attachment 491
17.4	GPCR Biosensing 491
	17.4.1 Level 1 Biosensing – Ligand Binding 492
	17.4.2 Level 2 Biosensing – Conformational Changes in the GPCR 496
	17.4.3 Level 3 Biosensing – GTP Binding ... 496
	17.4.4 Level 4 Biosensing – GPCR:G-Protein Dissociation 497
17.5	The Future of GPCRs 499
References	 499

which technologies based on GPCRs could be applied.

The Superfamily of GPCRs

G-protein coupled receptors (GPCRs) represent a superfamily of intramembrane proteins (polypeptides) which initiate many signal transduction pathways in virtually all eukaryotic cells. GPCRs are structurally characterized by their seven transmembrane (*serpentine*) spanning domains (Fig. 17.1). GPCR activation can be initiated by a wide variety of extracellular stimuli such as light, odorants, neurotransmitters, and hormones. In most cases the GPCR uses a transmembrane signaling system which involves three separate components (systems). Firstly, the extracellular ligand is specifically detected by a cell-surface GPCR. Once recognition takes place, the GPCR in turn triggers the activation of a heterotrimeric G-protein complex located on the peripheral intracellular (cytoplasmic) surface of the cell membrane (the term *G-protein* is used since these proteins bind guanine nucleotides such as guanosine di- and triphosphate present in cells, as discussed in detail later). Finally, the *signal transduction* cascade involves the activated G-protein altering the activity of some downstream *effector* protein(s), which can be enzymes or ion channels located in the cell membrane. This then leads to a change in the cellular concentration of cyclic adenosine monophosphate (cAMP), calcium ions or a metabolite such as phospho-

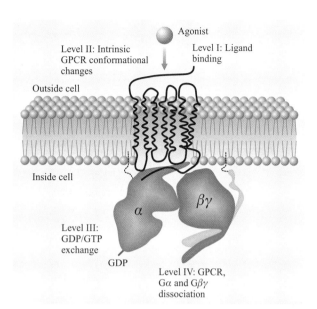

Fig. 17.1 Transmembrane topology of a typical *serpentine* G-protein coupled receptor (GPCR) and representation of the levels of biosensing. The receptor's amino terminal (N-terminal) is extracellular (outside of the cell), and its carboxy (C) terminal is within the cytoplasm (intracellular). The receptor polypeptide chain traverses the plane of the membrane phospholipid bilayer seven times. The hydrophobic transmembrane segments of the GPCR are indicated by spirals. The agonist approaches the receptor from the extracellular surface and binds, depending on the receptor type, to a site near the N-terminal or to a site deep within the receptor, surrounded by the transmembrane regions of the receptor protein. The G-proteins (Gα and G$\beta\gamma$) interact with cytoplasmic regions of the receptor, including the third intracellular loop between transmembrane regions V and VI. The Gα and Gγ subunits contain fatty acid modifications (myristate and palmitate on the Gα and isoprenylate on the Gγ) to help anchor the proteins at the lipid bilayer (shown as *dotted lines*). The *levels* of signaling that may be exploited for detection in a cell-free mode are shown (GDP, guanosine diphosphate) ◀

inositides within the cell, resulting in a physiological response such as stronger and faster contraction of the heart.

Since many disease processes involve aberrant or altered GPCR signaling dynamics, GPCRs represent a significant target for medicinal pharmaceuticals. Furthermore, more than 50% of all drugs currently marketed worldwide are directed against GPCRs [17.1], with this likely to increase with the recent elucidation of high-resolution structural data of the β_2-adrenergic

Table 17.1 Some examples of prescription drugs that target GPCRs for the indicated disease state

Brand name	Generic name	G-protein coupled receptor(s)	Indication
Zyprexa	Olanzapine	Serotonin 5-HT$_2$ and dopamine	Schizophrenia, antipsychotic
Risperdal	Risperidone	Serotonin 5-HT$_2$	Schizophrenia
Claritin	Loratidine	Histamine H$_1$	Rhinitis, allergies
Imigran	Sumatriptan	Serotonin 5-HT$_{1B/1D}$	Migraine
Cardura	Doxazosin	α-Adrenoceptor	Prostate hypertrophy
Tenormin	Atenolol	β_1-Adrenoceptor	Coronary heart disease
Serevent	Salmeterol	β_2-Adrenoceptor	Asthma
Duragesic	Fentanyl	Opioid	Pain
Imodium	Loperamide	Opioid	Diarrhea
Cozaar	Losartan	Angiotensin II	Hypertension
Zantac	Ranitidine	Histamine H$_2$	Peptic ulcer
Cytotec	Misoprostol	Prostaglandin PGE$_1$	Ulcer
Zoladex	Goserelin	Gonadotrophin-releasing factor	Prostate cancer
Requip	Ropinirole	Dopamine	Parkinson's disease
Atrovent	Ipratropium	Muscarinic	Chronic obstructive pulmonary disease (COPD)

Table 17.2 A partial list of some of the known endogenous and exogenous GPCR ligands

Acetylcholine	Ghrelin	Opioids
Adenosine	Glucagon	Orexin
Adrenaline	Glutamate	Oxytocin
Adrenocorticotropic hormone	Gonadotropin-releasing hormone	Parathyroid hormone
Angiotensin II	Growth hormone-releasing factor	Photons (light)
Bradykinin	Growth-hormone secretagogue	Platelet activating factor
Calcitonin	Histamine	Prolactin releasing peptide
Chemokines	Luteinising hormone	Prostaglandins
Cholecystokinin	Lymphotactin	Secretin
Corticotropin releasing factor	Lysophospholipids	Serotonin
Dopamine	Melanocortin	Somatostatin
Endorphins	Melanocyte-stimulating hormone	Substances P, K
Endothelin	Melatonin	Thrombin
Enkephalins	Neuromedin-K	Thromboxanes
Fatty acids	Neuromedin-U	Thyrotropin
Follitropin	Neuropeptide-FF	Thyrotropin releasing hormone
γ-Aminobutyric acid (GABA)	Neuropeptide-Y	Tyramine
Galanin	Neurotensin	Urotensin
Gastric inhibitory peptide	Noradrenaline	Vasoactive intestinal peptide
Gastrin	Odorants	Vasopressin

receptor [17.2]. Table 17.1 lists some commonly prescribed drugs acting on GPCRs. GPCRs are associated with almost every major therapeutic category or disease class, including pain, asthma, inflammation, obesity, cancer, as well as cardiovascular, metabolic, gastrointestinal, and central nervous system diseases [17.3]. It

Fig. 17.2 Future applications of GPCR platforms. The development of functional assay platforms for integral membrane proteins, particularly those of the GPCR class, which are compatible with future high-throughput microarray formats will offer significant opportunities in a number of areas. It is expected that such advances in assay technology will likely impact on drug discovery, diagnostics, and biosensors. There is a strong requirement for technologies that enable screening of multiple GPCR targets simultaneously (multiplexing). Therefore, it would be advantageous in the future to design new biosensor platforms using miniaturized nanotechnology approaches. Furthermore, to achieve this aim successfully, it will be an absolute requirement for cross-disciplinary fields of research (including biology, physics, and chemistry, as well as mathematics for molecular modeling and bioinformatics) to be highly integrated ▶

is this vitally important function of these cell-surface receptors, i.e., transduction of exogenous signals into an intracellular response, which makes GPCRs so physiologically significant. Indeed, there are reported to be ≈ 747 different human GPCRs as predicted from gene sequencing analyses, 380 of which are thought to be chemosensory receptors, whereas the remaining 367 GPCRs are predicted to bind endogenous ligands such as neurotransmitters, hormones, fatty acids, and peptides [17.4]. With about 230 of these GPCRs having been identified already (i.e., they have known ligands), this currently leaves about 140 *orphan* GPCRs with as yet undiscovered ligands. A summary of some of the known GPCR ligands is presented in Table 17.2.

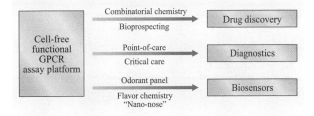

This chapter focuses firstly on possible cell-free approaches which could be used in biosensor applications, diagnostic platforms, and for high-throughput screening (HTS) of GPCR ligands, with particular emphasis on GPCR signaling complexes and associated enabling nanotechnologies (Fig. 17.2). Additionally, we include molecular biology approaches involving G-proteins and GPCRs with reference to biosensor and HTS applications. One of the most important breakthroughs permitting these developments for GPCR and G-protein signaling is the ability to produce these GPCRs and G-proteins in relatively high amounts and in purified form using recombinant DNA techniques. Also, it is becoming increasingly more routine to produce recombinant modifications of such proteins using basic molecular biological approaches. These modifications can include biotin tags, hexahistidine tags or fluorescent protein fusions which can allow site-specific interaction of the recombinant protein(s) with appropriately derivatized biosensor surfaces such as glass or gold or the generation of a biosensor signal.

17.1 The GPCR:G-Protein Activation Cycle

In order to understand how we measure the activation of GPCRs and their associated G-proteins, a first step is to revise the GPCR:G-protein activation cycle in more detail. At the cellular level, GPCRs are integral membrane proteins which reside within the cell membrane lipid bilayer and are closely associated with the peripheral G-protein heterotrimeric complex consisting of the Gα and the G$\beta\gamma$ dimer subunits (Fig. 17.1). Owing to the very high affinity between Gβ and Gγ, these two subunits are almost exclusively considered as the G$\beta\gamma$ dimer. (The Gα subunits are ≈ 41 kDa and have a theoretical diameter of ≈ 4.7 nm, whilst β subunits are ≈ 37 kDa and γ subunits are 8 kDa, giving G$\beta\gamma$ dimers an approximate diameter of 4.6 nm.) Figure 17.3 depicts the cycle of activation/inactivation of the heterotrimeric G-protein complex. In the resting inactive state (i.e., when there is no agonist bound to the receptor), the G-proteins Gα and G$\beta\gamma$ have high affinity for each other and remain tightly bound, forming the heterotrimeric G-protein complex. In this state, guanosine diphosphate (GDP) is tightly bound to the Gα subunit associated with the G$\beta\gamma$ dimer. Both Gα and G$\beta\gamma$ subunits can bind to the GPCR. When the agonist (a GPCR ligand which activates the GPCR signaling pathway) approaches the GPCR from the extracellular fluid and binds to the active site on the GPCR, the GPCR is in turn *activated*, possibly leading to a change in its conformation. The GDP-liganded Gα subunit responds with a conformational change which results in a decreased affinity, so that GDP is no longer bound to the Gα subunit. At this point guanosine triphosphate (GTP), which is in higher concentration in the cell than GDP, can rapidly bind to the Gα subunit, thus replacing the GDP. This replacement of GDP with GTP activates the Gα subunit, causing it to dissociate from the G$\beta\gamma$ subunit as well as from the receptor. This in effect results in exposure of new surfaces on the Gα and G$\beta\gamma$ subunits which can interact with cellular effectors such as the enzyme adenylate cyclase, which converts adenosine triphosphate (ATP) to cAMP. The activated state of the Gα subunit lasts until the GTP is hydrolyzed to GDP by the intrinsic GTPase

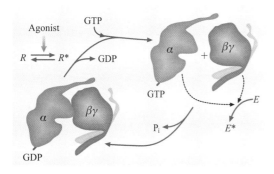

Fig. 17.3 Molecular switching: the regulatory cycle of agonist-induced (receptor-activated) heterotrimeric G-proteins. The binding of the agonist to the unoccupied receptor (R) causes a change in conformation, thus activating the receptor (R^*), which promotes the release of GDP from the heterotrimeric G-protein complex and rapid exchange with GTP into the nucleotide binding site on the Gα subunit. In its GTP-bound state, the G-protein heterotrimer dissociates into the Gα and G$\beta\gamma$ subunits, exposing new surfaces and allowing interaction with specific downstream effectors (E). The signal is terminated by hydrolysis of GTP to GDP (and P$_i$) by the intrinsic GTPase activity of the Gα subunit followed by return of the system to the basal unstimulated state. Asterisk indicates activated state of receptor (R) or effector (E); P$_i$, inorganic phosphate; GDP, guanosine diphosphate; GTP, guanosine triphosphate ◄

activity of the Gα subunit. The various families of Gα subunits, i.e., Gα$_s$, Gα$_{i/o}$, Gα$_{q/11}$, and Gα$_{12/13}$, are all GTPases, although the intrinsic rate of GTP hydrolysis varies greatly from one type of Gα subunit to another. Following the hydrolysis of GTP to GDP on the Gα subunit, the Gα and Gβγ subunits re-associate and are able to return to the receptor-associated state.

17.2 Preparation of GPCRs and G-Proteins

In cell-free assays, host cells are transfected with DNA, which allows high levels of expression of the GPCR of interest (in a similar manner to that of whole-cell assays). To date GPCRs have proven to be extremely

Table 17.3 Comparison of the main advantages and disadvantages of various commonly used expression systems to obtain GPCRs and/or G-proteins

Expression system	Advantage	Disadvantage
Bacteria, e.g., *Eschericia coli* spp.	• Many host species to chose from • Many DNA expression vectors available • Relatively cheap • Fast process and easy to scale up • Yield can be very high	• Prokaryotic, not eukaryotic • Truncated proteins can be produced • The expressed proteins often do not fold properly and so are biologically inactive • Insufficient posttranslational modifications made, e.g., GPCR glycosylation, G-protein palmitoylation • Overexpression can be toxic to the host cells
Yeast, e.g., *Saccharomyces cerevisiae*	• Eukaryotic • Fast process and relatively easy to scale up • Yield can be very high • Relatively cheap • Performs many of the post-translational modifications made to human proteins	• Cell wall may hinder recovery of expressed proteins • Presence of active proteases that degrade foreign (expressed) proteins, therefore may reduce yield
Insect, e.g., *Spodoptera frugiperda Sf*9, Hi-5	• High levels of expression • Correct folding • Posttranslational modifications similar to those in mammalian cells	• Expensive to scale up • Slow generation time • Difficult to work with
Mammalian, e.g., Chinese hamster ovary (CHO), human embryonic kidney 293 (HEK), CV-1 in origin with SV40 (COS)	• Good levels of expression • Correct folding and post-translational modifications	• Relatively low yields • Very expensive to scale up • Slow generation time • Difficult to work with • Health and safety implications involved

difficult to purify, primarily due to the lipophilic (hydrophobic) nature of these receptors and the fact that they are usually irreversibly denatured (inactivated) when they are removed from their native lipid environment using detergent treatment. However, partial purification of GPCRs is usually carried out in order to obtain a supply of them. This results in small (nanometer-scale) crude membrane fragments being produced. The GPCR membrane fragments will usually contain hydrophobic membrane lipids (which are required for functionality) as well as other native, *contaminating* proteins and can then be manipulated and immobilized by various means (discussed later) onto appropriate surfaces for use as *biosensors*. On the other hand, the G-proteins, which are classified as peripheral as opposed to integral membrane proteins and do not require an absolute lipid environment for activity, can be routinely purified in relatively large amounts (milligram quantities) when expressed using recombinant DNA-based technologies. Therefore, the first step in generating a GPCR biosensor technology is successfully obtaining functional proteins that may also have been engineered to provide new properties that enable surface attachment or a fluorescent signal.

17.2.1 Expression Systems for Recombinant GPCRs/G-Proteins

A prerequisite for molecular approaches to the design of cell-free GPCR assays is an expression system which produces recombinant proteins with the required activity and level of expression. Expression systems utilizing either bacteria, yeast, mammalian or insect cells are detailed in Table 17.3. These systems are generally well characterized and show the greatest promise in terms of their ability to produce large amounts of functional proteins which can be utilized in GPCR biosensing assay formats.

17.3 Protein Engineering in GPCR Signaling

Molecular engineering of proteins is likely to be of great importance for producing receptors and other associated signaling proteins which have a modified structure or function amenable for use in cell-free biosensing applications. Currently, many receptor or G-protein modifications are aimed at enhancing the purification of proteins, facilitating the attachment to a specific surface or to aid in generating the biosensor signal from GPCR activation (e.g., fluorescence). These modifications can range from the attachment of small tags to larger reporter proteins. These fusion proteins can be generated by engineering DNA sequences that encode the receptor and another protein or tag, joining them such that a single protein is expressed (Fig. 17.4).

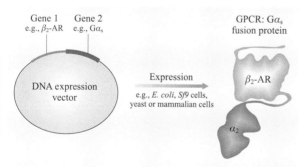

Fig. 17.4 Generation of a fusion protein. Two separate genes of interest are cloned and subsequently ligated into a DNA expression vector *in frame*. In this example, the DNA sequence encoding the GPCR (β_2-adrenergic receptor; β_2-AR) is incorporated into the expression vector within the multiple cloning site. The DNA sequence encoding the Gα_s protein is also cloned into this vector. The resultant recombinant expression vector contains the (carboxy) C-terminus of the β_2-AR fused in frame to the (amino) N-terminus of the Gα_s protein. The recombinant DNA expression vector is then transfected into an appropriate cell line and the fusion protein is expressed

17.3.1 Fluorescent Proteins

Green fluorescent protein (GFP) was first isolated from jellyfish and has been widely exploited in molecular/cell biology research applications due to its efficient fluorescence emission properties. GFPs are particularly useful as they do not require unusual substrates, external catalysis or accessory cofactors for fluorescence as do many other natural pigments [17.5].

Whilst fluorescent proteins provide many advantages, they are limited in their use as protein labels due to their property of being large, multimeric proteins. For this reason alternative methods of site-specific fluorescent labeling are emerging, including the use of lanthanide binding tags [17.6] and tetracysteine motifs (TCM) [17.7]. Each of these tags are significantly smaller than GFP variants and enable fluorescence through the binding of a lanthanide such as terbium or a fluorescent arsenic derivative such as FlAsH, respectively. In some cases the use of a smaller tag can prevent the loss of function of the proteins of interest. For example, when a TCM was used to label the adenosine A_{2A} receptor, aspects of the receptors function that had been lost when using yellow fluorescent protein (YFP) at the same site were restored [17.8]. In the future it is likely that fluorescence-based assay development involving compounds such as these will increase in efficiency and flexibility, allowing such methods to be at the forefront of technologies for determining molecular interactions using cell-free systems.

17.3.2 Engineering of Promiscuous Gα Proteins

A major impediment to the production of homogeneous, cell-free, GPCR-based screening systems is the coupling between a given GPCR and a subset of Gα subunits. For example, muscarinic receptor subtypes M_1, M_3, and M_5 typically couple to $G\alpha_{q/11}$, whilst M_2 and M_4 subtypes couple to G_i or G_o [17.9]. Biologically, this discrimination is the basis for correct cellular signaling but needs to be modified from the in vivo situation to allow production of a generic GPCR biosensing system. In this regard, recent attempts have been made to produce *promiscuous* Gα subunits capable of transducing signals resulting from extracellular interactions involving any GPCR [17.10, 11]. Many of the promiscuous subunits constructed thus far are based on variants of the human $G\alpha_{16}$ (a member of the $G\alpha_q$ subfamily). This protein was first isolated from hematopoietic cells [17.12] and was shown to couple to a wider range of receptors than other known alpha subunits, and to transduce ligand-mediated signaling through phospholipase C (PLC), resulting in the modification of intracellular calcium concentrations [17.13–17]. Molecular biology approaches have also been utilized to increase the promiscuity of various Gα subunits by altering the sequence of amino acids within the protein [17.11, 18–21]. Although cell-free applications have not been routinely used to date, it is expected that, within the near future, promiscuous G-proteins will be used in a similar manner to in whole-cell applications.

17.3.3 Protein Engineering for Surface Attachment

Ideally, proteins should be uniformly immobilized so that the protein remains functional and is orientated such that the required interaction can occur. To achieve these ends, protein engineering can be a powerful tool. Immobilized metal ion affinity chromatography (IMAC) used for protein purification has been extended to enable functional immobilization of proteins onto a surface. GPCRs and G-proteins are often fused to a 6 histidine tag [17.22–24] that has a high affinity for nitriloacetic acid (NTA) loaded with a divalent cation, often Ni^{2+}, on a surface. The length of the histidine tag can also be adjusted for higher affinity to Ni^{2+}, as has been demonstrated using $G\alpha_{i1}$ [17.25]. Utilizing histidine tags, it has also been possible to functionally reconstitute GPCRs with G-proteins on Ni-NTA beads and observe signaling upon ligand binding [17.24]. Surface immobilization has also been achieved by engineering short peptides such as the C9 peptide or Myc tags onto GPCRs of interest and using surfaces displaying the appropriate antibody to these tags to capture the receptors [17.23, 26].

17.4 GPCR Biosensing

The basic requirement of a biosensor is the use of a biological element, such as an immobilized protein, to act as a sensor for a specific binding analyte. This is coupled with a reporter system which amplifies the initial signal to produce some form of output. Depending on the type of output required for a given screening process, e.g., ligand binding to a GPCR or a functional assay such as G-protein activation, a number of protocols are available to target the site of interest. In this chapter we will refer to these as *levels* of GPCR activation (Fig. 17.1). Examples of each of these levels will be discussed below. Additionally, in this section, the levels of biosensing referred to represent those *cell-free* samples or biological preparations which are derived from cells and are used in the cell-free mode, i.e., the GPCRs and G-proteins have been either partially or fully purified from cells expressing the GPCRs or G-proteins, and then subsequently *reconstituted* at known concentrations, usually within the nanomolar range.

17.4.1 Level 1 Biosensing – Ligand Binding

We have defined level 1 biosensing as ligand binding to the receptor. This includes such techniques as radioligand binding (not discussed here) and fluorescent (and fluorescent polarization) ligand binding assays. Ligand binding can also be detected by techniques such as flow cytometry, two-photon excitation cross-correlation spectroscopy (TPE-FCCS), surface plasmon resonance (SPR), plasmon waveguide spectroscopy, and piezoelectric crystal sensing.

This level of biosensing does not discriminate between compounds which can be pharmacologically defined as agonists, antagonists, partial agonists or inverse agonists. Therefore its use in biosensing of the *activation* of a signaling pathway is somewhat limited. However it is still useful for some specific purposes such as screening for compounds which *interact* with a particular GPCR.

Fluorescence Polarization

Polarization is a general property of most fluorescent molecules. Polarization-based experiments are less dye dependent and less susceptible to environmental interferences (such as pH changes) than assays based on fluorescence intensity measurements. Fluorescence intensity variations due to the presence of samples which may be colored (e.g., in drug screening of compound libraries) tend to produce relatively minor interferences. The degree of polarization (or anisotropy) can be determined from measurements of fluorescence intensities parallel and perpendicular to the plane of linearly polarized excitation light [17.27].

Fluorescence and fluorescence polarization (FP) assays which are based on specific binding of the ligand to a GPCR can offer an alternative to traditional radioligand binding assays which utilize radionuclides (radioisotopes) [17.28]. FP assays usually take the form of a homogeneous or *mix and read* type of assay (and an example of level 1 assays), which indicates that they are readily transferable from assay development to high-throughput screening (HTS). FP allows for the development of protocols which are both real-time measurements (kinetic assays) and insensitive to variations in concentrations. One of the disadvantages of this assay format is the lack of adaptability to all GPCR ligands by virtue of the fact that only a small number of ligands can be chemically tagged with an appropriate fluorophore and still retain their intrinsic binding qualities. Finally, the choice of fluorophore is important, as the intensity of the fluorescent compound must be of sufficient magnitude as well as having good polarizing properties [17.29].

Two-Photon Excitation Fluorescence Cross-Correlation Spectroscopy

Two-photon excitation fluorescence cross-correlation spectroscopy (TPE-FCCS) is used to measure dynamic interactions between molecules, and in the bioscience field has applications in monitoring DNA, protein, and ligand interactions. The technique allows for small measurement volumes and low sample concentrations and has increased detection specificity over classical fluorescence techniques for monitoring molecular dynamics in solution. TPE-FCS is an extension of fluorescence correlation spectroscopy (FCS), which analyzes minute, spontaneous signal fluctuations arising from molecular diffusion. The term "two-photon" refers to the use of different fluorescent molecules with distinct emission properties, each of which can be excited by two photons of half the energy required for a transition to the excited state. Two-photon excitation spectra of many common fluorophores differ considerably from their one-photon spectra without a change in emission. This makes it possible to simultaneously excite two spectrally distinct dyes with a single infrared light source. A cross-correlation of the two fluorophores is only generated when the two detection channels measure synchronous fluorescence fluctuations, which suggests that the different colored species must be spectrally linked.

This technique is useful in the study of association and dissociation reactions such as that of a receptor–ligand pair. To obtain accurate kinetic information regarding the interaction of the human μ-opioid receptor (within nonpurified preparations that were termed *nanopatches*) with its ligand, *Swift* et al. have used TPE-FCCS [17.30]. A pentahistidine-tagged μ-opioid receptor was fluorescently tagged with an Alexa-conjugated antipentahistidine antibody and measured in the presence of fluorescein-labeled antagonists. Similarly to the FP previously mentioned, this fluorescence technique also enables a homogeneous assay platform amenable to HTS.

Flow Cytometry

Flow cytometry is a technique used to analyze the fluorescence of individual cells or particles (such as the dextran beads in the example below). Fluorescence can arise from intrinsic properties of the cell, but generally molecules/particles of interest are fluorescently labeled. Hydrodynamic focusing is used to force the cells or particles into a single file, after which they are passed through a laser beam where both scattered and emitted light are measured. A benefit of this method is that simultaneous measurements can be performed on individual particles.

Waller et al. [17.31] have conjugated dextran beads with the cognate ligand dihydroalprenolol, which allowed for capture of solubilized β_2-adrenergic receptors (β-AR) onto this immobilized surface ligand. To measure the specific binding of the receptor to the bead in this flow-cytometry-based assay system, the receptor was expressed as a fusion protein with GFP. It was then possible to screen for ligands (either agonists or antagonists) to β-AR using a competition assay. Another successful bead-based approach used paramagnetic beads [17.32]. In that study the authors built up a surface containing the captured CCR5 receptor from a cell lysate held within a lipid bilayer. In this instance the CCR5 receptor was not able to freely move laterally in the bilayer as it was tethered via an antibody (directed at the CCR5 receptor) conjugated to the paramagnetic beads (paramagnetic proteoliposome).

Total Internal Reflection Fluorescence

Total internal reflection fluorescence (TIRF) takes advantage of refractive index differences at a solid–liquid interface, with the solid surface being either glass or plastic, e.g., cell culture containers. At a critical angle, when total internal reflection occurs, an evanescent wave is produced in the liquid medium. This electromagnetic field decays exponentially with increasing distance from the surface. The range of this field limits background fluorescence, as only fluorophores in close proximity to the surface are excited. As such, the technique is used to examine interactions between the molecule of interest and the surface, for example, receptors binding to a surface.

Martinez et al. [17.33] used TIRF to demonstrate ligand binding to the neurokinin-1 GPCR by surface immobilization of membrane fragments containing this receptor protein. In this study, the GPCR expressed as a biotinylated protein using mammalian cells was selectively immobilized on a quartz sensor surface coated with streptavidin (streptavidin binds biotin with extremely high affinity). TIRF measurements were made using a fluorescence-labeled agonist (i. e., the cognate agonist substance-P labeled with fluorescein). Using this approach, it was not necessary to detergent-solubilize and reconstitute the neurokinin-1 receptors, thus avoiding the deleterious effect(s) associated with such processes. This receptor, in the form of a mammalian cell membrane homogenate, was then surface-immobilized without further purification. The selective, high-affinity interaction between biotin and streptavidin allowed template-directed and uniform orientation of the neurokinin-1 receptor on the support matrix. Additionally, the highly selective TIRF fluorescence detection methodology was able to resolve the binding of fluorescently tagged agonist to as little as 1 aM of receptor molecules.

Microspotting of GPCRs on Glass

The intrinsic difficulties in producing, purifying, and manipulating membrane proteins have delayed their introduction into microarray platforms. Hence there are no reports to date describing purified membrane protein (GPCR) microarrays and their use in functional screening or biosensor applications. However, as a first step towards such display technologies, researchers at Corning Inc. (Rochester, USA) have recently described the fabrication of GPCR membrane arrays for the screening of GPCR ligands [17.34–37]. The arraying of membrane GPCRs required appropriate surface chemistry for the immobilization of the lipid phase containing the GPCR of interest (Fig. 17.5). They reported that surface modification with γ-aminopropylsilane (an amine-presenting surface) provided the best combination of properties to allow surface capture of the GPCR:G-protein complex from crude membrane preparations, resulting in microspots of $\approx 100\,\mu$m diameter. Atomic force microscopy (AFM) demonstrated that the height of the supported lipid bilayer was ≈ 5 nm, corresponding to GPCRs confined in a single, supported lipid layer scaffold. Using these chemically derivatized surfaces, it was possible to demonstrate capture of the β_1, β_2, and α_{2A} subtypes of the adrenergic receptor, as well as neurotensin-1 receptors and D1-dopamine receptors. This was achieved by using ligands with fluorescent labels covalently attached and detecting fluorescence binding to the GPCRs with a fluorescence-based microarray scanner. Dose–response curves using the fluorescently labeled ligands gave 50% inhibition concentration (IC_{50}) values in the nanomolar range, suggesting that the GPCR:G-protein complex was largely preserved and biologically intact in the microspot. There was no change in the performance of the arrays over a 60 day period, indicating good long-term stability. Although the use of glass slides for the printing of the GPCR arrays was promoted by this research group, in some instances gold surfaces were required due to nonspecific binding of fluorescent ligands. A current limitation of this technology is the inability to carry out a functional (i. e., signaling) assay, which

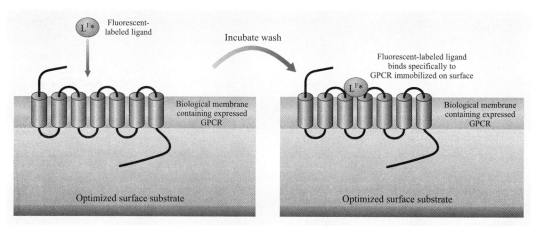

Fig. 17.5 Idealized schematic of an immobilized GPCR with associated G-proteins. The fabricated surface array is printed on a γ-aminopropylsilane (GAPS)-presenting surface. The height of the supported lipid bilayer is ≈ 5 nm. Fluorescently labeled (L^{F*}) ligands (such as 4,4,-difluoro-5,7-dimethyl-4-bora-3a,4a-diaza-s-indacene (BODIPY)-tetramethylrhodamine (TMR)) will bind specifically to the GPCR (for example, a neurotensin receptor) at nanomolar concentrations. The fluorescence is measured following an incubation/washing step to remove unbound fluorescent ligands. When compounds of unknown activity are added to the incubation step, as in drug screening programs, fluorescent-labeled ligand binding is blocked by agents that bind to the GPCR (for example, GPCR antagonists) (adapted from [17.13])

would allow test compounds to be classified as agonists or antagonists. Furthermore, although there are increasing numbers of commercially available fluorescently labeled ligands, the need to always structurally modify the ligand to accommodate some reporter moiety may limit the implementation of the technology. Nevertheless, the above-mentioned GPCR microarrays may have potential as *functional* GPCR assays when complexed with G-proteins and integrated with appropriate signal generation and detection methods.

Surface Plasmon Resonance

One of the most versatile techniques for measuring biospecific interactions in real time are biosensors based on the optical phenomenon of surface plasmon resonance (SPR, Fig. 17.6). Surface plasmon resonance occurs when light interacts with a conducting surface (plasmon interaction) which is positioned between two materials of different refractive index. At a specific angle the intensity of the reflected light decreases, with this angle being dependent on (among other things) the refractive index of the material on the opposite side to which the light is applied. Molecules associating with or disassociating from this material (e.g., receptor to surface or ligand to receptor) will change the refractive index of the material and can be detected by measuring the reflected light. The instrument detects the change in angle of the reflected light minimum. The technique can be used to study interactions between ligands, GPCRs, and G-proteins. SPR experiments do not require a large amount of sample, and detection does not require fluorescent or radioisotopic labeling. A variety of available surface chemistries allows for immobilization of many types of proteins using a range of strategies.

Ligand binding to a GPCR attached to a surface has been reported for the chemokine CCR5 receptor using SPR methodology [17.38]. For such display, purification of the GPCR has not always been necessary, and crude membrane preparations have either been fused with an alkylthiol monolayer (≈ 3 nm thickness) formed on a gold-coated glass surface or onto a carboxymethyl-modified dextran sensor surface [17.39]. One problem of surface-based assays is orientation of the receptor once attached to the surface. One means of overcoming this problem was to specifically select only those proteoliposomes (≈ 300 nm-diameter vesicles) in which the carboxy terminus of the receptor was orientated to the outside of the vesicle. This was performed using conformationally dependent antibodies [17.26]. In this biosensor application, SPR has a distinct advantage as a screening tool since it can detect the cognate ligand without requiring fluorescent or radio labeling. This allows SPR to be used in complex fluids of natural origin and simplifies, and potentially speeds up, the development of assay technologies.

Plasmon-Waveguide Resonance Spectroscopy

Plasmon-waveguide resonance (PWR) spectroscopy measures real-time binding of free molecules to immobilized molecules such as GPCRs without the application of specific labels (reviewed elsewhere [17.40]). PWR has several significant advantages compared with conventional surface plasmon resonance, including enhanced sensitivity and spectral resolution, as well as the ability to distinguish between mass and conformational changes. This latter property is a consequence of the use of both p- and s-polarized excitation to produce resonances. This allows for measurement of refractive index anisotropy, which reflects changes in mass distribution and, therefore, changes in molecular orientation and conformation.

In a recent study, ligand binding to the β_2-adrenergic receptor has been demonstrated using PWR [17.41]. Using this technique, changes in the refractive index upon ligand binding to surface-immobilized receptor results in a shift in the PWR spectra. The authors used ligands with similar molecular weight in order to study structural changes in the receptor caused by agonist, inverse agonist, and partial agonist binding. The technique was used to produce binding curves for five ligands using shifts in the PWR spectra (with both s- and p-polarized light) with increasing ligand concentration, with the results from PWR being compared with those obtained by traditional radioligand binding assays. Differences in s- and p-polarized light measurements demonstrated changes in receptor structure which varied depending on whether the ligand was a full, partial or inverse agonist. Previous work using PWR technology has been reported for the detection of conformational changes in a proteolipid membrane containing the human δ-opioid receptor following binding of nonpeptide agonists, partial agonists, antagonists, and inverse agonists [17.42]. Although the ligands in the above study were of similar molecular weight, there were distinctly different refractive index changes induced by ligand binding and these were too large to be accounted for by differences in mass alone. The infer-

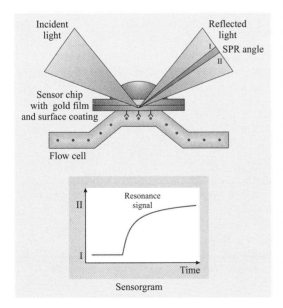

Fig. 17.6 Surface plasmon resonance (SPR) provides mass detection. Most importantly, this technique does not require labeling of the interacting components. Since it is the evanescent field wave and not the incident light which penetrates the sample, measurements can be made on turbid or even opaque samples. The detection principle of SPR relies on electron charge density wave phenomena arising at the surface of a metallic film when light is reflected at the film under specific conditions (surface plasmon resonance). The resonance is a result of energy and momentum being transformed from incident photons into surface plasmons, and is sensitive to the refractive index of the medium on the opposite side of the film from the reflected light. Quantitative measurements of the binding interaction between one or more molecules are dependent on the immobilization of a target molecule onto the sensor chip surface. Binding partners to the target can be captured from a complex mixture as they pass over the chip. Interactions between proteins, nucleic acids, lipids, carbohydrates, and even whole cells can be studied. The sensor chip consists of a glass surface coated with a thin layer of gold. This forms the basis for a range of specialized surfaces designed to optimize the binding of a variety of molecules. The gold layer in the sensor chip creates the physical conditions required for SPR. The *upper figure* shows a detector with sensor chip. When molecules in the test solution bind to a target molecule the mass increases; when they dissociate the mass falls. This simple principle forms the basis of the sensorgram for continuous, real-time monitoring of the association and dissociation of the interacting molecules (*lower figure*). The sensorgram provides quantitative information in real time on the specificity of binding, active concentration of molecules in a sample, kinetics, and affinity. Molecules as small as 100 Da can be studied ◄

ence from this finding was that a ligand-specific conformation change in the receptor protein may have been detected. Therefore this methodology may have use as a future biosensor, particularly with regard to GPCRs.

Piezoelectric Crystal Sensing

Piezoelectric crystal sensing measures a change in mass on molecule binding to the surface due to a change in resonance frequency of the crystal. The technique has been used in an *electronic nose* with olfactory receptors which are typically GPCRs [17.43], where an array of six sensor elements could be used to characterize each of six test compounds, emphasizing the potential for GPCR ligand screening in the sensory area. The use of an artificial nose (*bionose*) to mimic the properties of the human nose may find wide applications in the near future.

17.4.2 Level 2 Biosensing – Conformational Changes in the GPCR

Level 2 involves the detection of intrinsic conformational changes in the GPCR protein following agonist activation and may involve the use of fluorescence-based techniques. Cell-free measurements of conformational changes in the GPCR following ligand (usually agonist or partial agonist) binding have been limited to date.

A good example of a level 2 cell-free assay used β_2-adrenergic receptors immobilized onto glass and gold surfaces. In this study, the receptors were site-specifically labeled with the fluorophore tetramethyl-rhodamine-maleimide at cysteine 265 (Cys265) using a series of molecular biology approaches. It was then possible to show agonist (isoproterenol)-induced conformational changes within the vicinity of the fluorescent moiety (tetramethyl-rhodamine) at position Cys265 of the recombinant β_2-adrenergic receptors. Moreover, the agonist-induced signal was large enough to detect using a simple intensified charge-coupled device (ICCD) camera image. Thus, it was suggested that the technique may be useful for drug screening with GPCR arrays. Indeed this method did not require the formation of lipid bilayers and did not require the use of purified G-proteins or fluorescent ligands to detect receptor activation.

17.4.3 Level 3 Biosensing – GTP Binding

Measurements of GPCR activation further downstream from level 2 are considered for the purposes of this chapter to be truly functional assays since the transducer G-proteins are the first differentiated site of signalling initiated from the GPCR. This therefore means that the GPCR must be in a *functional* form, enabling it to interact and activate a G-protein signaling pathway. Level 3 biosensing involves the use of nonhydrolyzable GTP analogs such as radiolabeled $^{35}S\gamma GTP$ or fluorescent-tagged europium-GTP which bind to the receptor-activated form of the $G\alpha$ subunit targeting the site of guanine nucleotide exchange (GDP for GTP on the $G\alpha$ subunit of the $G\alpha\beta\gamma$ heterotrimer). The guanine nucleotide exchange process is generally considered the first major point of G-protein activation following GPCR stimulation (Figs. 17.1 and 17.3).

Guanine nucleotide exchange is a very early, generic event in the signal transduction process of GPCR activation and is therefore an attractive event to monitor as it is less subject to regulation by cellular processes further downstream (we denote this as level 3 biosensing, see Fig. 17.1). The radiolabeled $^{35}S\gamma GTP$ or fluorescent europium-GTP binding assays measure the level of G-protein activation following agonist activation of a GPCR by determining the binding of these nonhydrolyzable analogs of GTP to the $G\alpha$ subunit. Therefore, they are defined as *functional* assays of GPCR activation. Ligand regulation of the binding of $^{35}S\gamma GTP$ is one of the most widely used assay methods to measure receptor activation of heterotrimeric G-proteins, as discussed elsewhere in detail [17.44, 45]. This methodology also provides the basis for measurement of pharmacological characteristics such as potency, efficacy, and the antagonist affinity of compounds [17.45] in cell-free assays and artificial expression systems for GPCRs (an example of typical data is shown in Fig. 17.7). However, despite the highly desirable attributes of this methodology and its widespread use to date, ligand regulation of $^{35}S\gamma GTP$ binding has been largely restricted to those receptors which signal through the $G\alpha_{i/o}$ proteins (pertussis-toxin sensitive) and, to a lesser extent, the $G\alpha_s$ and $G\alpha_q$ families of G-proteins. As such, the use of these assay platforms can be problematic for high-throughput screening as they are not homogeneous (i.e., they require a separation step to remove bound from free $^{35}S\gamma GTP$). Additionally, the use of radioactive-based assays (including ligand binding assays) has led to safety, handling, waste disposal, and cost concerns. The newly developed, fluorescence-based europium-GTP assay partly overcomes some of the above limitations and has already been successfully used with the following GPCRs: motilin, neurotensin, muscarinic-M_1, and α_{2A}-adrenergic receptors.

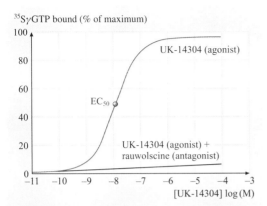

Fig. 17.7 Activation of GPCR-induced GTP binding. The data show results from an experiment which was conducted by incubating 20 nM purified G-proteins (Gα_{i1} and G$\beta_1\gamma_2$) reconstituted with 0.4 nM recombinant α_{2A}-adrenergic receptor-expressing membranes (these receptors normally bind adrenaline with high affinity). The assay also contained 0.2 nM ^{35}SγGTP (a radioactive nonhydrolyzable analog of GTP). An adrenaline analog (UK-14304) was then added to the reconstituted α_{2A}-adrenergic receptor membrane, at the concentrations indicated on the x-axis (0.01 nM to 100 μM) in the presence or absence of the α_{2A}-adrenergic receptor antagonist, rauwolscine (10 μM). Following a filtration step to remove the bound ^{35}SγGTP:Gα_{i1} complex from unbound ^{35}SγGTP, the filters were subsequently counted in a scintillation counter to measure the level of radioactivity. As the concentration of the agonist (UK-14304) was increased above 1 nM (10^{-9} M), the characteristic sigmoidal dose–response effect was seen. This result shows an increase in receptor-activated binding of ^{35}SγGTP to the Gα_{i1} subunits as the UK-14304 is increased in concentration, indicating functional signaling of the receptor through the G-protein complex. The concentration at which 50% (also the point of inflexion) of the signaling response (effective concentration) was observed (EC$_{50}$) was $\approx$ 12 nM. In the presence of excess α_{2A}-adrenergic receptor antagonist (rauwolscine), the signal was completely blocked at the receptor. Therefore, this type of biosensing application demonstrates sensitivity as well as specificity

17.4.4 Level 4 Biosensing – GPCR:G-Protein Dissociation

Procedures that utilize only ligand binding (level 1) do not distinguish between agonist (activates receptor), antagonist (blocks the action of the agonist at the receptor binding site) or inverse agonist [inhibits the intrinsic (nonagonist-stimulated) activity of the receptor signaling, often observed in overexpressed receptors]; however, if a functional GPCR assay is constructed in which G-protein *activation* is an endpoint, i.e., level 4 biosensing, then it is possible to distinguish between these functionally distinct ligands. For cell-free assays, both methodologies (levels 1 and 4) are important in HTS programs, for example, and may have differing extents of applicability. Indeed, novel nanotechnology approaches will be required to achieve level 4 biosensing, including suitable surface derivatization for immobilization of GPCR and G-protein complexes.

Level 4 biosensing encompasses those assays which measure the final stage of activation of the G-protein heterotrimeric complex, i.e., the putative dissociation or rearrangement of the subunits following GPCR-induced G-protein activation [17.46]. This level of GPCR activation has currently not been investigated in great detail but may prove to be extremely valuable in future functional biosensor applications. Assay methodologies which are examples of level 4 biosensing have been reported using surface plasmon resonance and flow cytometry technologies to demonstrate receptor dissociation from the G-protein complex.

SPR G-Protein Dissociation

Bieri et al. [17.47] used carbohydrate-specific biotinylation chemistry to achieve appropriate orientation and functional immobilization of the solubilized bovine rhodopsin receptor, with high-contrast micropatterns of the receptor being used to spatially separate protein regions. This reconstituted GPCR:G-protein system provided relatively stable results (over hours) with the added advantage of obtaining repeated activation/deactivation cycles of the GPCR:G-protein system. Measurements were made using SPR detection of G-protein dissociation from the receptor surface following the positioning of the biotinylated form of the rhodopsin receptor onto a self-assembled monolayer containing streptavidin. Using this approach, G-protein activation could be directly monitored, giving a functional output, as opposed to ligand–receptor binding interactions, which yield little information on the receptor-activated pathway when screening agonists and antagonists. Although SPR is useful for the study of G-protein interactions, it may not be well suited to detect binding of small ligand molecules directly due to its reliance on changes in mass concentration. An advantage of repeated activation/deactivation cycles of GPCRs is that different compounds can be tested serially with the same receptor preparation. The above approach appears promising for future applications of chip-based technologies in the area of GPCR biosensor applications.

Flow Cytometry – GPCR:G-Protein Interactions

Modifying the surface of epoxy-activated dextran beads by forming a Ni^{2+}-NTA conjugate was shown to produce beads with a surface capable of binding hexahistidine (his)-tagged $\beta_1\gamma_2$ subunits (Fig. 17.8). Tethered $\beta_1\gamma_2$ subunits were then used to capture $G\alpha_s$ subunits, which in turn were capable of binding membrane preparations with expressed β_2-adrenergic receptor containing a GFP fusion protein (see Sect. 17.3 for a detailed description of fusion proteins); alternatively, a fluorescence-labeled ligand could be detected binding to the tethered β_2-adrenergic receptor, the whole complex being measured using flow cytometry. Additionally, quantitative solubilization and re-assembly of the (hexahistidine-tagged) N-formyl peptide receptor (FPR) has been demonstrated on Ni^{2+}-silica particles using flow cytometry with dodecyl maltoside as the detergent [17.48]. Using such approaches, it may be possible to screen ligands for a known solubilized GPCR, or alternatively to test which G-proteins preferentially couple to a particular solubilized, reconstituted GPCR. The flow cytometry system used above had a sampling rate of $\approx 50-100$ samples per minute; however, flow cytometry's greatest advantage is its ability to be multiplexed, where different molecular assemblies can be made with one sample and yet be discriminated by their unique spectral characteristics [17.31, 49, 50]. In more detailed studies, the assembly and disassembly of the FPR and his-tagged G-proteins complexed on Ni^{2+}-silica particles provided insight into the activation kinetics of the ternary complex (i. e., receptors and heterotrimeric G-proteins) [17.49, 51]. The study by *Simons* et al. [17.49] extended the knowledge of ligand–GPCR interactions to involve G-protein–GPCR–ligand interactions assayed in a homogeneous format with a bead-based approach amenable to high-throughput flow cytometry. Indeed, HTS and proteomic applications could easily be based on such bead arrays with potential for color-coded particles and multiplexing (e.g., by using quantum-dot technology [17.52]).

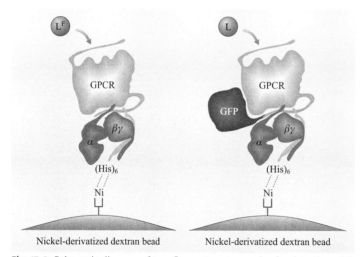

Fig. 17.8 Schematic diagram of two flow cytometry modes for detection of the ligand:receptor:G-protein assembly on nickel-coated beads. The G-proteins are immobilized on the bead surface containing a nickel (Ni) chelate. The exposed Ni binds to an engineered hexahistidine $(His)_6$ sequence on the N-terminal of the $G\gamma$ subunit and is able to capture the heterotrimeric G-protein complex (*left figure*). The fluorescent ligand (L^F) binds to the GPCR following capture of the GPCR with appropriate G-proteins complexed on the surface of dextran beads. This technique is useful for biosensing of the interaction of specific GPCR ligands (agonists and antagonists) and may be useful for demonstrating receptor:G-protein specificity and screening of ligands. In the *right figure*, the assembly uses a GPCR fusion protein containing enhanced green fluorescent protein (GFP). This technique demonstrates the requirement of the heterotrimeric complex for ligand activation and allows quantification of the receptor without the use of fluorescent ligands (adapted from [17.28])

Particle-based screening constitutes an enabling technology for the identification of agonists promoting assembly of G-protein–GPCR interactions as well as antagonists which inhibit such assembly.

17.5 The Future of GPCRs

Although this chapter has focused on the GPCR signaling system for biosensing applications, many other potential biological systems could equally be exploited for biosensing applications, including those involving antibodies, ion channels, and enzymes. We have emphasized that molecular biology, combined with nanobiotechnologies, provides important tools by which every facet of designing and investigating cell-free biosensing approaches can be improved. GPCR and G-protein engineering is a technique which has been employed not only to study GPCR interactions but to enhance the measurement of GPCR activation, which will interface with future biosensing applications. Fusion proteins, promiscuous and chimeric Gα proteins, and molecular tagging are some of the molecular attributes which have been described herein. Structural enhancements to GPCRs and G-protein subunits or effectors are only limited by the creativity of the researcher, and these enhancements will be imperative in the design of novel, cell-free assay technologies. Further research into microarray and chip-based technologies, recombinant protein design and production, assay automation, and new assay methodologies for studying GPCR signaling is rapidly developing. The involvement of GPCR signaling in such a multitude of cellular processes indicates that it is unlikely that the current interest in GPCRs will diminish in the foreseeable future.

References

17.1   A. Wise, K. Gearing, S. Rees: Target validation of G-protein coupled receptors, Drug Discov. Today **7**, 235–246 (2002)

17.2   S.G. Rasmussen, H.J. Choi, D.M. Rosenbaum, T.S. Kobilka, F.S. Thian, P.C. Edwards: Crystal structure of the human beta(2) adrenergic G-protein-coupled receptor, Nature **450**, 383–387 (2007)

17.3   K.L. Pierce, R.T. Premont, R.J. Lefkowitz: Seven-transmembrane receptors, Nat. Rev. Mol. Cell Biol. **3**, 639–650 (2002)

17.4   D.K. Vassilatis, J.G. Hohmann, H. Zeng, F. Li, J.E. Ranchalis, M.T. Mortrud: The G protein-coupled receptor repertoires of human and mouse, Proc. Natl. Acad. Sci. USA **100**, 4903–4908 (2003)

17.5   V.V. Verkhusha, K.A. Lukyanov: The molecular properties and applications of anthozoa fluorescent proteins and chromoproteins, Nat. Biotechnol. **22**, 289–296 (2004)

17.6   T.H. Cooper, W.R. Leifert, R.V. Glatz, E.J. McMurchie: Expression and characterisation of functional lanthanide binding tags fused to a Gα-protein and muscarinic (M2) receptor, J. Bionanosci. **2**(1), 27–34 (2008)

17.7   S.R. Adams, R.E. Campbell, L.A. Gross, B.R. Martin, G.K. Walkup, Y. Yao: New biarsenical ligands and tetracysteine motifs for protein labeling in vitro and in vivo: Synthesis and biological applications, J. Am. Chem. Soc. **124**, 6063–6076 (2002)

17.8   C. Hoffmann, G. Gaietta, M. Bunemann, S.R. Adams, S. Oberdorff-Maass, B. Behr: A FlAsH-based FRET approach to determine G protein-coupled receptor activation in living cells, Nat. Methods **2**, 171–176 (2005)

17.9   M.P. Caulfield, N.J. Birdsall: International union of pharmacology. XVII. classification of muscarinic acetylcholine receptors, Pharmacol. Rev. **50**, 279–290 (1998)

17.10 A.M. Liu, M.K. Ho, C.S. Wong, J.H. Chan, A.H. Pau, Y.H. Wong: Gα(16/z) chimeras efficiently link a wide range of G protein-coupled receptors to calcium mobilization, J. Biomol. Screen. **8**, 39–49 (2003)

17.11 A. Hazari, V. Lowes, J.H. Chan, C.S. Wong, M.K. Ho, Y.H. Wong: Replacement of the α_5 helix of Gα_{16} with gα_s-specific sequences enhances promiscuity of Gα_{16} toward Gs-coupled receptors, Cell Signal. **16**, 51–62 (2004)

17.12 T.T. Amatruda III, D.A. Steele, V.Z. Slepak, M.I. Simon: Gα_{16}, a G protein α subunit specifically expressed in hematopoietic cells, Proc. Natl. Acad. Sci. USA **88**, 5587–5591 (1991)

17.13 S. Offermanns, M.I. Simon: Gα_{15} and Gα_{16} couple a wide variety of receptors to phospholipase C, J. Biol. Chem. **270**, 15175–15180 (1995)

17.14 X. Zhu, L. Birnbaumer: G protein subunits and the stimulation of phospholipase C by Gs- and Gi-coupled receptors: Lack of receptor selectivity of Gα(16) and evidence for a synergic interaction

17.14 between G$\beta\gamma$ and the α subunit of a receptor activated G protein, Proc. Natl. Acad. Sci. USA **93**, 2827–2831 (1996)

17.15 G. Milligan, F. Marshall, S. Rees: G16 as a universal G protein adapter: Implications for agonist screening strategies, Trends Pharmacol. Sci. **17**, 235–237 (1996)

17.16 J.W. Lee, S. Joshi, J.S. Chan, Y.H. Wong: Differential coupling of μ-, δ-, and κ-opioid receptors to Gα_{16}-mediated stimulation of phospholipase C, J. Neurochem. **70**, 2203–2211 (1998)

17.17 E. Kostenis: Is Gα_{16} the optimal tool for fishing ligands of orphan G-protein-coupled receptors?, Trends Pharmacol. Sci. **22**, 560–564 (2001)

17.18 B.R. Conklin, Z. Farfel, K.D. Lustig, D. Julius, H.R. Bourne: Substitution of three amino acids switches receptor specificity of G$_q\alpha$ to that of G$_i\alpha$, Nature **363**, 274–276 (1993)

17.19 B.R. Conklin, P. Herzmark, S. Ishida, T.A. Voyno-Yasenetskaya, Y. Sun, Z. Farfel: Carboxyl-terminal mutations of G$_q\alpha$ and G$_s\alpha$ that alter the fidelity of receptor activation, Mol. Pharmacol. **50**, 885–890 (1996)

17.20 E. Kostenis, F.Y. Zeng, J. Wess: Functional characterization of a series of mutant G protein α_q subunits displaying promiscuous receptor coupling properties, J. Biol. Chem. **273**, 17886–17892 (1998)

17.21 S.M. Mody, M.K. Ho, S.A. Joshi, Y.H. Wong: Incorporation of Gα(z)-specific sequence at the carboxyl terminus increases the promiscuity of Gα(16) toward G(i)-coupled receptors, Mol. Pharmacol. **57**, 13–23 (2000)

17.22 C.H. Klaassen, P.H. Bovee-Geurts, G.L. Decaluwe, W.J. DeGrip: Large-scale production and purification of functional recombinant bovine rhodopsin with the use of the baculovirus expression system, Biochem. J. **342**, 293–300 (1999)

17.23 D. Ott, Y. Neldner, R. Cebe, I. Dodevski, A. Pluckthun: Engineering and functional immobilization of opioid receptors, Protein Eng. Design Sel. **18**, 153–160 (2005)

17.24 R.V. Glatz, W.R. Leifert, T.H. Cooper, K. Bailey, C.S. Barton, A.S. Martin: Molecular engineering of G protein-coupled receptors and G proteins for cell-free biosensing, Aust. J. Chem. **60**, 309 (2007)

17.25 C.S. Barton, R.V. Glatz, A.S. Martin, L. Waniganayake, E.J. McMurchie, W.R. Leifert: Interaction of self-assembled monolayers incorporating NTA disulfide with multilength histidine-tagged Gα_{i1} subunits, J. Bionanosci. **1**, 22–30 (2007)

17.26 P. Stenlund, G.J. Babcock, J. Sodroski, D.G. Myszka: Capture and reconstitution of G protein-coupled receptors on a biosensor surface, Anal. Biochem. **316**, 243–250 (2003)

17.27 J.C. Owicki: Fluorescence polarization and anisotropy in high throughput screening: Perspectives and primer, J. Biomol. Screen. **5**, 297–306 (2000)

17.28 C.J. Daly, J.C. McGrath: Fluorescent ligands, antibodies, and proteins for the study of receptors, Pharmacol. Ther. **100**, 101–118 (2003)

17.29 P. Banks, M. Harvey: Considerations for using fluorescence polarization in the screening of G protein-coupled receptors, J. Biomol. Screen. **7**, 111–117 (2002)

17.30 J.L. Swift, M.C. Burger, D. Massotte, T.E. Dahms, D.T. Cramb: Two-photon excitation fluorescence cross-correlation assay for ligand-receptor binding: Cell membrane nanopatches containing the human micro-opioid receptor, Anal. Chem. **79**, 6783–6791 (2007)

17.31 A. Waller, P. Simons, E.R. Prossnitz, B.S. Edwards, L.A. Sklar: High throughput screening of G-protein coupled receptors via flow cytometry, Comb. Chem. High Throughput Screen. **6**, 389–397 (2003)

17.32 T. Mirzabekov, H. Kontos, M. Farzan, W. Marasco, J. Sodroski: Paramagnetic proteoliposomes containing a pure, native, and oriented seven-transmembrane segment protein, CCR5, Nat. Biotechnol. **18**, 649–654 (2000)

17.33 K.L. Martinez, B.H. Meyer, R. Hovius, K. Lundstrom, H. Vogel: Ligand binding to G protein-coupled receptors in tethered cell membranes, Langmuir **19**, 10925–10929 (2003)

17.34 Y. Fang, A.G. Frutos, J. Lahiri: Membrane protein microarrays, J. Am. Chem. Soc. **124**, 2394–2395 (2002)

17.35 Y. Fang, A.G. Frutos, B. Webb, Y. Hong, A. Ferrie, F. Lai: Membrane biochips, BioTechniques **Supplement**, 62–65 (2002)

17.36 Y. Fang, A.G. Frutos, J. Lahiri: G-protein-coupled receptor microarrays, ChemBioChem **3**, 987–991 (2002)

17.37 Y. Fang, J. Peng, A.M. Ferrie, R.S. Burkhalter: Air-stable G protein-coupled receptor microarrays and ligand binding characteristics, Anal. Chem. **78**, 149–155 (2006)

17.38 N.M. Rao, V. Silin, K.D. Ridge, J.T. Woodward, A.L. Plant: Cell membrane hybrid bilayers containing the G-protein-coupled receptor CCR5, Anal. Biochem. **307**, 117–130 (2002)

17.39 O. Karlsson, L. Stefan: Flow-mediated on-surface reconsititution of G-protein coupled receptors for applications in surface plasmon resonance biosensors, Anal. Biochem. **300**, 132–138 (2002)

17.40 G. Tollin, Z. Salamon, V.J. Hruby: Techniques: Plasmon-waveguide resonance (PWR) spectroscopy as a tool to study ligand-GPCR interactions, Trends Pharmacol. Sci. **24**, 655–659 (2003)

17.41 S. Devanathan, Z. Yao, Z. Salamon, B. Kobilka, G. Tollin: Plasmon-waveguide resonance studies of ligand binding to the human beta 2-adrenergic receptor, Biochemistry **43**, 3280–3288 (2004)

17.42 I.D. Alves, S.M. Cowell, Z. Salamon, S. Devanathan, G. Tollin, V.J. Hruby: Different structural states of the proteolipid membrane are produced by ligand binding to the human delta-opioid receptor as shown by plasmon-waveguide resonance spectroscopy, Mol. Pharmacol. **65**, 1248–1257 (2004)

17.43 T.Z. Wu: A piezoelectric biosensor as an olfactory receptor for odour detection: Electronic nose, Biosens. Bioelectron. **14**, 9–18 (1999)

17.44 G. Milligan: Principles: Extending the utility of [^{35}S]GTPγS binding assays, Trends Pharmacol. Sci. **24**, 87–90 (2003)

17.45 C. Harrison, J.R. Traynor: The [^{35}S]GTPγS binding assay: Approaches and applications in pharmacology, Life Sci. **74**, 489–508 (2003)

17.46 M. Bunemann, M. Frank, M.J. Lohse: Gi protein activation in intact cells involves subunit rearrangement rather than dissociation, Proc. Natl. Acad. Sci. USA **100**, 16077–16082 (2003)

17.47 C. Bieri, O.P. Ernst, S. Heyse, K.P. Hofmann, H. Vogel: Micropatterned immobilization of a G protein-coupled receptor and direct detection of G protein activation, Nat. Biotechnol. **17**, 1105–1108 (1999)

17.48 L.A. Sklar, J. Vilven, E. Lynam, D. Neldon, T.A. Bennett, E. Prossnitz: Solubilization and display of G protein-coupled receptors on beads for real-time fluorescence and flow cytometric analysis, BioTechniques **28**, 975–976 (2000)

17.49 P.C. Simons, M. Shi, T. Foutz, D.F. Cimino, J. Lewis, T. Buranda: Ligand-receptor-G-protein molecular assemblies on beads for mechanistic studies and screening by flow cytometry, Mol. Pharmacol. **64**, 1227–1238 (2003)

17.50 A. Waller, P.C. Simons, S.M. Biggs, B.S. Edwards, E.R. Prossnitz, L.A. Sklar: Techniques: GPCR assembly, pharmacology and screening by flow cytometry, Trends Pharmacol. Sci. **25**, 663–669 (2004)

17.51 T.A. Bennett, T.A. Key, V.V. Gurevich, R. Neubig, E.R. Prossnitz, L.A. Sklar: Real-time analysis of G protein-coupled receptor reconstitution in a solubilized system, J. Biol. Chem. **276**, 22453–22460 (2001)

17.52 X. Michalet, F.F. Pinaud, L.A. Bentolila, J.M. Tsay, S. Doose, J.J. Li: Quantum dots for live cells, in vivo imaging, and diagnostics, Science **307**, 538–544 (2005)

18. Microfluidic Devices and Their Applications to Lab-on-a-Chip

Chong H. Ahn, Jin-Woo Choi

Various microfluidic components and their characteristics, along with the demonstration of two recent achievements of lab-on-chip systems are reviewed and discussed. Many microfluidic devices and components have been developed during the past few decades, as introduced earlier for various applications. The design and development of microfluidic devices still depend on the specific purposes of the devices (actuation and sensing) due to a wide variety of application areas, which encourages researchers to develop novel, purpose-specific microfluidic devices and systems. Microfluidics is the multidisciplinary research field that requires basic knowledge in fluidics, micromachining, electromagnetics, materials, and chemistry for various applications.

Among the various application areas of microfluidics, one of the most important is the lab-on-a-chip system. Lab-on-a-chip is becoming a revolutionary tool for many different applications in chemical and biological analyses due to its fascinating advantages (fast speed and low cost) over conventional chemical or biological laboratories. Furthermore, the simplicity of lab-on-a-chip systems will enable self-testing

18.1	**Materials for Microfluidic Devices and Micro/Nanofabrication Techniques** ... 504
	18.1.1 Silicon 504
	18.1.2 Glass 505
	18.1.3 Polymer 505
18.2	**Active Microfluidic Devices** 507
	18.2.1 Microvalves 508
	18.2.2 Micropumps 510
18.3	**Smart Passive Microfluidic Devices** 513
	18.3.1 Passive Microvalves 513
	18.3.2 Passive Micromixers..................... 515
	18.3.3 Passive Microdispensers 517
	18.3.4 Microfluidic Multiplexer Integrated with Passive Microdispenser 517
	18.3.5 Passive Micropumps 519
	18.3.6 Advantages and Disadvantages of the Passive Microfluidic Approach . 520
18.4	**Lab-on-a-Chip for Biochemical Analysis**......................... 520
	18.4.1 Magnetic Micro/Nano-Bead-Based Biochemical Detection System....... 521
	18.4.2 Disposable Smart Lab-on-a-Chip for Blood Analysis 523
References 527

capability for patients or health consumers by overcoming space limitations.

Microfluidics covers the science of fluidic behaviors on the micro/nanoscales and the design engineering, simulation, and fabrication of fluidic devices for the transport, delivery, and handling of fluids on the order of microliters or smaller volumes. It is the backbone of biological or biomedical microelectromechanical systems (bioMEMS) and lab-on-a-chip concept, as most biological analyses involve fluid transport and reaction. Biological or chemical reactions on the micro/nanoscale are usually rapid since small amounts of samples and reagents are used, which offers quick and low-cost analysis.

A fluidic volume of 1 nl can be understood as the volume in a cube surrounded by 100 μm in each direction. It is much smaller than the size of a grain of table salt. Microfluidic devices and systems handle sample fluids in this range for various applications, including inkjet printing, blood analysis, biochemical detection, chemical synthesis, drug screening/delivery, protein analysis, DNA sequencing, and so on.

Microfluidic systems consist of microfluidic platforms or devices for fluidic sampling, control, monitoring, transport, mixing, reaction, incubation, and analysis. To construct microfluidic systems, or labs-on-a-chip, microfluidic devices must be functionally integrated on a microfluidic platform using proper micro/nanofabrication techniques. In this chapter, the basics of microfluidic devices and their applications to lab-on-a-chip are briefly reviewed and summarized. Basic materials and fabrication techniques for microfluidic devices will be introduced first and various active and passive microfluidic components will be described. Then, their applications to lab-on-a-chip, or biochemical analysis, will be discussed.

18.1 Materials for Microfluidic Devices and Micro/Nanofabrication Techniques

Various materials are being used for the fabrication of microfluidic devices and systems. Silicon is one of the most popular materials in micro/nanofabrication because its micromachining has been well established over a period of decades. In general, the advantages of using silicon as a substrate or structural material include good mechanical properties, excellent chemical resistance, well-characterized processing techniques, and the capability for integration of control/sensing circuitry in the semiconductor. Other materials such as glass, quartz, ceramics, metals, and polymers are also being used for substrates and structures in micro/nanofabrication, depending on the application. Among these materials, polymers or plastics have recently become one of the more promising materials for lab-on-a-chip applications, due to their excellent material properties for biochemical fluids and their low-cost manufacturability. The main issues in the fabrication techniques of microfluidic devices and systems usually lie in forming microfluidic channels, which are key micro/nanostructures of lab-on-a-chip. In this section, the basic micro/nanofabrication techniques for silicon, glass, and polymers are described.

18.1.1 Silicon

Microfluidic channels on silicon substrates are usually formed either by wet (chemical) etching or by dry (plasma) etching. Crystalline silicon has a preferential etch direction, depending on which crystalline plane is exposed to an etchant. Etch rate is slowest in the (111) crystalline planes – approximately 100 : 1 anisotropic etch rate compared with (100) : (111) or (110) : (111). Potassium hydroxide (KOH), tetramethyl ammonium hydroxide (TMAH), and ethylene diamine pyrocatechol (EDP) are commonly used silicon anisotropic etchants. In most cases, silicon dioxide (SiO_2) or silicon nitride (Si_3N_4) is used as a masking material during the etching process. Anisotropic etchants and basic etching mechanisms are summarized by *Ristic* et al. [18.1]. There is also an isotropic wet-etching process available using a mixture of hydrofluoric acid (HF), nitric acid (HNO_3), and acetic acid (CH_3COOH): the so-called hydrofluoric-nitric-acetic (HNA) etch. HNA etches in all directions with almost the same etch rate regardless of crystalline directions. Figure 18.1 illustrates wet anisotropic and isotropic etching profiles.

Reactive ion etching (RIE) is also one of the most commonly used dry-etching processes to generate microfluidic channels or deep trench structures on silicon substrate. In this dry-etching technique, radio-frequency (RF) energy is used to excite ions in a gas to an energetic state. The energized ions supply the necessary energy to generate physical and chemical reactions on the exposed area of the substrate, which starts the etching process. RIE can generate strong anisotropic, as well as isotropic profiles, depending on the gases used, the condition of plasma, and the applied power. Further information on reactive ion etching process, including deep reactive ion etching (DRIE), on silicon substrates can be found in the literature [18.2–5].

Many microfluidic devices have been realized using silicon as a substrate material, including microvalves and micropumps, which are covered in the next section.

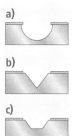

Fig. 18.1a–c Wet etching of silicon substrate for anisotropic and isotropic etching: (a) isotropic etching profile, (b) anisotropic etching profile (long term), and (c) anisotropic etching profile (short term)

18.1.2 Glass

Glass substrate has been widely used for the fabrication of microfluidic systems and lab-on-a-chip due to its excellent optical transparency and ease of electroosmotic flow (EOF). Chemical wet etching and thermal fusion bonding are the common fabrication techniques for glass substrate. Chemical wet etching and the bonding technique have also been widely reported [18.6, 7]. The most commonly used etchants are hydrofluoric acid (HF), buffered hydrofluoric acid, and a mixture of hydrofluoric acid, nitric acid, and deionized water (HF, HNO_3, H_2O). Gold with an adhesion layer of chrome is most often used as an etch mask for wet etching of a glass substrate. Since glass has no crystalline structure, only isotropic etch profiles are obtained, such as forming a hemispherical-shaped channel. Often the problem is that stresses within the surface layers of the glass cause preferential etching, and scratches created by polishing or handling errors cause spikes to be etched in the channels. Pre-etching is one method to release stress, which causes defects in channels after etching. Another way to improve the channel-etching process is to anneal the glass wafers before etching. Annealing can be done close to the glass-transition temperature for at least a couple of hours. Figure 18.2 shows two examples of poorly etched channels without the pre-etching and annealing steps compared with an etched channel without defects. Other fabrication techniques for glass substrate include photoimageable glass, as reported by *Dietrich* et al. [18.8]. Anisotropy is introduced into glass by making the glass photosensitive using lithium/aluminum/silicates in its composition.

One of the most successful examples of using glass as a substrate material in lab-on-a-chip applications is the capillary electrophoresis (CE) chip, which is fabricated using the glass etching and fusion bonding techniques, since the most advantageous property of glass is its excellent optical transparency, which is required for most lab-on-a-chip applications that use optical detection, including capillary electrophoresis microchips [18.6, 7, 9].

18.1.3 Polymer

Among the various substrates available for lab-on-a-chip, polymers, or plastics, have recently become one of the most popular and promising substrates due to their low cost, ease of fabrication, and favorable biochemical reliability and compatibility. Plastic substrates, such as polyimide, poly(methyl methacrylate) (PMMA), poly(dimethyl siloxane) (PDMS), polyethylene or polycarbonate, offer a wide range of physical and chemical material parameters for the applications of lab-on-a-chip, generally at low cost using replication approaches. Polymers and plastics are promising materials in microfluidic and lab-on-a-chip applications because they can be used for mass production using casting, hot embossing, and injection-molding techniques. This mass-production capability allows the successful commercialization of lab-on-a-chip technology, including disposable lab-on-a-chip. While several fabrication methods have recently been developed, three fabrication techniques – casting, hot embossing, and injection molding – are major techniques of great interest. Figure 18.3 shows schematic illustrations of these polymer microfabrication techniques.

For polymer or plastic micro/nanofabrication, a mold master is essential for replication. Mold masters are fabricated using photolithography, silicon/glass bulk etching, and metal electroplating, depending on the application. Figure 18.4 summarizes the mold masters from different fabrication techniques.

Photolithography, including LIGA (from the German words Lithografie, Galvanoformung, Abformung, meaning lithography, electroforming and molding) [18.10] and UV-LIGA (ultraviolet LIGA) [18.11, 12], is used to fabricate mold masters for casting or

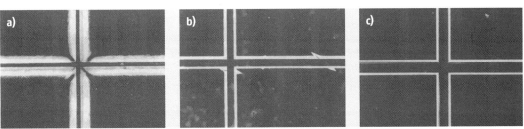

Fig. 18.2a–c Isotropically etched microfluidic channels on glass substrate: (**a**) poorly etched channel with large underetching, (**b**) poorly etched channel with spikes, and (**c**) well-etched channel without any defects

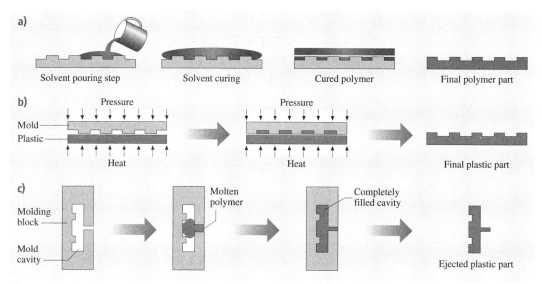

Fig. 18.3a–c Concept of polymer micro/nanofabrication techniques: (**a**) casting, (**b**) hot embossing, and (**c**) injection molding

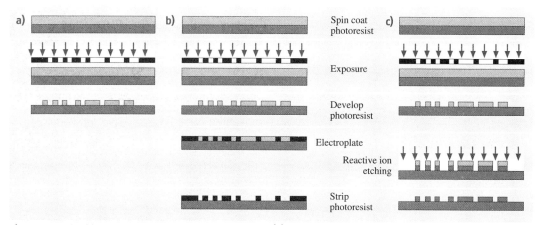

Fig. 18.4a–c Mold masters in polymer/plastic fabrication: (**a**) photolithography-based mold master, (**b**) silicon-based mold master, and (**c**) mold master by electroplating

soft lithography replication, while silicon-based and electroplated mold masters are used for hot-embossing replication. For injection molding, electroplated metallic mold masters are preferable.

Casting or soft lithography [18.13, 14] usually offers flexible access to microfluidic structures using mostly poly(dimethyl siloxane) (PDMS) as a casting material. A mixture of the elastomer precursor and curing agent is poured over a master mold structure. After curing, the replicated elastomer is released from the mold master, having transferred a reverse structure of the mold master. Patterns of a few nm can be achieved using this technique.

While casting can be carried out at room temperature, hot embossing requires a slightly higher temperature – up to the glass-transition temperature of the plastic substrate to be replicated. The hot-embossing technique has been developed by several

Table 18.1 Overview of the different polymer micro/nanofabrication techniques

Fabrication type	Casting	Hot embossing	Injection molding
Investment	Low	Moderate	High
Manufacturability	Low	Moderate	High
Cycle time	8–10 h	1 h	1 min
Polymer choices	Low	Moderate	Moderate
Mold replication	Good	Good	Good
Reusability of mold	No (photolithography-based molds)	Yes	Yes

research groups [18.15–17]. A mold master is placed in the chamber of a hot-embossing system with the plastic substrate, then heated plates press both the plastic substrate and the mold master, as illustrated in Fig. 18.3b. After a certain amount of time (typically 5–20 min, depending on the plastic substrate), the plates are cooled down to release the replicated plastic substrate. Hot embossing offers mass production of polymer microstructures, as its cycle time is less than an hour.

Injection molding is a technique to fabricate polymer microstructures at low cost and high volumes [18.18]. A micromachined mold master is placed in the molding block of the injection molding machine, as illustrated in Fig. 18.3c. The plastic in granular form is melted and then injected into the cavity of a closed mold block, where the mold master is located. The molten plastic continues to flow into and fill the mold cavity until the plastic cools down to a highly viscous melt, and a cooled plastic part is ejected. In order to ensure good flow properties during the injection molding process, thermoplastics with low or medium viscosity are desirable. The filling of the mold cavity, and subsequently the microstructures, depend on the viscosity of the plastic melt, injection speed, molding-block temperature, and nozzle temperature of the injection unit. This technique allows very rapid replication and high-volume mass production. The typical cycle time is several seconds for most applications. However, due to the high shear force on the mold master inside the mold cavity during injection molding, metallic mold masters are highly recommended. Poly(methyl methacrylate) (PMMA), polyethylene (PE), polystyrene (PS), polycarbonate (PC), and cyclic olefin copolymers (COC) are common polymer/plastic materials for both hot embossing and injection molding replication.

All of these polymer/plastic replication techniques are summarized in Table 18.1. Since each technique differs from the others, fabrication techniques and materials have to be selected according to the application.

18.2 Active Microfluidic Devices

Microfluidic devices are essential for the development of lab-on-a-chip or micro-total analysis systems (mTAS). A number of different microfluidic devices have been developed with basic structures analogous to macroscale fluidic devices. Such devices include microfluidic valves, microfluidic pumps, microfluidic mixers, etc. The devices listed above have been developed both as active and passive devices. While passive microfluidic devices are generally easy to fabricate, they do not offer the same functional diversity that the active microfluidic devices provide. For example, passive microvalves based on the surface tension effect [18.19] can operate a few times to hold the liquid, which is acceptable for a disposable format. Once the air–liquid interface passes over the valve mechanism, the operation characteristics of the valve will differ, due to the change of surface energy over the channel. Similarly, passive check valves [18.20–22] are dependent on the pressure of the fluid for operation. Since the active microvalves can be triggered on/off depending on an external signal regardless of the status of the fluid system, there has been considerable research effort to develop active microfluidic devices. However, active devices are usually more expensive due to their desired functional and fabrication complexity.

This section reviews some of the active microfluidic devices, such as microvalves, micropumps, and active microfluidic mixers. Passive counterparts of these microfluidic devices will be discussed in the next section.

18.2.1 Microvalves

Active microvalves have been an area of intense research over the past decade, and a number of novel design and actuation schemes have been developed. This makes the categorization of active microvalves a confusing enterprize. Classification schemes for active microvalves include:

1. Fluidics handled: liquid/gas/liquid and gas
2. Materials used for the structure: silicon/polysilicon/glass/polymer
3. Actuation mechanisms: electrostatic/pneumatic/thermopneumatic, etc.
4. Physical actuating microstructures: membrane-type, flap-type, ball valve, etc.

All of the classification schemes listed above are valid, but the most commonly used method [18.23, 24] is the classification based on the actuation mechanisms.

In this section, the various microvalves are discussed in terms of their actuation mechanisms and their relevance to the valve mechanism, as well as fluid handled and special design criteria.

Pneumatically/Thermopneumatically Actuated Microvalves

Pneumatic actuation uses an external air line (or pneumatic source) to actuate a flexible diaphragm. Pneumatic actuation offers such attractive features as high force, high displacement, and rapid response time. Figure 18.5 illustrates a schematic concept of pneumatically actuated microvalves. *Schomburg* et al. [18.25] demonstrated pneumatically actuated microvalves.

Pneumatically actuated microvalves have also been demonstrated using polymeric substrate. *Hosokawa* and *Maeda* [18.26] demonstrated a pneumatically actuated three-way microvalve system using a PDMS platform. The microfluidic lines and the pneumatic lines are fabricated on separate layers.

Thermopneumatic actuation is typically performed by heating a fluid (usually a gas) in a confined cavity, as illustrated in Fig. 18.6. The increase in temperature leads to a rise in the pressure of the gas, and this pressure is used to deflect a membrane for valve operation. Thermopneumatic actuation is an inher-

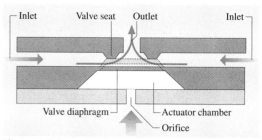

Fig. 18.5 Schematic concept of pneumatically actuated microvalve

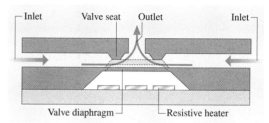

Fig. 18.6 Schematic concept of thermopneumatically actuated microvalve

ently slow technique but offers very high forces when compared to other techniques [18.23]. Thermopneumatically actuated microvalves have been realized by many researchers using various substrates and diaphragm materials [18.27–29].

Electrostatically Actuated Microvalves

Electrostatic actuation has been widely explored for a number of applications, including pressure sensors, comb drives, active mirror arrays, etc. Electrostatically actuated devices typically have a fairly simple structure and are easy to fabricate. A number of fabrication issues, such as stiction and release problems of membranes and valve flaps, need to be addressed to realize practical electrostatic microvalves. *Sato* and *Shikida* [18.30] have developed a novel membrane design in which the deflection *propagates* through the membrane, rather than deforming it entirely. Figure 18.7 shows a schematic sketch of the actuation mechanism and valve design. The use of this S-shaped design allows them to have relatively large gaps across the two surfaces, as the electrostatic force need only be concentrated at the edges of the S-shape where the membrane is deflected. *Robertson* and *Wise* [18.31] have developed an array of electrostatically actuated valves using a flap design (rather than a membrane) to

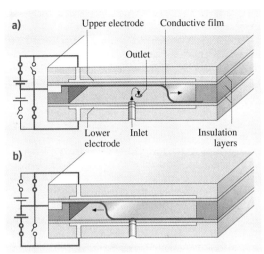

Fig. 18.7a,b Electrostatically actuated microvalve with an S-shaped film element: propagation of bend in the film as (a) open and (b) closed (after [18.30], © IOP Publishing Limited)

seal the fluid flow. The demonstrated system is suitable for very-low-pressure gas control systems such as those needed in a clean-room environment.

Wijngaart et al. [18.33] have developed a high-stroke high-pressure electrostatic actuator for valve applications. This reference provides a good overview of the theoretical design parameters used to design and analyze an electrostatically actuated microvalve.

Piezoelectrically Actuated Microvalve

Piezoelectric actuation schemes offer a significant advantage in terms of operating speed; they are typically the fastest actuation scheme at the expense of a reduced actuator stroke. Also, piezoelectric materials are more challenging to incorporate into fully integrated MEMS devices. *Watanabe* and *Kuwano* [18.34] and *Stehr* et al. [18.35] demonstrated piezoelectric actuators for valve applications. A film of piezoelectric material is deposited on the movable membrane, and upon application of an electric potential, a small deformation occurs in the piezo film that is transmitted to the valve membrane.

Electromagnetically Actuated Microvalves

Electromagnetic actuators are typically capable of delivering high force and range of motion. A significant advantage of electromagnetic microvalves is that they are relatively insensitive to external interference. How-ever, electromagnetic actuators involve a fairly complex fabrication process. Usually, a soft electromagnetic material such as NiFe (nickel iron, also known as permalloy) is used as a membrane layer, and an external electromagnet is used to actuate this layer. *Sadler* et al. [18.32] have developed a microvalve using the electromagnetic actuation scheme shown in Fig. 18.8. They have demonstrated a fully integrated magnetic actuator with magnetic interconnection vias to guide the magnetic flux. The valve seat design, also shown in Fig. 18.8, allows for very intimate contact between the NiFe valve membrane and the valve seat, hence, achieving an ultra-low leak rate when the valve is closed.

Jackson et al. [18.36] demonstrate an electromagnetic microvalve using *magnetic* PDMS as the membrane material. For this application, the PDMS pre-polymer is loaded with soft magnetic particles and then cured to form the valve membrane. The PDMS membrane is then assembled over the valve body, and miniature electromagnets are used to actuate the membrane.

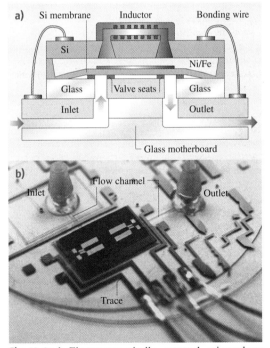

Fig. 18.8a,b Electromagnetically actuated microvalves: (a) schematic illustration and (b) photograph of the electromagnetically actuated microvalve as a part of lab-on-a-chip (after [18.32])

Other Microvalve Actuation Schemes

Microvalves have most commonly been implemented with one of the actuation schemes listed above. However, these are not the only actuation schemes that are used for microvalves. Some others include the use of (shape memory alloys) (SMA) [18.37], electrochemical actuation [18.38], etc. SMA actuation schemes offer the advantage of generating very large forces when the SMA material is heated to its original state. *Neagu* et al. [18.38] presented an electrochemically actuated microvalve. In their device, an electrolysis reaction is used to generate oxygen in a confined chamber. This chamber is sealed by a deformable membrane that is deflected due to increased pressure. The reported microvalve has a relatively fast actuation time and can generate very high pressures. *Yoshida* et al. [18.39] present a novel approach to the microvalve design: a micro-electrorheological valve (ER valve). An electrorheological fluid is loaded into the microchannel, and, depending on the strength of an applied electric field, the viscosity of the ER fluid changes considerably. A higher viscosity is achieved when an electric field is applied perpendicular to the flow direction. This increased viscosity leads to a drop in the flow rate, allowing the ER fluid to act as a valve. Of course, this technique is limited to fluids that can exhibit such properties; nevertheless, it provides a novel idea to generate on-chip microvalves.

The ideal characteristics for a microvalve are listed by *Kovacs* [18.24]. However, of all the microvalves listed above, none can satisfy all the criteria. Thus, microvalve design, fabrication, and utility are highly application-specific and most microvalves try to generate the performance characteristics that are most useful for the intended application.

18.2.2 Micropumps

One of the most challenging tasks in developing a fully integrated microfluidic system has been the development of efficient and reliable micropumps. On the macroscale, a number of pumping techniques exist, such as peristaltic pumps, vacuum-driven pumps, Venturi-effect pumps, etc. However, in microscale, most mechanical pumps rely on pressurizing the working fluid and forcing it to flow through the system. Practical vacuum pumps are not available on the microscale. There are also some effects such as electroosmotic pumping, that are only possible on the microscale. Consequently, electrokinetic driving mechanisms have also been widely studied for microfluidic pumping applications.

In the previous section, various microvalves that are used for microfluidic control were discussed. They were presented based on the actuation schemes that they employ. A similar classification can also be adopted for micropumps. However, rather than using the same classification scheme, the micropumps are categorized based on the type of microvalve mechanism used as part of the pumping mechanism. Broadly, the mechanical micropumps can be classified as check-valve-controlled microvalves or diffuser pumps. Either mechanism can use various actuation schemes such as electrostatic, electromagnetic, piezoelectric, etc. Pneumatic control is another actuation scheme for micropumping. A peristaltic micropump based on multilayer soft lithography of elastomers is a remarkable example of pneumatically controlled micropumps. Micropumps driven by direct electrical control form a separate category and are discussed following the mechanical micropumps.

Micropumps Using a Check-Valve Design

Figure 18.9 shows a typical mechanical micropump with check valves. This pump consists of an inlet and an outlet check valve with a pumping chamber in between. A membrane is deflected upwards, and a low-pressure zone is created in the pumping chamber. This forces the inlet check valve open, and fluid is sucked into the pumping chamber. As the membrane returns to its original state and continues to travel downwards, a positive pressure builds up, which seals the inlet valve while simultaneously opening the outlet valve. The fluid is then ejected and the pump is ready for another cycle.

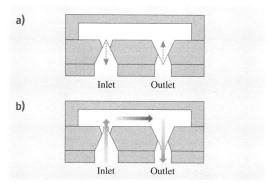

Fig. 18.9a,b Operation of a micropump using a check-valve design: (**a**) check valves are closed and (**b**) check valves are open

A number of other techniques have been used to realize mechanical micropumps with check valves and a pumping chamber. *Jeon* et al. [18.40] present a micropump that uses PDMS flap valves to control the pumping mechanism, *Koch* et al. [18.41], *Cao* et al. [18.42], *Park* et al. [18.43], and *Koch* et al. [18.44] present piezoelectrically driven micropumps. *Xu* et al. [18.45] and *Makino* et al. [18.46] present SMA-driven micropumps. *Chou* et al. [18.47] present a novel rotary pump using a soft-lithography approach. As explained in the active valve section, valves can be created on a PDMS layer using separate liquid and air layers. When a pressure is applied to the air lines, the membranes deflect to seal the fluidic path. *Chou* et al. [18.47] have implemented a series of such valves in a loop. When they are deflected in a set sequence, the liquid within the ring is pumped by peristaltic motion.

An interesting approach toward the development of a bidirectional micropump has been used by *Zengerle* et al. [18.48]. Their design has two flap valves at the inlet and outlet of the micropump, which work in the forward mode at low actuation frequencies and in the reverse direction at higher frequencies. *Zengerle* et al. [18.48] attribute the change in pumping direction to the phase shift between the response of the valves and the pressure difference that drives the fluid. *Carrozza* et al. [18.49] use a different approach to generate the check valves. Rather than using the conventional membrane or flap-type valves, they use ball valves by employing a stereolithographic approach. The developed pump is actuated using a piezoelectric actuation scheme.

Diffuser Micropumps

The use of nozzle-diffuser sections, or pumps with fixed valves, or pumps with dynamic valves has been extensively researched. The basic principle of these pumps is based on the idea that the geometrical structure used as an inlet valve has a preferential flow direction toward the pumping chamber, and the outlet valve structure has a preferential flow direction away from the pumping chamber. An illustrative example of this concept is shown in Fig. 18.10. As can be readily seen, these pumps are designed to work with liquids only. When the pump is in suction mode, the flow in the inlet diffuser structure is primarily directed toward the pump, and a slight back-flow occurs from the outlet diffuser (acting as a nozzle) section. When the pump is in pressure mode, the outlet diffuser sec-

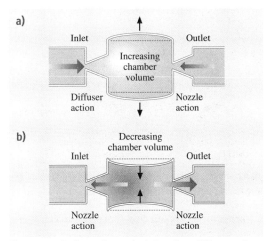

Fig. 18.10a,b Operating principle of a diffuser micropump: (**a**) suction mode and (**b**) pumping mode

tion allows most of the flow out of the pump, whereas the inlet diffuser section allows a slight back-flow. The net effect is that liquid pumping occurs from left to right.

Diffuser micropumps are simple and valveless structures that improve pumping reliability [18.50], but they cannot eliminate back-flow problems.

These micropumps can also be described as flow rectifiers, analogously to diodes in electrical systems. *Forster* et al. [18.51] have used the Tesla-valve geometry, instead of the diffuser section, and the reference presents a detailed discussion of the design parameters and operational characteristics of their fixed-valve micropumps.

Peristaltic Micropumps
Using Multilayer Soft Lithography

A flexible substrate and actuating diaphragm can be utilized to develop a micropump to reduce power consumption and to increase actuating ranges. An interesting micropumping device was recently reported using multilayer monolithic soft lithography and assembly [18.52]. Multilayer soft lithography utilizes sequence soft lithography cast-molding and bonding steps to generate a microfluidic system. In realizing micropumps, the dead volume and sealing issues have been the most cumbersome tasks. The reported micropumping device based on multilayer soft lithography shown successfully addressed the dead volume and sealing issues as illustrated in Fig. 18.11. There are a fluidic layer and a control layer. When air pressure

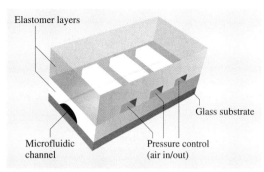

Fig. 18.11 Schematic illustration of a peristaltic micropump based on multilayer soft lithography

is applied to the control line, a thin elastomer membrane collapses and blocks the flow line working as a microvalve. Integrating three microvalves, a peristaltic micropump can be formed with a proper pressure control on the three microvalves. Multichannel pressure control is necessary to operate the microvalve and micropump. The technique provides rapid prototyping, ease of fabrication, minimized dead volume, and excellent sealing. The concept was applied to a large-scale microfluidic platform [18.53], a parallel microfluidic analysis system [18.54], a three-dimensional microfluidic system [18.55], and so on.

Electric/Magnetic-Field-Driven Micropumps

An electric field can be used to directly pump liquids in microchannels using such techniques as electroosmosis (EO), electrohydrodynamic (EHD) pumping, magnetohydrodynamic (MHD) pumping, etc. These pumping techniques rely on creating an attractive force for some of the ions in the liquid, and the remaining liquid is dragged along to form a bulk flow.

When a liquid is introduced into a microchannel, a double-layer charge exists at the interface of the liquid and the microchannel wall. The magnitude of this charge is governed by the zeta potential of the channel–liquid pair. Figure 18.12 shows a schematic view of the electroosmotic transport phenomenon.

As shown in Fig. 18.12, the channel wall is negatively charged, which attracts the positive ions in the solution. When a strong electric field is applied along the length of the microchannel, the ions at the interface experience an attractive force toward the cathode. As the positive ions move toward the cathode, they exert a drag force on the bulk fluid, and net fluid transport occurs from the anode to the cathode. It is interesting to note that the flow profile of the liquid plug is significantly different from pressure-driven flow. Unlike the parabolic flow profile of pressure-driven flow, electroosmotic transport leads to an almost vertical flow profile. The electroosmotic transport phenomenon is only effective across very narrow channels.

EHD can be broadly broken down into two subcategories: injection type and induction type. In injection-type EHD, a strong electric field ($\approx$ 100 kV/cm) is applied across a dielectric liquid. This induces charge formation in the liquid, and these induced (or injected) charges are then acted upon by the electrical field for pumping. Induction-type EHD relies on generating a gradient/discontinuity in the conductivity and/or permittivity of the liquid. *Fuhr* et al. [18.56] explain various techniques that can be used to generate gradients for noninjection-type EHD pumping.

MHD pumps rely on creating a Lorentz force on the liquid particles in the presence of an externally applied electric field. MHD has been demonstrated using both direct-current (DC) [18.57] and alternating-current (AC) [18.58] excitations. MHD, EHD, and electroosmotic transport share one feature in common that makes them very appealing for microfluidic systems, namely that none of them requires microvalves to regulate the flow. This makes these pumping techniques very reliable, as there is no concern about wear and tear on the microvalves, or any other moving parts of the micropump. However, it has been difficult to implement these actuation schemes fully on the microscale, owing to the high voltages, electromagnets, etc. required for these actuation schemes.

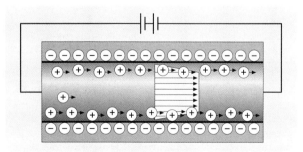

Fig. 18.12 Schematic sketch explaining the principle of electroosmotic fluid transport

18.3 Smart Passive Microfluidic Devices

Passive microfluidics is a powerful technique for the rapidly evolving discipline of bioMEMS. It is a fluid control topology in which the physical configuration of the microfabricated system primarily determines the functional characteristics of the device/system. Typically, passive microfluidic devices do not require an external power source, and the control exerted by the devices is based, in part, on energy drawn from the working fluid, or based purely on surface effects, such as surface tension, selective hydrophobic/hydrophilic control, etc. Most passive microfluidic devices exploit various physical properties such as shape, contact angle, and flow characteristics to achieve the desired function. Passive microfluidic systems are usually easier to implement and allow for a simple microfluidic system with little or no control circuitry. A further list of advantages and disadvantages of passive microfluidic systems (or devices) is considered toward the end of this section.

Passive microfluidic devices can be categorized based on:

- Function: microvalves, micromixers, filters, reactors, etc.
- Fluidic medium: gas or liquid
- Application: biological, chemical, or other
- Substrate material: silicon, glass, polysilicon, polymer, or others.

In this section, we will study various passive microfluidic devices that are categorized based on their function. Passive microfluidic devices include, but are not limited to, microvalves, micromixers, filters, dispensers, etc. [18.24].

18.3.1 Passive Microvalves

Passive microvalves have been a subject of great interest ever since the inception of the lab-on-a-chip concept. Microvalves are a key component of any microfluidic system and are essential for fluidic sequencing operations. Since most chemical and biochemical reactions require about five to six reaction steps, passive microvalves with limited functionality are ideally suited for such simple tasks. Passive microvalves can be broadly categorized as follows:

- Silicon/polysilicon or polymer-based check valves
- Passive valves based on surface tension effects
- Hydrogel-based biomimetic valves.

Passive Check Valves

Shoji and *Esashi* [18.23] provide an excellent review of check-type passive microvalves. Some of the valves, shown in Fig. 18.13, illustrate the various techniques that can be used to fabricate check valves.

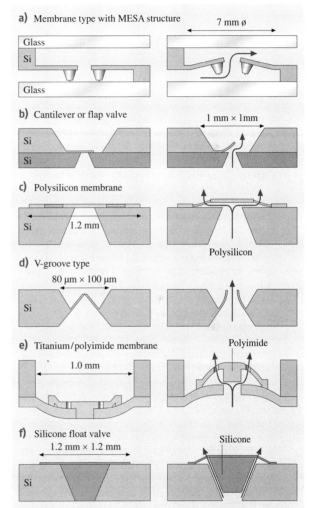

Fig. 18.13a–f Various types of passive microvalve designs: (**a**) membrane type with a mesa structure; (**b**) cantilever or flap valve; (**c**) polysilicon membrane; (**d**) V-groove type; (**e**) titanium/polyimide membrane; and (**f**) silicone float valve (after [18.23], © IOP Publishing Limited)

Figure 18.13a shows a microvalve fabricated using silicon bulk-etching techniques. A through-hole (pyramidal cavity) is etched through a silicon wafer that is sandwiched between two glass wafers. The normally closed valve is held in position by the spring effect of the silicon membrane. Upon applying pressure to the lower fluidic port, the membrane deflects upwards, allowing fluid flow through the check valve. The same working principle is employed by the microvalve shown in Fig. 18.13b. However, instead of using a membrane supported on all sides, a cantilever structure is used for the flap. This reduces the burst pressure, i.e., the minimum pressure required to open the microvalve. Figure 18.13c shows that the membrane structure can also be realized using a polysilicon layer deposited on a bulk-etched silicon wafer. Polysilicon processes typically allow tighter control over dimensions and, consequently, offer more repeatable operating characteristics. Figure 18.13d shows the simplest type of check valve where the V-groove etched in a bulk silicon substrate acts as a check valve. However, the low contact area between the flaps of the microvalve leads to nontrivial leakage rates in the forward direction. Figure 18.13e and f show check-valve designs that are realized using polymer/metal films, in addition to the traditional glass/silicon platform. This technique offers a significant advantage in terms of biocompatibility characteristics and controllable operating characteristics. The surface properties of polymers can be easily tailored using a wide variety of techniques such as plasma treatment and surface adsorption [18.59]. Thus, in applications for which the biocompatibility requirements are very stringent, it is preferable to have polymers as the fluid-contacting material. Furthermore, polymer properties such as stiffness can also be controlled in some cases based on the composition and/or processing conditions. Thus, it may be possible to fabricate microvalves with different burst pressures by using different processing conditions for the same polymer.

In addition to the passive check valves reviewed by *Shoji* and *Esashi* [18.23], other designs include check valves using composite titanium/polyimide membranes [18.20, 21], polymeric membranes such as Mylar or KAPTON [18.60], or PDMS [18.40], and metallic membranes such as [18.22].

Terray et al. [18.61] present an interesting approach to fabricating ultrasmall passive valve structures. They have demonstrated a technique to polymerize colloidal particles into linear structures using an optical trap to form microscale particulate valves.

Passive Valves Based on Surface Tension Effects

The passive valves listed in the previous section use the forced motion of the membrane or flap to control the flow of fluids. These valves are prone to such problems as clogging and mechanical wear and tear. Passive valves based on surface tension effects, on the other hand, have no moving parts and control the fluid motion based on their physical structure and the surface property of the substrate. Figure 18.14 shows a schematic sketch of a passive microvalve on a hydrophobic substrate [18.19].

The Hagen–Poiseuille equation for laminar flow governs the pressure drops in microfluidic systems with laminar flow. For a rectangular channel with a high width-to-height ratio, the pressure drop is governed by the equation

$$\Delta P = \frac{12L\mu Q}{wh^3}, \quad (18.1)$$

where L is the length of the microchannel, μ is the dynamic viscosity of the fluid, Q is the flow rate, and w and h are the width and height of the microchannel, respectively. Varying L or Q can control the pressure drop for a given set of w and h.

An abrupt change in the width of the channel causes a pressure drop at the point of restriction. For a hydrophobic channel material, an abrupt decrease in channel width causes a positive pressure drop

$$\Delta P_2 = 2\sigma_l \cos(\theta_c)\left[\left(\frac{1}{w_1}+\frac{1}{h_1}\right)-\left(\frac{1}{w_2}+\frac{1}{h_2}\right)\right], \quad (18.2)$$

where σ_l is the surface tension of the liquid, θ_c is the contact angle, and w_1, h_1 and w_2, h_2 are the width and height of the two sections before and during the restriction, respectively. Setting h as constant through the

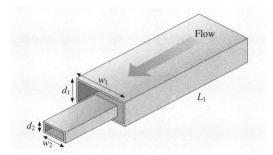

Fig. 18.14 Structure of a passive microvalve based on surface tension effects (after [18.19])

system, ΔP_2 can be varied by adjusting the ratio of w_1 and w_2. *Ahn* et al. [18.19] have proposed and implemented a novel structurally programmable microfluidic system (sPROMs) based on the passive microfluidic approach. In short, the sPROMs system consists of a network of microchannels with passive valves of the type shown in the passive valve section. If the pressure drop of the passive valves is set to be significantly higher than the pressure drop of the microchannel network, then the position of the liquid in the microchannel network can be controlled accurately. By applying sequentially higher pressure pulses, the liquid is forced to move from one passive valve to another. Thus, the movement of the fluid within the microfluidic channels is *programmed* using the physical structure of the microfluidic system, and this forms the basic idea behind sPROMs.

The abrupt transition from a wide channel to a narrow channel can also be affected along the height of the microchannel. Furthermore, the passive valve geometry shown in Fig. 18.14 is not exclusive. *Puntambekar* et al. [18.62] have demonstrated different geometries of passive valves, as shown in Fig. 18.15. The graph shows that the various geometries of the passive valves in Fig. 18.15 can act as effective passive valves without having an abrupt transition. This is important in order to avoid the dead volume that is commonly encountered across an abrupt step junction.

The use of surface tension to control the operation of passive valves is not limited to hydrophobic substrates. *Madou* et al. [18.63] demonstrate a capillarity-driven stop valve on a hydrophilic substrate. On a hydrophilic substrate, the fluid can easily *wick through* in the narrow region. However, at the abrupt transition to a larger channel section, the surface tension effects will not allow the fluid to leave the narrow channel. Thus, in this case, the fluid is held at the transition from the narrow capillary to the wide outlet channel.

Another mechanism to implement passive valves is the use of hydrophobic patches on a normally hydrophilic channel. *Handique* et al. [18.64] demonstrate the use of this technique to implement passive valves for a DNA analysis system. The fluid is sucked into the microfluidic channels via capillary suction force. The hydrophobic patch exerts a negative capillary pressure that stops further flow of the fluid. The use of hydrophobic patches as passive valves is reported by *Andersson* et al. [18.65].

Other Passive Microvalves

A novel approach to realizing passive valves that are responsive to their surrounding environment is demonstrated by *Low* et al. [18.66] and *Yu* et al. [18.67]. *Yu* et al. [18.67] have developed customized polymer *cocktails* that are polymerized in situ around prefabricated posts. The specialized polymer is selectively responsive to stimuli such as pH, temperature, electric fields, light, carbohydrates, and antigens.

Another interesting approach in developing passive valves has been adopted by *Forster* et al. [18.51]. They have developed the so-called no-moving-part (NMP) valves, which are based on a physical configuration that allows a higher flow rate along one direction compared to the reverse direction.

18.3.2 Passive Micromixers

The successful implementation of microfluidic systems for many lab-on-a-chip systems is partly owing to the significant reduction in volumes handled by such sys-

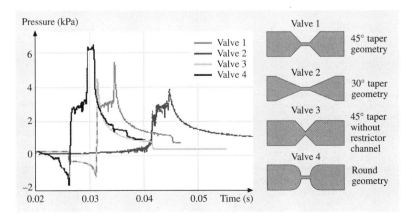

Fig. 18.15 Analysis of different geometries of passive valves (after [18.62], © The Royal Society of Chemistry)

tems. This reduction in volume is made possible by the use of microfabricated features and channel dimensions ranging from a few to several hundred μm. Despite the advantage offered by the μm-sized channels, one of the significant challenges has been the implementation of effective microfluidic mixers on the microscale. Mixing on the macroscale is a turbulent-flow-regime process. However, on the microscale, because of the low Reynolds numbers as a result of the small channel dimensions, most flow streams are laminar in nature, which does not allow for efficient mixing. On the other hand, diffusion is an important factor in mixing because of the short diffusion lengths.

There have been numerous attempts to realize micromixers using both active and passive techniques. Active micromixers rely on creating localized turbulence to enhance the mixing process, whereas passive micromixers usually enhance the diffusion process. The diffusion process can be modeled by the following equation

$$\tau = \frac{d^2}{D},\qquad(18.3)$$

where τ is the mixing time, d is the distance, and D is the diffusion coefficient. Equation (18.3) illustrates the diffusion-dominated mixing at the microscale. Because of the small diffusion lengths (d), the mixing times can be made very short. The simplest category of micromixers is illustrated in Fig. 18.16. In these mixers, creating a convoluted path increases the path length that the two fluids share, leading to higher diffusion and more complete mixing. However, these mixers only exhibit good mixing performance at low flow rates, in the range of a few μl/min.

Mitchell et al. [18.69] have demonstrated three-dimensional micromixers that can achieve better performance by alternately laminating the two fluid streams to be mixed. *Beebe* et al. [18.70] have created a chaotic mixer that has the convoluted channel along three dimensions. This micromixer works on the principle of forced advection resulting from repeated turns in three dimensions. Furthermore, at each turn, eddies are generated because of the difference in flow velocities along the inner and outer radii, which enhances the mixing.

Stroock et al. [18.71] demonstrate a passive micromixer that uses chaotic mixing by superimposing a transverse flow component on the axial flow. Ridges are fabricated at the bottom of microchannels. The flow resistance is lower along the ridges (peak/valley) and higher in the axial direction. This generates a helical flow pattern that is superimposed on the laminar flow. The demonstrated mixer shows good mixing performance over a wide range of flow velocities.

Hong et al. [18.68] have demonstrated a passive micromixer based on the Coanda effect. Their design uses the effects of diffusion mixing at low flow velocities; at high flow velocities, a convective component is added perpendicular to the flow direction, allowing for rapid mixing. This mixer shows excellent mixing performances across a wide range of flow rates because of

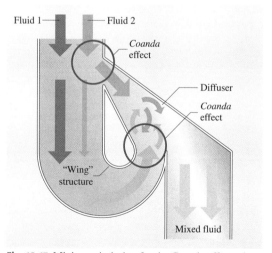

Fig. 18.17 Mixing unit design for the Coanda effect mixer. The actual mixer has mixing unit pairs in series (after [18.68])

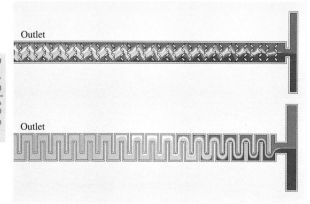

Fig. 18.16 Diffusion-enhanced mixers based on a long, convoluted flow path

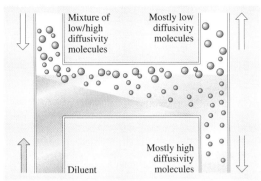

Fig. 18.18 Conceptual illustration of the H-sensor (after [18.72])

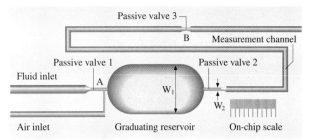

Fig. 18.19 Schematic sketch of the microdispenser (after [18.62], © The Royal Society of Chemistry)

the dual mixing effects. Figure 18.17 shows a schematic sketch of the mixer structure.

The mixer works on the principle of superimposing a parabolic flow profile in a direction perpendicular to the flow direction. The parabolic profile creates a Taylor dispersion pattern across the cross section of the flow path. The dispersion is directly proportional to the flow velocity, and higher flow rates generate more dispersion mixing.

Brody and *Yager* [18.72] have used the laminar flow characteristics in a microchannel to develop a diffusion-based extractor. When two fluid streams, where fluid 1 is loaded with particles of different diffusivity and fluid 2 is a diluent, are forced to flow together in a microchannel, they form two laminar streams with little mixing. If the length that the two streams are in contact is carefully adjusted, only particles with high diffusivity (usually small molecules) can diffuse across into the diluent stream, as shown in Fig. 18.18. The same idea can be extended to a T-filter. *Weigl* and *Yager* [18.73] have demonstrated a rapid diffusion immunoassay using the T-filter.

18.3.3 Passive Microdispensers

The principle of the structurally programmable microfluidic systems (sPROMs) was introduced earlier in the passive valve section. One of the key components of the sPROMs system is the microdispenser, which is designed to accurately and repeatedly dispense fluidic volumes in the micro- to nanoliter range. This would allow the dispensing of a controlled amount of the analyte into the system that could be used for further biochemical analysis. Figure 18.19 shows a schematic sketch of the microdispenser design [18.62].

The microdispenser works on the principle of graduated volume measurement. The fluid fills up the exact fixed volume of the reservoir, and the second passive valve at the other end of the microdispenser stops further motion. When the reservoir is filled with fluid, the fluidic actuation is stopped, and, simultaneously, pneumatic actuation from the air line causes a split in the fluid column at point A (Fig. 18.19). Thus, the accuracy of the reservoir decides the accuracy of the dispensed volume. Since the device is manufactured using UV-LIGA lithography techniques, highly accurate and reproducible volumes can be defined. The precisely measured volume of fluid is expelled to the right from the reservoir. The expelled fluid then starts to fill up the measuring channel. When the fluid reaches point B (Fig. 18.19), the third microvalve holds the fluid column.

Figure 18.20 shows an actual operation sequence of the microdispenser. Figure 18.20f shows that the dispensed volume is held by passive valve 3, and at this stage, the length (and hence volume) can be calculated using the on-chip scale. In experiments only, the region in the immediate vicinity of the scale was viewed using a stereomicroscope to measure the length of the fluid column. The microdispenser demonstrated above is reported to have dispensing variation of less than 1% between multiple dispensing cycles.

18.3.4 Microfluidic Multiplexer Integrated with Passive Microdispenser

Ahn et al. [18.19] have demonstrated the sPROMs technology to be an innovative method of controlling liquid movement in a programmed fashion in a microfluidic network. By integrating this technique with the microdispensers, a more functionally useful microfluidic system can be realized. Figure 18.21a shows a schematic illustration of the microfluidic multiplexer

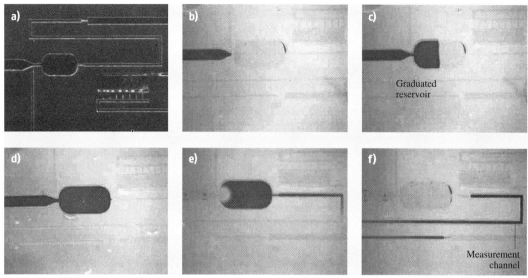

Fig. 18.20a–f Microphotographs of the microdispenser sequence: (**a**) fabricated device; (**b**) fluid at reservoir inlet; (**c**) reservoir filling; (**d**) reservoir filled; (**e**) split in liquid column due to pneumatic actuation; and (**f**) fluid ejected to measurement channel and locked in by passive valve (after [18.62], © The Royal Society of Chemistry)

with the integrated dispenser, and Fig. 18.21b shows an actual device fabricated using rapid prototyping techniques [18.74].

The operation of the microdispenser has been explained earlier in this section. Briefly, the fluid is loaded in the fixed-volume metering reservoir via a syringe pump. The fluid is locked in the reservoir by the passive valve at the outlet of the reservoir. When a higher pressure is applied via the air inlet line, the liquid column is split and the fluid is dispensed into the graduating channel.

The microfluidic multiplexer is designed to have a programmed delivery sequence, as shown in Fig. 18.21a, where the numbers on each branch of the multiplexer indicates the filling sequence. This sequential filling is achieved by using different ratios of passive valves along the multiplexer section. For instance, at the first branching point, the passive valve at the upper branch offers less resistance than the passive valve at the beginning of the lower branch. Thus, the dispensed fluid will first fill the top branch. After filling the top branch, the liquid encounters another passive

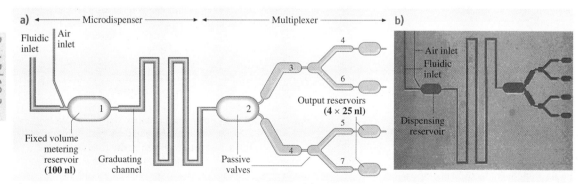

Fig. 18.21a,b Microfluidic multiplexer with integrated dispenser: (**a**) schematic sketch and (**b**) fabricated device filled with dye (after [18.62])

valve at the end of the top branch. The pressure needed to push beyond this valve is higher than the pressure needed to push liquid into the lower branch of the first split. The liquid will then fill the lower branch. This sequence of nonsymmetrical passive valves continues along all the branches of the multiplexer, as shown in Fig. 18.22.

The ability to sequentially divide and deliver liquid volumes was demonstrated for the first time using a passive microfluidic system. This approach has the potential to deliver very simple microfluidic control systems that are capable of a number of sequential microfluidic manipulation steps required in a biochemical analysis system.

18.3.5 Passive Micropumps

A passive system is defined inherently as one that does not require an external energy source. Thus, the term passive micropump might seem a misnomer upon initial inspection. However, there have been some efforts dedicated to realizing a passive micropump that essentially does not draw energy from an external source, but stores the required actuation energy in some form and converts it to mechanical energy on demand.

Passive Micropumps Based on Osmotic Pressure

Nagakura et al. [18.75] have demonstrated a mesoscale osmotic actuator that converts chemical energy to mechanical displacement. Osmosis is a well-known phenomenon by which liquid is transported across a semipermeable membrane to achieve a uniform concentration distribution across the membrane. If the membrane is flexible, such as the one used by *Nagakura* et al. [18.75], then the transfer of liquid would cause the membrane to deform and act as an actuator. The inherent drawback of using osmosis as an actuation mechanism is that it is a very slow process: typical response times (on a macroscale) are on the order of several hours. However, osmotic transport scales favorably to the microscale, and it is expected that these devices will have response times on the order of several minutes, rather than hours. Based on this idea, *Nagakura* et al. [18.75] are developing a miniature insulin pump. *Su* et al. [18.76] have demonstrated a microscale osmotic actuator that is capable of developing pressures as high as 35 MPa. This is still a relatively unexplored realm in bioMEMS actuation, and it has good potential for applications such as sustained drug delivery.

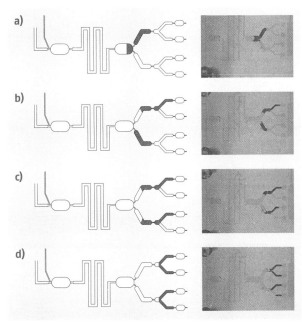

Fig. 18.22a–d Microphotographs showing operation of sequential multiplexer: (**a**) first-level division; (**b**) second-level division; (**c**) continued multiplexing; and (**d**) end of sequential multiplexing sequence (after [18.62])

Passive Pumping Based on Surface Tension

Walker and *Beebe* [18.77] have demonstrated pumping action using the difference between the surface tension pressure at the inlet and outlet of a microfluidic channel. In the simplest case, a small drop of a fluid is placed at one end of a straight microchannel, and a much larger drop of fluid is placed at the opposite end of the microchannel. The pressure within the small drop is significantly higher than the pressure within the large drop, due the difference in the surface tension effects across the two drops. Consequently, the liquid will flow from the small drop and add to the larger drop. The flow rate can be varied by changing various parameters such as the volume of the pumping drop, the surface free energy of the liquid, or the resistance of the microchannel, etc. This pumping scheme is very easy to realize and can be used for a wide variety of fluids.

Evaporation-Based Continuous Micropumps

Effenhauser et al. [18.78] have demonstrated a continuous-flow micropump based on a controlled evaporation approach. Their concept is based on the controlled evaporation of a liquid through a membrane into a gas reservoir. The reservoir contains a suitable adsorption agent that draws out the liquid vapors and maintains a low vapor pressure conducive to further evaporation. If the liquid being pumped is replenished from a reservoir, capillary forces will ensure that the fluid is continuously pumped through the microchannels as it evaporates at the other end into the adsorption reservoir. Though the pump suffers from inherent disadvantages such as strong temperature dependence and operation only in suction mode, it offers a very simple technique for fluidic transport.

18.3.6 Advantages and Disadvantages of the Passive Microfluidic Approach

This chapter has covered a number of different passive microfluidic devices and systems. Passive microfluidic devices have only recently been a subject of considerable research effort. One of the reasons for this interest is the long list of advantages that passive microfluidic devices offer. However, since most microfluidic devices are very application-specific (and even more so for passive microfluidic devices), the advantages are not to be considered universally applicable for all the devices/systems. Some of the advantages that are commonly found are:

- Avoiding the need for an *active* control system.
- They are usually very easy to fabricate.
- Passive microfluidic systems with no moving parts are inherently more reliable because of the lack of mechanical wear and tear.
- They offer very repeatable performance once the underlying phenomena are well understood and characterized.
- They are highly suited for bioMEMS applications; they can easily handle a limited number of microfluidic manipulation sequences.
- Well suited for low-cost mass production.
- Their low cost offers the possibility of having disposable microfluidic systems for specific applications, such as working with blood.
- They can offer other interesting possibilities, such as biomimetic responses.

However, like all MEMS devices, passive microfluidic devices or systems are not the solution to the microfluidic handling problem. Usually they are very application-specific; they cannot be reconfigured for another application easily. Other disadvantages are listed below:

- They are suited for well-understood, niche applications for which the fluidic sequencing steps are well decided.
- They are strongly dependent on variations in the fabrication process.
- They are usually not very suitable for a wide range of fluidic mediums.

18.4 Lab-on-a-Chip for Biochemical Analysis

Recent development in MEMS (microelectromechanical systems) has brought a new and revolutionary tool in biological or chemical applications: *lab-on-a-chip*. New terminology, such as micro-total analysis systems and lab-on-a-chip, was introduced in the last decade, and several prototype systems have been reported.

The idea of lab-on-a-chip is basically to reduce biological or chemical laboratories to a microscale system, hand-held size or smaller. Lab-on-a-chip systems can be made out of silicon, glass, and polymeric materials, and the typical microfluidic channel dimensions are in the range of several tens to hundreds of μm. Liquid samples or reagents can be transported through the microchannels from reservoirs to reactors using electrokinetic, magnetic, or hydrodynamic pumping methods. Fluidic motion or biochemical reactions can also be monitored using various sensors, which are often used for biochemical detection of products.

There are many advantages to using lab-on-a-chip over conventional chemical or biological laboratories. One of the important advantages lies in its low cost. Many reagents and chemicals used in biological and chemical reactions are expensive, so the prospect of using very small amounts (in the micro- to nanoliter ranges) of reagents and chemicals for an application is very appealing. Another advantage is that lab-on-a-chip requires very small amounts of reagents/chemicals (which enables rapid mixing and reaction) because biochemical reaction is mainly involved in the diffusion of two chemical or biological reagents, and microscale fluidics reduces diffusion time as it increases reaction probabilities. In practical terms, reaction products can be produced in a matter of seconds/minutes, whereas laboratory-scale reactions can take hours, or even days. In addition, lab-on-a-chip systems minimize harmful by-products since their volume is so small.

Complex reactions with many reagents could happen on a lab-on-a-chip, ultimately with potential in DNA analysis, biochemical warfare-agent detection, biological cell/molecule sorting, blood analysis, drug screening/development, combinatorial chemistry, and protein analysis. In this section, three recent developments of microfluidic systems for lab-on-a-chip applications will be introduced: (a) a magnetic micro/nano-bead-based biochemical detection system; (b) a disposable smart lab-on-a-chip for blood analysis, and (c) a disposable lab-on-a-chip for magnetic immunoassay.

18.4.1 Magnetic Micro/Nano-Bead-Based Biochemical Detection System

In the past few years, a large number of microfluidic prototype devices and systems have been developed, specifically for biochemical warfare detection systems and portable diagnostic applications. The bioMEMS team at the University of Cincinnati has been working on the development of a remotely accessible generic microfluidic system for biochemical detection and biomedical analysis, based on the concepts of surface-mountable microfluidic motherboards, sandwich immunoassays, and electrochemical detection techniques [18.79, 81]. The limited goal of this work is to develop a generic MEMS-based microfluidic system and to apply the fluidic system to detect biomolecules, such as specific proteins and/or antigens, in liquid samples. Figure 18.23 illustrates the schematic diagram of a generic microfluidic system for biochemical detection using a magnetic-bead approach for both sampling and manipulating the target biomolecules [18.80, 82].

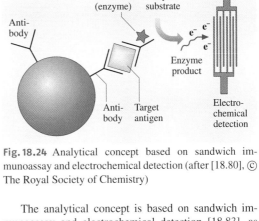

Fig. 18.24 Analytical concept based on sandwich immunoassay and electrochemical detection (after [18.80], © The Royal Society of Chemistry)

The analytical concept is based on sandwich immunoassay and electrochemical detection [18.83], as illustrated in Fig. 18.24. Magnetic beads are used as both substrates for the antibodies and carriers for the target antigens. A simple concept of magnetic-bead-based bio-sampling with an electromagnet for the case of sandwich immunoassay is shown in Fig. 18.25.

Antibody-coated beads are introduced on the electromagnet and separated by applying magnetic fields. While holding the antibody-coated beads, antigens are injected into the channel. Only target antigens are immobilized and, thus, separated on the magnetic bead surface due to antibody/antigen reaction. Other antigens get washed out with the flow. Next, enzyme-labeled secondary antibodies are introduced and incubated, along with the immobilized antigens. The chamber is then rinsed to remove all unbound secondary antibodies.

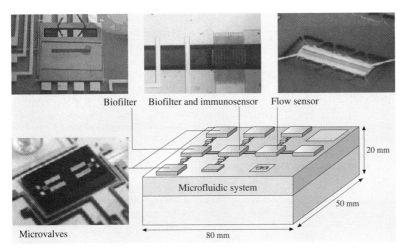

Fig. 18.23 Schematic diagram of a generic microfluidic system for biochemical detection (after [18.79])

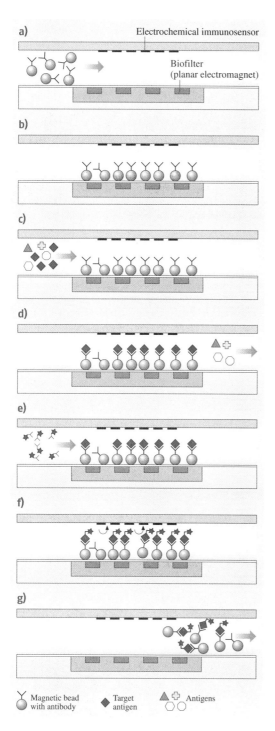

Fig. 18.25a–g Conceptual illustration of biosampling and immunoassay procedure: (**a**) injection of magnetic beads; (**b**) separation and holding of beads; (**c**) flowing samples; (**d**) immobilization of target antigen; (**e**) flowing labeled antibody; (**f**) electrochemical detection; and (**g**) washing out magnetic beads and ready for another immunoassay (after [18.80], © The Royal Society of Chemistry) ◄

A substrate solution, which will react with the enzyme, is injected into the channel, and the electrochemical detection is performed. Finally, the magnetic beads are released to the waste chamber, and the bio-separator is ready for another immunoassay. Alkaline phosphatase (AP) and p-aminophenyl phosphate (PAPP) were chosen as the enzyme and electrochemical substrate, respectively. Alkaline phosphatase makes PAPP turn into its electrochemical product, p-aminophenol (PAP). By applying a potential, PAP gives up electrons and turns into 4-quinoneimine (4QI), which is the oxidant form of PAP.

For a successful immunoassay, the biofilter [18.82] and the immunosensor were fabricated separately and integrated together. The integrated biofilter and immunosensor were surface-mounted using a fluoropolymer bonding technique [18.84] on a microfluidic motherboard, which contains microchannels fabricated using the glass-etching and glass-to-glass direct-bonding technique. Each the inlet and outlet were connected to sample reservoirs through custom-designed microvalves. Figure 18.26 shows the integrated microfluidic biochemical detection system for the magnetic-bead-based immunoassay.

After a fluidic sequencing test, full immunoassays were performed in the integrated microfluidic system to prove magnetic-bead-based biochemical detection and sampling function. Magnetic beads (Dynabeads M-280, Dynal Biotech Inc.) coated with biotinylated sheep anti-mouse immunoglobulin G (IgG) were injected into the reaction chamber and separated on the surface of the biofilter by applying magnetic fields. While holding the magnetic beads, antigen (mouse IgG) was injected into the chamber and incubated. Then secondary antibody with label (rat anti-mouse IgG conjugated alkaline phosphatase) and electrochemical substrate (PAPP) to alkaline phosphatase were sequentially injected and incubated to ensure production of PAP. Electrochemical detection using an amperometric time-based detection method was performed during incubation. After detection, magnetic beads with all the reagents were washed away, and the system was ready for another immunoassay. This sequence was repeated for every

Fig. 18.26 Photograph of the fabricated lab-on-a-chip for magnetic-bead-based immunoassay

new immunoassay. The flow rate was set to 20 μl/min in every step. After calibration of the electrochemical immunosensor, full immunoassays were performed following the sequence stated above for different antigen concentrations: 50, 75, 100, 250, and 500 ng/ml. Concentration of the primary antibody-coated magnetic beads and conjugated secondary antibody was 1.02×10^7 beads/ml and 0.7 μg/ml, respectively. Immunoassay results for different antigen concentrations are shown in Fig. 18.27.

Immunoreactant consumed during one immunoassay was 10 μl (20 μl/min × 30 s), and total assay time was less than 20 min, including all incubation and detection steps.

The integrated microfluidic biochemical detection system has been successfully developed and fully tested for fast and low-volume immunoassays using magnetic beads, which are used as both immobilization surfaces and biomolecule carriers. Magnetic-bead-based immunoassay, as a typical example of biochemical detection and analysis, has been performed on the integrated microfluidic biochemical analysis system that includes a surface-mounted biofilter and immunosensor on a glass microfluidic motherboard. Protein-sampling capability has been demonstrated by capturing target antigens.

The methodology and system can also be applied to generic biomolecule detection and analysis systems by replacing the antibody/antigen with appropriate bioreceptors/reagents, such as DNA fragments or oligonucleotides, for application to DNA analysis and/or high-throughput protein analysis.

18.4.2 Disposable Smart Lab-on-a-Chip for Blood Analysis

One of several substrates available for biofluidic chips, plastics have recently become one of the most pop-

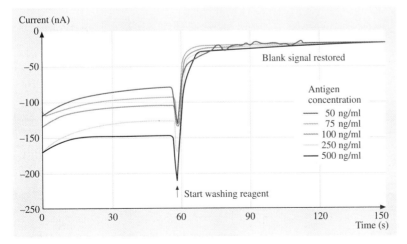

Fig. 18.27 Immunoassay results measured by amperometric time-based detection method (after [18.80], © The Royal Society of Chemistry)

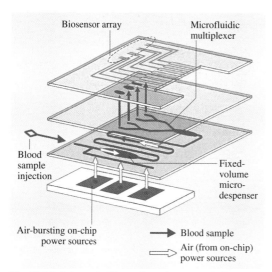

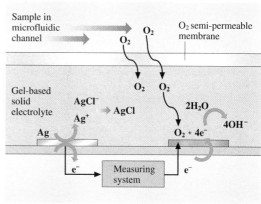

Fig. 18.30 Electrochemical and analytical principle of the developed disposable biosensor for partial oxygen concentration sensing (after [18.86], © 2002 IEEE)

Fig. 18.28 Schematic illustration of smart and disposable plastic lab-on-a-chip by *Ahn* et al. (after [18.87], © 2004 IEEE)

ular and promising substrates due to their low cost, ease of fabrication, and favorable biochemical reliability and compatibility. Plastic substrates, such as polyimide, PMMA, PDMS, polyethylene, or polycarbonate, offer a wide range of physical and chemical material parameters for the applications of biofluidic chips, generally at low cost using replication approaches. The disposable smart plastic biochip is composed of integrated modules of plastic fluidic chips for fluid regulation, chemical and biological sensors, and electronic controllers. As a demonstration vehicle, the biochip has the specific goal of detecting and identifying three metabolic parameters such as p_{O_2} (partial pressure of oxygen), lactate, and glucose from a blood sample. The schematic concept of the cartridge-type disposable lab-on-a-chip for blood analysis is illustrated in Fig. 18.28. The disposable lab-on-a-chip cartridge has been fabricated using plastic micro-injection molding and plastic-to-plastic direct-bonding techniques. The biochip cartridge consists of a fixed-volume microdispenser based on the structurally programmable microfluidic system (sPROMs) technique [18.74], an air-bursting on-chip pressure source [18.85], and electrochemical biosensors [18.86].

A passive microfluidic dispenser measures exact amounts of sample to be analyzed, and then the air-bursting on-chip power source is detonated to push the graduated sample fluid from the dispenser reservoir. Upon air-bursting, the graduated sample fluid travels through the microfluidic channel into sensing reservoirs, under which the biosensor array is located, as shown in Fig. 18.29.

An array of disposable biosensors consisting of an oxygen sensor, a glucose sensor and a lactate sensor has been fabricated using screen-printing technology [18.86]. Measurements from the developed biosensor array can be done based on tiny amounts of

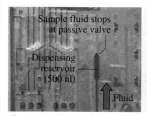 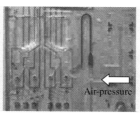 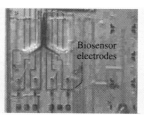 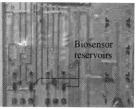

Fig. 18.29 Upon air-bursting, sample fluid travels through the microfluidic channel into the biosensor detection chamber. In sequence: loading the dispenser; dispensing; multiplexing, and delivered volume to biosensor array

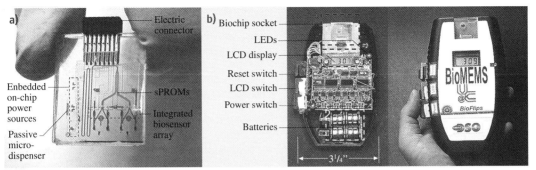

Fig. 18.31a,b Disposable biochip and hand-held analyzer: (**a**) developed smart and disposable biochip cartridge and (**b**) hand-held analyzer developed at the University of Cincinnati

sample (as low as 100 nl). One of the most fundamental sensor designs is the oxygen sensor, which is the basic sensing structure for many other metabolic products such as glucose and lactate. The principle of the oxygen sensor is based on amperometric detection. Figure 18.30 shows a schematic representation of an oxygen sensor. When the diffusion profile for oxygen from the sample to the electrode surface is saturated, a constant oxygen gradient profile is generated. Under these circumstances the detection current is proportional only to the oxygen concentration in the sample.

The gel-based electrolyte is essential for the ion-exchange reactions at the anode of the electrochemical pair. The oxygen semipermeable membrane ensures that mainly oxygen molecules permeate through this layer and that the electrochemical cell is not exposed to other ions. A silicone layer was spin-coated and utilized as an oxygen semipermeable membrane because of its high permeability and low signal-to-noise ratio. Water molecules pass through the silicone membrane and reconstitute the gel-based electrolyte so the Cl^- ions can move close to the anode to coalesce with Ag^+ ions. The number of electrons in this reaction is counted by the measuring system.

For the glucose sensor, additional layers – a glucose semipermeable membrane (polyurethane) and immobilized glucose oxidase (GOD) in a polyacrylamide gel for the glucose sensor – allow direct conversion of the oxygen sensor into a glucose sensor. A similar modification is made for the lactate sensor by replacing the immobilized glucose oxidase with lactate oxidase. The glucose molecules will pass through the semipermeable layer and be oxidized immediately. The oxygen sensor will measure hydrogen peroxide, which is a by-

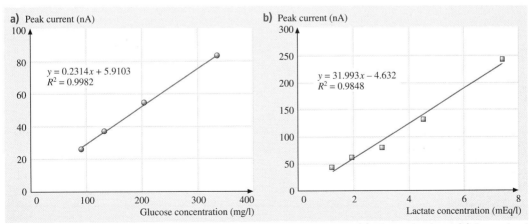

Fig. 18.32a,b Measurement results from the biochip cartridge and analyzer: (**a**) glucose level and (**b**) lactate level (after [18.87], © 2004 IEEE)

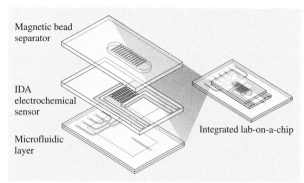

Fig. 18.33 Schematic illustration of a disposable lab-on-a-chip system for magnetic immunoassay (after [18.88], © The Royal Society of Chemistry)

product of glucose oxidation. The level of hydrogen peroxide is proportional to the glucose level in the sample. The fabricated disposable plastic lab-on-a-chip cartridge was inserted into a hand-held biochip analyzer for analysis of human blood samples, as shown in Fig. 18.31. The prototype biochip analyzer consists of biosensor detection circuitry, timing/sequence circuitry for the air-bursting, on-chip power source, and a display unit. The hand-held biochip analyzer initiated the sensing sequence and displayed readings in one minute. The measured glucose and lactate levels in human blood samples are also shown in Fig. 18.32.

The development of disposable smart microfluidic-based biochips is of immediate relevance to several patient-monitoring systems, specifically for point-of-care health monitors. Since the developed biochip is a low-cost plastic-based system, we envision a disposable application for monitoring clinically significant parameters such as p_{O_2}, glucose, lactate, hematocrit, and pH. These health indicators provide an early warning system for the detection of patient status and can also serve as markers for disease and toxicity monitoring.

Disposable Lab-on-a-Chip for Magnetic Immunoassay

There has been a large demand for inexpensive and smart lab-on-a-chip systems for immunoassay. Section 18.4.1 showed feasibility of magnetic bead-based immunoassay in a microfluidic lab-on-a-chip system. The development of a disposable lab-on-a-chip system with a magnetic bead-based immunoassay capability is very desirable for point-of-care testing (POCT) applications. Recently, a polymer lab-on-a-chip has been developed and reported for magnetic immunoassay [18.88] as illustrated in Fig. 18.33. The polymer lab-on-a-chip is composed of three components: (i) a microfluidic module for the introduction of sample fluid and immunoassay reagents; (ii) a sampling/detection module comprised of a magnetic separator, and (iii) an interdigitated array (IDA) microelectrodes as an electrochemical sensor. The key components of the polymer lab-on-a-chip system are the magnetic separator and the IDA electrochemical sensor.

Magnetic separation is achieved with electroplated permalloy microarray on a polymer substrate where magnetic field is applied by external permanent magnets as shown in Fig. 18.34. External permanent magnets (NdFeB) are located at both sides of the magnetic separator creating the focused magnetic flux along with the permalloy microarray and trapping magnetic particles on the magnetic array. The IDA electrochemical sensor is similar to that of in previous Sect. 18.4.1,

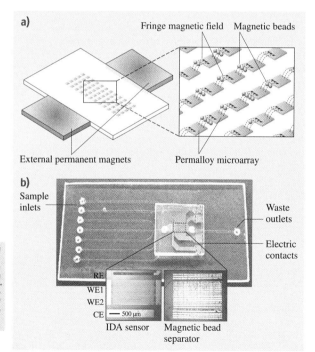

Fig. 18.34a,b Magnetic bead separator and disposable lab-on-a-chip: (**a**) schematic illustration of magnetic bead separator with electroplated permalloy microarray and (**b**) photograph of a fabricated lab-on-a-chip (after [18.88], © The Royal Society of Chemistry)

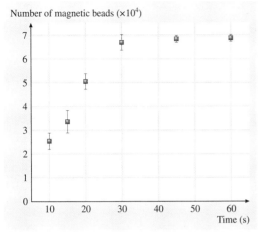

Fig. 18.35 Separation efficiency of magnetic beads. Time period for flowing magnetic beads varied from 10 to 60 s at flow rate of 20 μl/min (after [18.88], © The Royal Society of Chemistry)

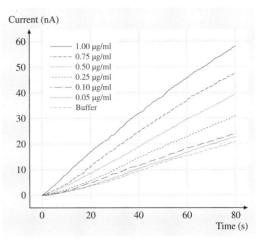

Fig. 18.36 Immunoassay result for the various mouse IgG concentrations detected in the disposable lab-on-a-chip system for magnetic immunoassay (after [18.88], © The Royal Society of Chemistry)

which has two working electrodes for signal amplification by redox cycling [18.87]. All three modules are assembled using a room temperature UV bonding technique [18.89] and a new metal pattern embedding technique [18.90].

The polymer lab-on-a-chip for magnetic bead-based immunoassay was characterized to verify its functionality of sampling and detection. The steps are described in Sect. 18.4.1 using mouse IgG as target antigen. The dynamic range of the IDA electrochemical sensor was investigated by measuring the signal from known concentrations of bead-conjugated alkaline phosphatase (AP) before the magnetic immunoassay. Bead-conjugated alkaline phosphatase was externally prepared at known concentrations. The separation efficiency was obtained by measuring the signal from the bead-conjugated alkaline phosphatase and comparing it with a calibration curve as shown in Fig. 18.35.

After verification of each components of the polymer lab-on-a-chip, a full immunoassay was carried out with the lab-on-a-chip system. Same sequences described in Sect. 18.4.1 were applied. Figure 18.36 shows the immunoassay results for different antigen concentration. A low detection limit of 16.4 ng/ml was achieved using the described method with a disposable polymer lab-on-a-chip.

The immunoassay results have proved that the disposable polymer lab-on-a-chip system promises a potential in fast and small volume biochemical detection and analysis.

References

18.1 L. Ristic, H. Hughes, F. Shemansky: Bulk micromachining technology. In: *Sensor Technology and Devices*, ed. by L. Ristic (Artech House, Norwood 1994) pp. 49–93

18.2 M. Esashi, M. Takinami, Y. Wakabayashi, K. Minami: High-rate directional deep dry etching for bulk silicon micromachining, J. Micromech. Microeng. **5**, 5–10 (1995)

18.3 H. Jansen, M. de Boer, R. Legtenberg, M. Elwenspoek: The black silicon method: A universal method for determining the parameter setting of a fluorine-based reactive ion etcher in deep silicon trench etching with profile control, J. Micromech. Microeng. **5**, 115–120 (1995)

18.4 Z.L. Zhang, N.C. MacDonald: A RIE process for submicron silicon electromechanical structures, J. Micromech. Microeng. **2**, 31–38 (1992)

18.5 C. Linder, T. Tschan, N.F. de Rooij: Deep dry etching techniques as a new IC compatible tool for silicon micromachining, Technical Digest 6th Int. Conf. Solid-State Sens. Actuators (Transducers '91), San Francisco June 24–27, 1991 (IEEE, Piscataway 1991) 524–527

18.6 D.J. Harrison, A. Manz, Z. Fan, H. Ludi, H.M. Widmer: Capillary electrophoresis and sample injection systems integrated on a planar glass chip, Anal. Chem. **64**, 1926–1932 (1992)

18.7 A. Manz, D.J. Harrison, E.M.J. Verpoorte, J.C. Fettinger, A. Paulus, H. Ludi, H.M. Widmer: Planar chips technology for miniaturization and integration of separation technique into monitoring systems, J. Chromatogr. **593**, 253–258 (1992)

18.8 T.R. Dietrich, W. Ehrfeld, M. Lacher, M. Krämer, B. Speit: Fabrication technologies for microsystems utilizing photoetchable glass, Microelectron. Eng. **30**, 497–504 (1996)

18.9 S.C. Jacobson, R. Hergenroder, L.B. Koutny, J.M. Ramsey: Open channel electrochromatography on a microchip, Anal. Chem. **66**, 2369–2373 (1994)

18.10 E.W. Becker, W. Ehrfeld, P. Hagmann, A. Maner, D. Munchmeyer: Fabrication of microstructures with high aspect ratios and great structural heights by synchrotron radiation lithography, galvanoforming, and plastic moulding (LIGA process), Microelectron. Eng. **4**, 35–36 (1986)

18.11 M.G. Allen: Polyimide-based process for the fabrication of thick electroplated microstructures, Proc. 7th Int. Conf Solid-State Sens. Actuators (Transducers '93), Yokohama June 7–10, 1993 (IEE of Japan, Tokyo 1993) 60–65

18.12 C.H. Ahn, M.G. Allen: A fully integrated surface micromachined magnetic microactuator with a multilevel meander magnetic core, IEEE/ASME J. Microelectromech. Syst. (MEMS) **2**, 15–22 (1993)

18.13 C.S. Effenhauser, G.J.M. Bruin, A. Paulus, M. Ehrat: Integrated capillary electrophoresis on flexible silicone microdevices: Analysis of DNA restriction fragments and detection of single DNA molecules on microchips, Anal. Chem. **69**, 3451–3457 (1997)

18.14 D.C. Duffy, J.C. McDonald, O.J.A. Schueller, G.M. Whitesides: Rapid prototyping of microfluidic systems in poly(dimethylsiloxane), Anal. Chem. **70**, 4974–4984 (1998)

18.15 L. Martynova, L. Locascio, M. Gaitan, G.W. Kramer, R.G. Christensen, W.A. MacCrehan: Fabrication of plastic microfluid channels by imprinting methods, Anal. Chem. **69**, 4783–4789 (1997)

18.16 H. Becker, W. Dietz, P. Dannberg: Microfluidic manifolds by polymer hot embossing for micro-TAS applications, Proc. Micro-Total Anal. Syst. '98, Banff, Canada 1998, ed. by D.J. Harrison, A. van den Berg (Kluwer, Dordrecht 1998) 253–256

18.17 H. Becker, U. Heim: Hot embossing as a method for the fabrication of polymer high aspect ratio structures, Sens. Actuators A **83**, 130–135 (2000)

18.18 R.M. McCormick, R.J. Nelson, M.G. Alonso-Amigo, J. Benvegnu, H.H. Hooper: Microchannel electrophoretic separations of DNA in injection-molded plastic substrates, Anal. Chem. **69**, 2626–2630 (1997)

18.19 C.H. Ahn, A. Puntambekar, S.M. Lee, H.J. Cho, C.-C. Hong: Structurally programmable microfluidic systems, Proc. Micro-Total Anal. Syst. 2000, Enschede 2000, ed. by A. van den Berg, W. Olthius, P. Bergveld (Kluwer, Dordrecht 2000) 205–208

18.20 W. Schomburg, B. Scherrer: 3.5 mm thin valves in titanium membranes, J. Micromech. Microeng. **2**, 184–186 (1992)

18.21 A. Ilzofer, B. Ritter, C. Tsakmasis: Development of passive microvalves by the finite element method, J. Micromech. Microeng. **5**, 226–230 (1995)

18.22 B. Paul, T. Terharr: Comparison of two passive microvalve designs for microlamination architectures, J. Micromech. Microeng. **10**, 15–20 (2000)

18.23 S. Shoji, M. Esashi: Microflow device and systems, J. Micromech. Microeng. **4**, 157–171 (1994)

18.24 G.T.A. Kovacs: *Micromachined Transducers Sourcebook* (McGraw-Hill, New York 1998)

18.25 W. Schomburg, J. Fahrenberg, D. Maas, R. Rapp: Active valves and pumps for microfluidics, J. Micromech. Microeng. **3**, 216–218 (1993)

18.26 K. Hosokawa, R. Maeda: A pneumatically-actuated three way microvalve fabricated with polydimethylsiloxane using the membrane transfer technique, J. Micromech. Microeng. **10**, 415–420 (2000)

18.27 J. Fahrenberg, W. Bier, D. Maas, W. Menz, R. Ruprecht, W. Schomburg: A microvalve system fabricated by thermoplastic molding, J. Micromech. Microeng. **5**, 169–171 (1995)

18.28 C. Goll, W. Bacher, B. Bustgens, D. Maas, W. Menz, W. Schomburg: Microvalves with bistable buckled diaphragms, J. Micromech. Microeng. **6**, 77–79 (1996)

18.29 A. Ruzzu, J. Fahrenberg, M. Heckele, T. Schaller: Multi-functional valve components fabricated by combination of LIGA process and high precision mechanical engineering, Microsyst. Technol. **4**, 128–131 (1998)

18.30 K. Sato, M. Shikida: An electrostatically actuated gas valve with S-shaped film element, J. Micromech. Microeng. **4**, 205–209 (1994)

18.31 J. Robertson, K.D. Wise: A low pressure micromachined flow modulator, Sens. Actuators A **71**, 98–106 (1998)

18.32 D.J. Sadler, K.W. Oh, C.H. Ahn, S. Bhansali, H.T. Henderson: A new magnetically actuated microvalve for liquid and gas control applications, Proc. 6th Int. Conf. Solid-State Sens. Actuators (Transducers '99), Sendai 1999 (IEE of Japan, Tokyo 1999) 1812–1815

18.33 W. Wijngaart, H. Ask, P. Enoksson, G. Stemme: A high-stroke, high-pressure electrostatic actuator for valve applications, Sens. Actuators A **100**, 264–271 (2002)

18.34 T. Watanabe, H. Kuwano: A microvalve matrix using piezoelectric actuators, Microsyst. Technol. **3**, 107–111 (1997)

18.35 M. Stehr, S. Messner, H. Sandmaier, R. Zangerle: The VAMP – A new device for handling liquids and gases, Sens. Actuators A **57**, 153–157 (1996)

18.36 W. Jackson, H. Tran, M. O'Brien, E. Rabinovich, G. Lopez: Rapid prototyping of active microfluidic components based on magnetically modified elastomeric materials, J. Vac. Sci. Technol. B **19**, 596 (2001)

18.37 M. Kohl, K. Skrobanek, S. Miyazaki: Development of stress-optimized shape memory microvalves, Sens. Actuators A **72**, 243–250 (1999)

18.38 C. Neagu, J. Gardeniers, M. Elwenspoek, J. Kelly: An electrochemical active valve, Electrochem. Acta **42**, 3367–3373 (1997)

18.39 K. Yoshida, M. Kikuchi, J. Park, S. Yokota: Fabrication of micro-electro-rheological valves (ER valves) by micromachining and experiments, Sens. Actuators A **95**, 227–233 (2002)

18.40 N. Jeon, D. Chiu, C. Wargo, H. Wu, I. Choi, J. Anderson, G.M. Whitesides: Design and fabrication of integrated passive valves and pumps for flexible polymer 3-dimensional microfluidic systems, Biomed. Microdevices **4**, 117–121 (2002)

18.41 M. Koch, N. Harris, A. Evans, N. White, A. Brunnschweiler: A novel micropump design with thick-film piezoelectric actuation, Meas. Sci. Technol. **8**, 49–57 (1997)

18.42 L. Cao, S. Mantell, D. Polla: Design and simulation of an implantable medical drug delivery system using microelectromechanical systems technology, Sens. Actuators A **94**, 117–125 (2001)

18.43 J. Park, K. Yoshida, S. Yokota: Resonantly driven piezoelectric micropump fabrication of a micropump having high power density, Mechatronics **9**, 687–702 (1999)

18.44 M. Koch, A. Evans, A. Brunnschweiler: The dynamic micropump driven with a screen printed PZT actuator, J. Micromech. Microeng. **8**, 119–122 (1998)

18.45 D. Xu, L. Wang, G. Ding, Y. Zhou, A. Yu, B. Cai: Characteristics and fabrication of NiTi/Si diaphragm micropump, Sens. Actuators A **93**, 87–92 (2001)

18.46 E. Makino, T. Mitsuya, T. Shibata: Fabrication of TiNi shape memory micropump, Sens. Actuators A **88**, 256–262 (2001)

18.47 H. Chou, M. Unger, S. Quake: A microfabricated rotary pump, Biomed. Microdevices **3**, 323–330 (2001)

18.48 R. Zengerle, J. Ulrich, S. Kulge, M. Richter, A. Richter: A bidirectional silicon micropump, Sens. Actuators A **50**, 81–86 (1995)

18.49 M. Carrozza, N. Croce, B. Magnani, P. Dario: A piezoelectric driven stereolithography-fabricated micropump, J. Micromech. Microeng. **5**, 177–179 (1995)

18.50 H. Andersson, W. Wijngaart, P. Nilsson, P. Enoksson, G. Stemme: A valveless diffuser micropump for microfluidic analytical systems, Sens. Actuators B **72**, 259–265 (2001)

18.51 F. Forster, R. Bardell, M. Afromowitz, N. Sharma: Design, fabrication and testing of fixed-valve micropumps, Proc. ASME Fluids Engineering Division, ASME Int. Mech. Eng. Congr. Expo. **234**, 39–44 (1995)

18.52 T. Thorsen, S.J. Maerkl, S.R. Quake: Monolithic microfabricated valves and pumps by multilayer soft lithography, Science **288**, 113–116 (2000)

18.53 T. Thorsen, S.J. Maerkl, S.R. Quake: Microfluidic large-scale integration, Science **298**, 580–584 (2002)

18.54 J.W. Hong, V. Studer, G. Hang, W.F. Anderson, S.R. Quake: A nanoliter-scale nucleic acid processor with parallel architecture, Nature Biotechnol. **22**, 435–439 (2004)

18.55 J.C. Love, J.R. Anderson, G.M. Whitesides: Fabrication of three-dimensional microfluidic systems by soft lithography, MRS Bulletin **7**, 523–528 (2001)

18.56 G. Fuhr, T. Schnelle, B. Wagner: Travelling wave driven microfabricated electrohydrodynamic pumps for liquids, J. Micromech. Microeng. **4**, 217–226 (1994)

18.57 J. Jang, S. Lee: Theoretical and experimental study of MHD micropump, Sens. Actuators A **80**, 84–89 (2000)

18.58 A. Lemoff, A.P. Lee: An AC magnetohydrodynamic micropump, Sens. Actuators B **63**, 178–185 (2000)

18.59 J. DeSimone, G. York, J. McGrath: Synthesis and bulk, surface and microlithographic characterization of poly(1-butene sulfone)-g-poly(dimethylsiloxane), Macromolecules **24**, 5330–5339 (1994)

18.60 A. Wego, L. Pagel: A self-filling micropump based on PCB technology, Sens. Actuators A **88**, 220–226 (2001)

18.61 A. Terray, J. Oakey, D. Marr: Fabrication of colloidal structures for microfluidic applications, Appl. Phys. Lett. **81**, 1555–1557 (2002)

18.62 A. Puntambekar, J.-W. Choi, C.H. Ahn, S. Kim, V.B. Makhijani: Fixed-volume metering microdispenser module, Lab Chip **2**, 213–218 (2002)

18.63 M. Madou, Y. Lu, S. Lai, C. Koh, L. Lee, B. Wenner: A novel design on a CD disc for 2-point calibration measurement, Sens. Actuators A **91**, 301–306 (2001)

18.64 K. Handique, D. Burke, C. Mastrangelo, C. Burns: On-chip thermopneumatic pressure for discrete drop pumping, Anal. Chem. **73**, 1831–1838 (2001)

18.65 H. Andersson, W. Wijngaart, P. Griss, F. Niklaus, G. Stemme: Hydrophobic valves of plasma deposited octafluorocyclobutane in DRIE channels, Sens. Actuators B **75**, 136–141 (2001)

18.66 L. Low, S. Seetharaman, K. He, M. Madou: Microactuators toward microvalves for responsive controlled drug delivery, Sens. Actuators B **67**, 149–160 (2000)

18.67 Q. Yu, J. Bauer, J. Moore, D. Beebe: Responsive biomimetic hydrogel valve for microfluidics, Appl. Phys. Lett. **78**, 2589–2591 (2001)

18.68 C.-C. Hong, J.-W. Choi, C.H. Ahn: A novel in-plane passive microfluidic mixer with modified Tesla structures, Lab. Chip **4**, 109–113 (2004)

18.69 M. Mitchell, V. Spikmans, F. Bessoth, A. Manz, A. de Mello: Towards organic synthesis in microfluidic devices: Multicomponent reactions for the construction of compound libraries, Proc. Micro-Total Anal. Syst. 2000, Enschede 2000, ed. by A. van den Berg, W. Olthius, P. Bergveld (Kluwer, Dordrecht 2000) 463–465

18.70 D. Beebe, R. Adrian, M. Olsen, M. Stremler, H. Aref, B. Jo: Passive mixing in microchannels: Fabrication and flow experiments, Mec. Ind. **2**, 343–348 (2001)

18.71 A. Stroock, S. Dertinger, A. Ajdari, I. Mezic, H. Stone, G.M. Whitesides: Chaotic mixer for microchannels, Science **295**, 647–651 (2002)

18.72 J. Brody, P. Yager: Diffusion-based extraction in a microfabricated device, Sens. Actuators A **54**, 704–708 (1996)

18.73 B. Weigl, P. Yager: Microfluidic diffusion-based separation and detection, Science **283**, 346–347 (1999)

18.74 A. Puntambekar, J.-W. Choi, C.H. Ahn, S. Kim, S. Bayyuk, V.B. Makhijani: An air-driven fluidic multiplexer integrated with microdispensers, Proc. μTAS 2001 Symp., Monterey 2001, ed. by J.M. Ramsey, A. van den Berg (Kluwer, Dordrecht 2001) 78–80

18.75 T. Nagakura, K. Ishihara, T. Furukawa, K. Masuda, T. Tsuda: Auto-regulated osmotic pump for insulin therapy by sensing glucose concentration without energy supply, Sens. Actuators B **34**, 229–233 (1996)

18.76 Y. Su, L. Lin, A. Pisano: Water-powered osmotic microactuator, Proc. of the 14th IEEE MEMS Workshop (MEMS 2001), Interlaken 2001 (IEEE, Piscataway) pp. 393–396

18.77 G.M. Walker, D.J. Beebe: A passive pumping method for microfluidic devices, Lab. Chip **2**, 131–134 (2002)

18.78 C. Effenhauser, H. Harttig, P. Kramer: An evaporation-based disposable micropump concept for continuous monitoring applications, Biomed. Microdevices **4**, 27–32 (2002)

18.79 J.-W. Choi, K.W. Oh, A. Han, N. Okulan, C.A. Wijayawardhana, C. Lannes, S. Bhansali, K.T. Schlueter, W.R. Heineman, H.B. Halsall, J.H. Nevin, A.J. Helmicki, H.T. Henderson, C.H. Ahn: Development and characterization of microfluidic devices and systems for magnetic bead-based biochemical detection, Biomed. Microdevices **3**, 191–200 (2001)

18.80 J.-W. Choi, K.W. Oh, J.H. Thomas, W.R. Heineman, H.B. Halsall, J.H. Nevin, A.J. Helmicki, H.T. Henderson, C.H. Ahn: An integrated microfluidic biochemical detection system for protein analysis with magnetic bead-based sampling capabilities, Lab Chip **2**, 27–30 (2002)

18.81 C.H. Ahn, H.T. Henderson, W.R. Heineman, H.B. Halsall: Development of a generic microfluidic system for electrochemical immunoassay-based remote bio/chemical sensors, Proc. Micro-Total Anal. Syst. '98, Banff, Canada 1998, ed. by D.J. Harrison, A. van den Berg (Kluwer, Dordrecht 1998) pp. 225–230

18.82 J.-W. Choi, C.H. Ahn, S. Bhansali, H.T. Henderson: A new magnetic bead-based, filterless bio-separator with planar electromagnet surfaces for integrated bio-detection systems, Sens. Actuators B **68**, 34–39 (2000)

18.83 O. Niwa, Y. Xu, H.B. Halsall, W.R. Heineman: Small-volume voltammetric detection of 4-aminophenol with interdigitated array electrodes and its application to electrochemical enzyme immunoassay, Anal. Chem. **65**, 1559–1563 (1993)

18.84 K.W. Oh, A. Han, S. Bhansali, H.T. Henderson, C.H. Ahn: A low-temperature bonding technique using spin-on fluorocarbon polymers to assemble microsystems, J. Micromech. Microeng. **12**, 187–191 (2002)

18.85 C.-C. Hong, J.-W. Choi, C.H. Ahn: An on-chip air-bursting detonator for driving fluids on disposable lab-on-a-chip systems, J. Micromech. Microeng. **17**, 410–417 (2007)

18.86 C. Gao, J.-W. Choi, M. Dutta, S. Chilukuru, J.H. Nevin, J.Y. Lee, M.G. Bissell, C.H. Ahn: A fully integrated biosensor array for measurement of metabolic parameters in human blood, Proc. 2nd Annu. Int. IEEE–EMBS Spec. Top. Conf. Microtechnol. Med. Biol., Madison May 2–4, 2002, ed. by A. Dittmar, D. Beebe (IEEE, New York 2002) pp. 223–226

18.87 C.H. Ahn, J.-W. Choi, G. Beaucage, J.H. Nevin, J.-B. Lee, A. Puntambekar, J.Y. Lee: Disposable Smart Lab on a Chip for Point-of-Care Clinical Diagnostics, Proc. IEEE (2004) pp. 154–173

18.88 J. Do, C.H. Ahn: A polymer lab-on-a-chip for magnetic immunoassay with on-chip sampling and detection capabilities, Lab. Chip **8**, 542–549 (2008)

18.89 J. Han, S.H. Lee, A. Puntambekar, S. Murugesan, J.-W. Choi, G. Beaucage, C.H. Ahn: UV adhesive bonding techniques at room temperature for plastic lab-on-a-chip, Proc. 7th Int. Conf. Micro Total Anal. Syst. (μ-TAS), Squaw Valley 2003, ed. by M.A. Northrup, K.F. Jensen, D.J. Harrison (Transducers Research Foundation, Cleveland 2003) pp. 1113–1116

18.90 J. Do, J.-W. Choi, C.H. Ahn: Low-cost magnetic interdigitated array on a plastic wafer, IEEE Trans. Magn. **40**, 3009–3011 (2004)

# 19. Centrifuge-Based Fluidic Platforms

Jim V. Zoval, Guangyao Jia, Horacio Kido, Jitae Kim, Nahui Kim, Marc J. Madou

In this chapter centrifuge-based microfluidic platforms are reviewed and compared with other popular microfluidic propulsion methods. The underlying physical principles of centrifugal pumping in microfluidic systems are presented and the various centrifuge fluidic functions such as valving, decanting, calibration, mixing, metering, heating, sample splitting, and separation are introduced. Those fluidic functions have been combined with analytical measurements techniques such as optical imaging, absorbance and fluorescence spectroscopy and mass spectrometry to make the centrifugal platform a powerful solution for medical and clinical diagnostics and high-throughput screening (HTS) in drug discovery. Applications of a compact disc (CD)-based centrifuge platform analyzed in this review include: two-point calibration of an optode-based ion sensor, an automated immunoassay platform, multiple parallel screening assays and cellular-based assays. The use of modified commercial CD drives for high-resolution optical imaging is discussed as well. From a broader perspective, we compare the technical barriers involved in applying microfluidics for sensing and diagnostic as opposed to applying such techniques to HTS. The latter poses less challenges and explains why HTS products based on a CD fluidic platform are already commercially available, while we might have to wait longer to see commercial CD-based diagnostics.

19.1	Why Centripetal Force for Fluid Propulsion?	532
19.2	Compact Disc or Microcentrifuge Fluidics	534
	19.2.1 How It Works	534
	19.2.2 Some Simple Fluidic Functions Demonstrated on a CD	535
19.3	CD Applications	538
	19.3.1 Two-Point Calibration of an Optode-Based Detection System	538
	19.3.2 CD Platform for Enzyme-Linked Immunosorbant Assays (ELISA)	539
	19.3.3 Multiple Parallel Assays	539
	19.3.4 Cellular-Based Assays on CD Platform	540
	19.3.5 Integrated Nucleic-Acid Sample Preparation and PCR Amplification	542
	19.3.6 Sample Preparation for MALDI MS Analysis	543
	19.3.7 Modified Commercial CD/DVD Drives in Analytical Measurements	544
	19.3.8 Microarray Hybridization for Molecular Diagnosis of Infectious Diseases	546
	19.3.9 Cell Lysis on CD	547
	19.3.10 CD Automated Culture of *C. Elegans* for Gene Expression Studies	548
19.4	Conclusion	549
References		550

Once it became apparent that individual chemical or biological sensors used in complex samples would not attain the hoped for sensitivity or selectivity, wide commercial use became severely hampered and sensor arrays and sensor instrumentation were proposed instead. It was projected that by using orthogonal sensor array elements (e.g., in electronic noses and tongues) selectivity would be improved dramatically [19.1]. It was envisioned that instrumentation would reduce matrix complexities through filtration, separation, and concentration of the target compound while at the same time ameliorating selectivity and sensitivity of the overall system by frequent recalibration and washing of the sensors. With microfluidics, the miniaturization of analytical equipment may potentially alleviate the shortcomings associated with large and expensive instrumentation through the reduction in reagent volumes, favorable scaling properties of several important instrument processes (basic theory of hydrodynamics and diffusion predicts faster heating and cooling and more efficient chromatographic and electrophoretic separations in miniaturized equipment) and batch fabrication which may enable low-cost disposable instruments to be used once and then thrown away to prevent sample contamination [19.2]. Micromachining (microelectromechanical systems (MEMS)) might also allow cofabrication of many integrated functional instrument blocks. Tasks that are now performed in a series of conventional bench top instruments could then be combined into one unit, reducing labor and minimizing the risk of sample contamination.

Today it appears that sensor-array development in electronic noses and tongues has slowed down because of the lack of highly stable chemical and biological sensors: too frequent recalibration of the sensors and relearning of the pattern recognition software is putting a damper on the original enthusiasm for this sensor approach. In the case of miniaturization of instrumentation through the application of microfluidics, progress was made in the development of platforms for high-throughput screening (HTS) as evidenced by new products introduced by, for example, Caliper and Tecan Boston [19.3, 4]. In contrast, progress with miniaturized analytical equipment remains limited; platforms have been developed for a limited amount of human and veterinary diagnostic tests that do not require complex fluidic design, see for example Abaxis [19.5]. In this review paper we are, in a narrow sense, summarizing the state of the art of compact disc (CD)-based microfluidics and in a broader sense we are comparing the technical barriers involved in applying microfluidics to sensing and diagnostic as opposed to applying such techniques to HTS. It will quickly become apparent that the former poses the more severe technical challenges and as a result the promise of lab-on-a-chip has not been fulfilled yet.

19.1 Why Centripetal Force for Fluid Propulsion?

There are various technologies for moving small quantities of fluids or suspended particles from reservoirs to mixing and reaction sites, to detectors, and eventually to waste or to a next instrument. Methods to accomplish this include syringe and peristaltic pumps, electrochemical bubble generation, acoustics, magnet-

Table 19.1 Comparison of microfluidics propulsion techniques

Comparison	Fluid propulsion mechanism			
	Centrifuge	Pressure	Acoustic	Electrokinetic
Valving solved?	Yes for liquids, no for vapor	Yes for liquids and vapor	No solution shown yet for liquid or vapor	Yes for liquids, no for vapor
Maturity	Products available	Products available	Research	Products available
Propulsion force influenced by	Density and viscosity	Generic	Generic	pH, ionic strength
Power source	Rotary motor	Pump, mechanical roller	5 to 40 V	10 kV
Materials	Plastics	Plastics	Piezoelectrics	Glass, plastics
Scaling	L^3	L^3	L^2	L^2
Flow rate	From less than 1 nL/s to greater than 100 μl/s	Very wide range (less than nL/s to L/s)	20 μl/s	0.001–1 μl/s
General remarks	Inexpensive CD drive, mixing is easy, most samples possible (including cells) Better for diagnostics	Standard technique Difficult to miniaturize and multiplex	Least mature of the four techniques Might be too expensive Better for smallest samples	Mixing difficult High-voltage source is dangerous and many parameters influence propulsion, better for smallest samples (HTS)

ics, direct-current (DC) and alternating-current (AC) electrokinetics, centrifuge, etc. In Table 19.1 we compare four of the more important and promising fluid propulsion means [19.6]. The pressure that mechanical pumps have to generate to propel fluids through capillaries is higher the narrower the conduit. Pressure and centripetal force are both volume-dependent forces, which scale as L^3 (in this case L is the characteristic length corresponding to the capillary diameter). Piezoelectric, electroosmotic, electrowetting and electrohydrodynamic (EHD) pumping (the latter two are not shown in Table 19.1) all scale as surface forces (L^2), which represent more favorable scaling behavior in the microdomain (propulsion forces scaling with a lower power of the critical dimension become more attractive in the microdomain) and lend themselves better to pumping in smaller and longer channels. In principle, this should make pressure- and centrifuge-based systems less favorable but other factors turn out to be more decisive; despite better scaling of the nonmechanical pumping approaches in Table 19.1, almost all biotechnology equipment today remain based on traditional external syringe or peristaltic pumps. The advantages of these approach are that they rely on well-developed, commercially available components and that a very wide range of flow rates is attainable. Although integrated micromachined pumps based on two one-way valves may achieve precise flow control on the order of 1 μl/min with fast response, high sensitivity, and negligible dead volume, these pumps generate only modest flow rates and low pressures, and consume a large amount of chip area and considerable power.

Acoustic streaming is a constant (DC) fluid motion induced by an oscillating sound field at a solid/fluid boundary. A disposable fluidic manifold with capillary flow channels can simply be laid on top of the acoustic pump network in the reader instrument. The method is considerably more complex to implement than electroosmosis (see next paragraph) but the insensitivity of acoustic streaming to the chemical nature of the fluids inside the fluidic channels and its ability to mix fluids make it a potentially viable approach. A typical flow rate measured for water in a small metal pipe lying on a piezoelectric plate is 0.02 cm³/s at 40 V, peak to peak [19.7]. Today acoustic streaming as a propulsion mechanism remains in the research stage.

Electro-osmotic pumping (DC electrokinetics) in a capillary does not involve any moving parts and is easily implemented. All that is needed is a metal electrode in some type of a reservoir at each end of a small flow channel. Typical electroosmotic flow velocities are on the order of 1 mm/s with a 1200 V/cm applied electric field. For example, in free-flow capillary electrophoresis work by *Jorgenson* and *Guthrie*, electroosmotic flow of 1.7 mm/s was reported [19.8]. This is fast enough for most analytical purposes. *Harrison* et al. achieved electroosmotic pumping with flow rates up to 1 cm/s in 20 μm capillaries that were micromachined in glass [19.9]. They also demonstrated the injection, mixing and reaction of fluids in a manifold of micromachined flow channels without the use of valves. The key aspect for tight valving of liquids at intersecting capillaries in such a manifold is the suppression of convective and diffusion effects. The authors demonstrated that these effects can be controlled by the appropriate application of voltages to the intersecting channels simultaneously. Some disadvantages of electroosmosis are the high voltage required (1–30 kV power supply) and direct electrical–fluid contact with resulting sensitivity of flow rate to the charge of the capillary wall

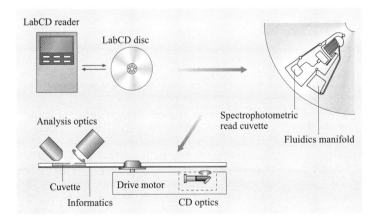

Fig. 19.1 LabCD instrument and disposable disc. Here, the analytical result is obtained through reflection spectrophotometry

and the ionic strength and pH of the solution. It is consequently more difficult to make it into a generic propulsion method. For example, liquids with high ionic strength cause excessive Joule heating; it is therefore difficult or impossible to pump biological fluids such as blood and urine.

Using a rotating disc, centrifugal pumping provides flow rates ranging from less than 10 nL/s to greater than 100 μL/s depending on disc geometry, rotational rate (revolutions per minute (RPM)), and fluid properties (Fig. 19.1) [19.10]. Pumping is relatively insensitive to physicochemical properties such as pH, ionic strength, or chemical composition (in contrast to AC and DC electrokinetic means of pumping). Aqueous solutions, solvents (e.g., dimethyl sulfoxide (DMSO)), surfactants, and biological fluids (blood, milk, and urine) have all been pumped successfully. Fluid gating, as we will describe in more detail further below, is accomplished using capillary valves in which capillary forces pin fluids at an enlargement in a channel until rotationally induced pressure is sufficient to overcome the capillary pressure (at the so-called burst frequency) or by hydrophobic methods. Since the types and the amounts of fluids one can pump on a centrifugal platform spans a greater dynamic range than for electrokinetic and acoustic pumps, this approach seems more amenable to sample preparation tasks than electrokinetic and acoustic approaches. Moreover miniaturization and multiplexing are quite easily implemented. A whole range of fluidic functions including valving, decanting, calibration, mixing, metering, sample splitting, and separation can be implemented on this platform and analytical measurements may be electrochemical, fluorescent or absorption based, and informatics embedded on the same disc could provide test-specific information.

An important deciding factor in choosing a fluidic system is the ease of implementing valves; the method that solves the valving issue most elegantly (traditional pumps) is already commercially accepted, even if it is not the most easily scaled method. In traditional pumps two one-way valves form a barrier for both liquids and vapors. In the case of the microcentrifuges, valving is accomplished by varying the rotation speed and capillary diameter. Thus, no real physical valve is required to stop water flow, but as in the case of acoustic and electrokinetic pumping, there is no simple means to stop vapors from spreading over the whole fluidic platform. If the liquids need to be stored for a long time, the valves, which are often disposable in sensing and diagnostics applications, must be barriers to both liquid and vapor. Some initial attempts at implementing vapor barriers on CDs will be reported in this review.

From the preceding comparison of fluidic propulsion methods for sensing and diagnostic applications, centrifugation in fluidic channels and reservoirs crafted in a CD-like plastic substrate as shown in Fig. 19.1 constitute an attractive fluidic platform.

19.2 Compact Disc or Microcentrifuge Fluidics

19.2.1 How It Works

CD fluid propulsion is achieved through centrifugally induced pressure and depends on rotation rate, geometry and the location of channels and reservoirs, and fluid properties. *Madou* and *Kellogg* [19.11] and *Duffy* et al. [19.10] characterized the flow rate of aqueous solutions in fluidic CD structures and compared the results to simple centrifuge theory. The average velocity of the liquid (U) from centrifugal theory is given as

$$U = D_h^2 \rho \omega^2 \bar{r} \Delta r / 32 L \mu , \tag{19.1}$$

and the volumetric flow rate (Q) as

$$Q = UA , \tag{19.2}$$

where D_h is the hydraulic diameter of the channel (defined as $4A/P$, where A is the cross-sectional area and P is the wetted perimeter of the channel), ρ is the density of the liquid, ω is the angular velocity of the CD, $\bar{r}$ is the average distance of the liquid in the channels to the center of the disc, Δr is the radial extent of the fluid, μ is the viscosity of the solution, and L is the length of the liquid in the capillary channel (Fig. 19.2). Flow rates ranging from 5 nL/s to > 0.1 mL/s have been achieved by various combinations of rotational speeds (400–1600 rpm), channel widths (20–500 μm), and channel depths (16–340 μm). The experimental flow rates were compared to rates predicted by the theoretical model and exhibited an 18.5% coefficient of variation. The authors note that experimental errors in measuring the highest and lowest flow rates made the largest contribution to this coefficient of variation. The absence of systematic deviation from the theory validates the model for describing flow in microfluidic channels under centripetal force. *Duffy* et al. [19.10]

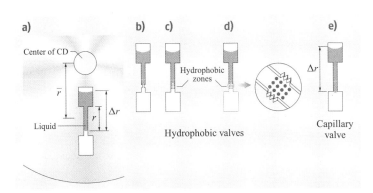

Fig. 19.2a–e Schematic illustrations for the description of CD microfluidics. (**a**) Two reservoirs connected by a microfluidic chamber. (**b**) Hydrophobic valve made by a constriction in a chamber made of hydrophobic material. (**c**) Hydrophobic valve made by the application of hydrophobic material to a zone in the channel. (**d**) Hydrophobic channel made by the application of hydrophobic material to a zone in a channel made with structured vertical walls (*inset*). (**e**) Capillary valve made by a sudden expansion in channel diameter such as when a channel meets a reservoir

measured flow rates of water, plasma, bovine blood, three concentrations of hematocrit, urine, dimethyl sulfoxide (DMSO), and polymerase chain reaction (PCR) products and report that centrifugal pumping is relatively insensitive to such physicochemical properties as ionic strength, pH, conductivity, and the presence of various analytes, noting good agreement between experiment and theory for all the liquids.

19.2.2 Some Simple Fluidic Functions Demonstrated on a CD

Fluid Mixing
In the work by *Madou* and *Kellogg* [19.11] and *Duffy* et al. [19.10], different means to mix liquids were designed, implemented, and tested. Observations of flow velocities in narrow channels on the CD enabled Reynolds numbers (Re) calculations that established that the flow remained laminar in all cases. Even in the largest fluidic channels tested Re was smaller than 100, well below the transition regime from laminar to turbulent flow (Re ≈ 2300) [19.12]. The laminar flow condition necessitates mixing by simple diffusion or by creating special features on the CD that enable advection or turbulence. In one scenario, fluidic diffusional mixing was implemented by emptying two microfluidic channels together into a single long meandering fluidic channel. Proper design of channel length and reagent reservoirs allowed for stoichiometric mixing in the meandering channel by maintaining equal flow rates of the two streams joining in the mixing channel. Concentration profiles may be calculated from the diffusion rates of the reagents and the time required for the liquids to flow through the tortuous path. Mixing can also be achieved by chaotic advection [19.6]. Chaotic advection is a result of the rapid distortion and elongation of the fluid–fluid interface, increasing the interfacial area where diffusion occurs, which increases the mean values of the diffusion gradients that drive the diffusion process; one may call this process an enhanced diffusional process. In addition to the simple and enhanced diffusional processes, one can create turbulence on the CD by emptying two narrow streams to be mixed into a common chamber. The streams violently splash against a common chamber wall, causing their effective mixing (no continuity of the liquid columns is required on the CD-based system, in contrast to the case of electrokinetics platforms where a broken liquid column would cause a voltage overload).

Valving
Valving is an important function in any type of fluidic platform. Both hydrophobic and capillary valves have been integrated into the CD platform [19.10, 11, 13–23]. Hydrophobic valves feature an abrupt decrease in the hydrophobic channel cross section, i.e., a hydrophobic surface prevents further fluid flow (Fig. 19.2b–d). In contrast, in capillary valves (Fig. 19.2e), liquid flow is stopped by a capillary pressure barrier at junctions where the channel diameter suddenly expands.

Hydrophobic Valving. The pressure drop in a channel with laminar flow is given by the Hagen–Poiseuille equation [19.12]

$$\Delta P = \frac{12L\mu Q}{wh^3}, \quad (19.3)$$

where L is the microchannel length, μ is the dynamic viscosity, Q is the flow rate, and w and h are the channel width and height, respectively. The pressure required to overcome a sudden narrowing in a rectangular channel is given by [19.6]

$$\Delta P = 2\sigma_l \cos\theta_c \left(\frac{1}{w_1} + \frac{1}{h_1}\right) - \left(\frac{1}{w_2} + \frac{1}{h_2}\right), \quad (19.4)$$

where σ_l is the liquid's surface tension, θ_c is the contact angle, w_1 and h_1 are the width and height of the channel before the restriction, and w_2 and h_2 are the width and height after the restriction, respectively. In hydrophobic valving, in order for liquid to move beyond these pressure barriers, the CD must be rotated above a critical speed, at which point the centripetal forces exerted on the liquid column overcome the pressure needed to move past the valve.

Ekstrand et al. [19.13] used hydrophobic valving on a CD to control discrete sample volumes in the nanoliter range with centripetal force. Capillary forces draw liquid into the fluidic channel until there is a change in the surface properties at the hydrophobic valve region. The valving was implemented as described schematically in Fig. 19.2c. *Tiensuu* et al. [19.14] introduced localized hydrophobic areas in CD microfluidic channels by inkjet printing of hydrophobic polymers onto hydrophilic channels. In this work, hydrophobic lines were printed onto the bottom wall of channels with both unstructured (Fig. 19.2c) and structured (Fig. 19.2d) vertical channel walls. Several channel width-to-depth ratios were investigated. The CDs were made by injection molding of polycarbonate and were subsequently rendered hydrophilic by oxygen plasma treatment. Ink-jet printing was used for the introduction of the hydrophobic polymeric material at the valve position. The parts were capped with polydimethylsiloxane (PDMS) to form the fourth wall of the channel. In testing of unstructured channels (without the sawtooth pattern) there were no valve failures for 300 and 500 μm wide channels but some failures for the 100 μm channels, however, in structured vertical walls (with sawtooth patterns), there were no valve failures. The authors attribute the better results of the structured vertical walls to both the favorable distribution of hydrophobic polymer within the channel and the sharper sidewall geometry to be wetted (the side walls are hydrophilic since the printed hydrophobic material is only on the bottom of the channel) compared to the unstructured vertical channel walls.

Capillary Valving. Capillary valves have been implemented frequently on CD fluidic platforms [19.10, 11, 15–18, 21, 22]. The physical principle involved is based on the surface tension, which develops when the cross section of a hydrophilic capillary expands abruptly as illustrated in Fig. 19.2e. As shown in this figure, a capillary channel connects two reservoirs, and the top reservoir (the one closest to the center of the CD) and the connecting capillary is filled with liquid. For capillaries with axisymmetric cross sections, the maximum pressure at the capillary barrier expressed in terms of the interfacial free energy [19.16] is given by

$$P_{cb} = 4\gamma_{al} \sin\theta_c / D_h , \quad (19.5)$$

where γ_{al} is the surface energy per unit area of the liquid–air interface, θ_c is the equilibrium contact angle, and D_h is the hydraulic diameter. Assuming low liquid velocities, the flow dynamics may be modeled by balancing the centripetal force and the capillary barrier pressure (19.5). The liquid pressure at the meniscus, from the centripetal force acting on the liquid, can be described as

$$P_m = \rho\omega^2 \bar{r} \Delta r , \quad (19.6)$$

where ρ is the density of the liquid, ω is the angular velocity, $\bar{r}$ is the average distance from the liquid element to the center of the CD, and Δr is the radial length of the liquid sample (Fig. 19.2a,e). Liquid will not pass a capillary valve as long as the pressure at the meniscus P_m is less than or equal to the capillary barrier pressure P_{cb}. *Zeng* and coworkers [19.16] named the point at which P_m equals P_{cb}, the critical burst condition and the rotational frequency at which it occurs they called the burst frequency. Experimental values of critical burst frequencies versus channel geometry, for rectangular cross sections over a range of channel sizes, show good agreement with simulation over the entire range of diameters studied. Since these simulations did not assume an axisymmetric capillary with a circular contact line and a diameter D_h, the meniscus contact line may be a complex shape. Burst frequencies were shown to be cross-section dependent for equal hydraulic diameters. The theoretical burst-frequency equation was modified as follows to account for variation of the channel cross section

$$\rho\omega^2 \bar{r} \Delta r < 4\gamma_{al} \sin\theta_c / (D_h)^n , \quad (19.7)$$

where $n = 1.08$ for an equilateral triangular cross section and $n = 1.14$ for a rectangular cross section. For *pipe flow* (circular cross section) an additional term is used in the burst-frequency expression

$$\rho\omega^2 \bar{r} \Delta r < 4\gamma_{al} \sin\theta_c / D_h \\ + \gamma_{al} \sin\theta_c (1/D_h - 1/D_0) , \quad (19.8)$$

where the empirically determined constant $D_0 = 40$ μm. The physical reason for the additional pipe-flow term, used to get a fit to the simulation results, is not well understood at this time.

Duffy et al. [19.10] modeled capillary valving by balancing the pressure induced by the centripetal force ($\rho\omega^2 \bar{r} \Delta r$) at the exit of the capillary with the pressure inside the liquid droplet being formed at the capillary outlet and the pressure required to wet the chamber beyond the valve. The pressure inside a droplet is given by the Young–Laplace equation [19.24]

$$\Delta P = \gamma(1/R_1 + 1/R_2),\qquad(19.9)$$

where γ is the surface tension of the liquid and R_1 and R_2 are the meniscus radii of curvature in the x- and y-dimensions of the capillary cross section. In the case of small circular capillary cross sections with spherical droplet shapes, $R_1 = R_2 \cong$ channel cross-section radius and (19.9) can be rewritten as

$$\Delta P = 4\gamma/D_h .\qquad(19.10)$$

On this basis *Duffy* et al. [19.10] derived a simplified expression for the critical burst frequency (ω_c) as

$$\rho\omega_c^2 \bar{r} \Delta r = a(4\gamma/D_h) + b ,\qquad(19.11)$$

with the first term on the right representing the pressure inside the liquid droplet being formed at the capillary outlet scaled by a factor a (for nonspherical droplet shapes) and the second term on the right b representing the pressure required to wet the chamber beyond the valve. The b term depends on the geometry of the chamber to be filled and the wettability of its walls.

A plot of the centripetal pressure ($\rho\omega_c^2 \bar{r} \Delta r$) at which the burst occurs verses $1/D_h$ was linear, as expected from (19.11), with a 4.3% coefficient of variation. The authors note a potential limitation with capillary valves due to the fact that liquids with low surface tension tend to wet the walls of the chamber at the capillary valve opening, resulting in the inability to gate the flow. The b term in (19.11) is beneficial in gating flow unless the surface walls at the abrupt enlargement of the capillary valve are so hydrophilic that the liquid is drawn past the valve and into the reservoir.

Badr et al. [19.17] and *Johnson* et al. [19.18] have designed a CD to sequentially valve fluids through a monotonic increase of rotational rate with progressively higher burst frequencies. The CD, shown in Fig. 19.3, was designed to carry out an assay for ions based on an optode-based detection scheme. The CD design employed five serial capillary valves opening at different times as actuated by rotational speed. Results showed good agreement between the observed and the calculated burst frequencies (see later).

It is very important to realize that the valves we mentioned thus far constitute liquid barriers and that they are not barriers for vapors. Vapor barriers must be implemented in any fluidic platform where reagents need to be stored for long periods of time. This is especially important for a disposable diagnostic assay platform. A multimonth, perhaps multiyear, shelf life would require vapor locks in order to prevent reagent solutions from drying or liquid evaporation and condensation in undesirable areas of the fluidic pathway. Tecan Boston have investigated vapor-resistant valves made of wax that was melted to actuate valve opening [19.25].

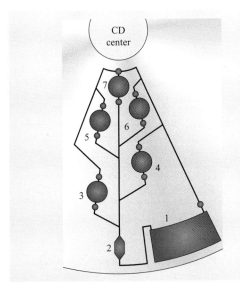

Fig. 19.3 Schematic illustration of the microfluidic structure employed for the ion-selective optode CD platform. The fluidic structure contains five solution reservoirs (numbered 1–5), a detection chamber (6), and a waste reservoir (7). Reservoir (1) and (3) contain the first and second calibrant, respectively, reservoirs (2) and (4) contain wash solutions, and reservoir (5) contains the sample. Upon increasing rotation rates, calibrant 1, wash 1, calibrant 2, rinse 2, and then sample were serially gated into the optical detection chamber. Absorption of the calibrants and sample was measured

Volume Definition (Metering) and Common Distribution Channels

The CD centrifugal microfluidic platform enables very fine volume control (or metering) of liquids. Precise volume definition is one of the important functions, necessary in many analytical sample-processing protocols, which has been added, for example, to the fluidic design in the Gyrolab MALDI SP1 CD [19.20]. In this CD, developed for matrix assisted laser desorption ionization (MALDI) sample preparation, a common distribution channel feeds several parallel individual sample-preparation fluidic structures (Fig. 19.4).

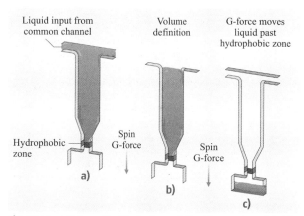

Fig. 19.4a–c Schematic illustration of liquid metering. (**a**) The common distribution channel and liquid metering reservoirs are filled (by capillary forces) with a reagent to be metered. Liquid entering the reservoir does not pass the hydrophobic zone (valve) because of surface tension forces. (**b**) The CD is rotated at a rate that supplies enough centripetal force to empty the common distribution but not enough to force the liquid through the hydrophobic zone. The volume of the fluid metered is determined by the volume of the reservoir. (**c**) A further increase in the rotational speed provides enough force to move the well-defined volume of solution past the hydrophobic valve (after [19.20])

Reagents are introduced by the capillary force exerted by the hydrophilic surfaces into the common channel and defined volume (200 nl) chambers until a hydrophobic valve stops the flow. When all of the defined-volume chambers are filled, the CD is spun at a velocity large enough to move the excess liquid from the common channel into the waste. Although there is sufficient centripetal force to empty the common channel, the velocity is not high enough to allow liquid to move past the hydrophobic valve and the well-defined-volume chambers remain filled. These precisely defined volumes can be introduced into the subsequent fluidic structures by increasing the CD angular momentum until the centripetal force allows the liquid to move past the hydrophobic barriers.

Packed Columns

Many commercial products are now available that use conventional centrifuges to move liquid, in a controlled manner, through a chromatographic column. One example is the Quick Spin protein desalting column (Roche Diagnostics Corp., Indianapolis), based on the size-exclusion principle. There is an obvious fit for this same type of separation experiment to be carried out on a CD fluidic device (we sometimes refer to the CD platform as a smart, miniaturized centrifuge). Affinity chromatography has been implemented in the fluidic design of the Gyrolab MALDI SP1 CD [19.20]. A reverse phase chromatography column material (SOURCE15 RPC) is packed into a microfluidic channel and protein is adsorbed on the column from an aqueous sample as it passes through the column under centrifugally controlled flow rates. A rinse solution is subsequently passed through the column and finally an elution buffer is flown through to remove the protein and carry it into the fluidic system for further processing. The complete Gyrolab MALDI SP1 CD is discussed in a later section of this review.

19.3 CD Applications

19.3.1 Two-Point Calibration of an Optode-Based Detection System

A CD based system with ion-selective optode detection and a two-point-calibration structure for the accurate detection of a wide variety of ions has been developed [19.15, 17, 18]. The microfluidic architecture, depicted in Fig. 19.3, is comprised of channels, five solution reservoirs, a chamber for colorimetric measurement of the optode membrane, and a waste reservoir, all manufactured onto a poly(methyl methacrylate) disc. Ion-selective optode membranes, composed of plasticized poly(vinyl chloride) impregnated with an ionophore, a chromoionophore, and a lipophilic anionic additive, were cast, with a spin-on device, onto a support layer and then immobilized on the disc. With this system, it is possible to deliver calibrant solutions, washing buffers, and unknown solutions (e.g., saliva, blood, urine, etc.) to the measuring chamber where the optode membrane is located. Absorbance measurements on a potassium optode indicate that optodes immobilized on the platform exhibit the theoretical absorbance response. Samples of unknown concentration can be quantified to within 3% error by fitting the response curve for a given optode membrane using an acid (for measuring the signal for a fully protonated chromoionophore), a base (for fully deprotonated chromoionophore), and two standard solutions. Further, the ability to measure ion concentrations employing one standard solution in conjunction with an acid and base, and with two standards alone were studied to delineate whether the current architecture could be simplified. Finally, the efficacy of incorporating washing steps into the calibration protocol was investigated.

This work was further extended to include anion-selective optodes and fluorescence rather than absorbance detection [19.17]. Furthermore, in addition to employing a standard excitation source where a fiber optic probe is coupled to a lamp, laser diodes were evaluated as excitation sources to enhance the fluorescence signal.

19.3.2 CD Platform for Enzyme-Linked Immunosorbant Assays (ELISA)

The automation of immunoassays on microfluidic platforms presents multiple challenges because of the high number of fluidic processes and the many different liquid reagents involved. Often there is also the need for highly accurate quantitative results at extremely low concentration and care must be taken to prevent nonspecific binding of reporter enzymes and to deliver well-defined volumes of reagents consistently. An enzyme-linked immunosorbant assay (i.e., ELISA) is one of the most common immunoassay methods and is often carried out in microtiter plates using labor-intensive manual pipetting techniques. Recently, *Lai* et al. [19.21] have implemented an automated enzyme-linked immunosorbant assay on the CD platform. This group used a five-step flow sequence in the same CD design illustrated in Fig. 19.3. A capture antibody (anti-rat IgG) was applied to the detection reservoir (reservoir 2 in Fig. 19.3) by adsorption to the PMMA CD surface, then the surface was blocked to prevent nonspecific binding. Antigen/sample (rat IgG), wash solution, second antibody, and substrate solutions were loaded into reservoirs 3–7 (Fig. 19.3) respectively. Using capillary valving techniques, the sample and reagents were pumped, one at a time, through the detection chamber. First, the sample was introduced for antibody antigen binding (reservoir 3), then a wash solution (reservoir 4), then an enzyme-labeled secondary antibody (reservoir 5), then another wash solution (reservoir 6), and finally the substrate was added (reservoir 7). The U-shaped bend in the fluidic path allows the solutions to incubate in the capture zone/detection chamber until the next solution is released into the chamber. Detection of the fluorescence is performed after the substrate is introduced into the detection reservoir (reservoir 2). Endpoint measurements (completion of enzyme–substrate reaction) were made and compared to conventional microtiter plate methods using similar protocols. The CD ELISA platform was shown to have advantages such as lower reagent consumption, and shorter assay times, explained in terms of larger surface-to-volume ratios, which favor diffusion-limited processes. Since the reagents were all loaded into the CD at the same time, there was no need for manual operator interventions in between fluidic assay steps. The consistent control and repeatability of liquid propulsion removes experimental errors associated with inconsistent manual pipetting methods, for example, rinsing/washing can be carried out not only with equal volumes, but equal flow conditions.

19.3.3 Multiple Parallel Assays

The ability to obtain simultaneous and identical flow rates, incubation times, mixing dynamics, and detection makes the CD an attractive platform for multiple parallel assays. *Kellogg* et al. [19.22] have reported on a CD system that performs multiple (48) enzymatic assays simultaneously by combining centrifugal pumping in microfluidic channels with capillary valving and colorimetric detection. The investigation of multiplexed parallel enzyme inhibitor assays are needed for high-throughput screening in diagnostics and in screening of drug libraries. For example, enzymatic dephosphorylation of colorless *p*-nitrophenol phosphate by alkaline phosphatase results in the formation of the yellow-colored *p*-nitrophenol and inhibition of this reaction may be quantified by light absorption measurement. Theophylline, a known inhibitor of the reaction, was used as the model inhibitory compound in *Kellogg* et al.'s [19.22] feasibility study. A single assay element on the CD contains three reservoirs: one for the enzyme, one for the inhibitor, and one for the substrate. Rotation of the CD allows the enzyme and inhibitor to pass capillary valves, mix in a meandering 100 μm-wide channel, and then move to a point where flow is stopped by another capillary valve. A further increase in the rotational speed allows the enzyme/inhibitor mixture and substrate to pass through the next set of capillary valves where they are mixed in a second meandering channel and emptied into an on-disc planar cuvette. The CD is slowed and absorption through each of the 48 parallel assay cuvettes is measured by reflectance, all in a period of 60 s, the entire fluidic process including measurement took about 3 min. The CDs were fabricated using

PDMS replication techniques [19.26], with the addition of a white pigment to the PDMS polymerization for enhanced reflectivity in the colorimetric measurements. The flow rates and meandering channel widths were selected such that the diffusion rate would allow 90% mixing of the solutions.

The variation in performance between the individual fluidic CD structures was quantified by carrying out the same assay 45 times simultaneously on a CD. The background-corrected absorbance was measured and the coefficient of variation in the assay was $\approx 3.2\%$. When the experiment was repeated on different discs the coefficient of variation was 3–3.5%. Furthermore, variation of absorption across a single cuvette was less than 1%, confirming complete mixing. In experiments to show enzyme inhibition, 45 simultaneous reactions were carried out on the CD using fixed concentrations of enzyme and substrate and 15 concentrations of theophylline in triplicate and a complete isotherm was generated for the inhibition of alkaline phosphatase. The three remaining structures were used for calibration with known concentrations of p-nitrophenol. A dose response was seen over three logs of theophylline concentration in the range of 0.1–100 mM. The authors concluded that a large number of identical assays, with applications in rapid, high-throughput screening, can be carried out on the CD platform simultaneously because of the symmetric force acting on the fluids in high-quality identical microfluidic structures and that detection was simplified by rotating all the reaction mixtures under a fixed detector. In later work [19.22], the same group has extended the number of assays to 96 per CD and has investigated fluorescent enzymatic assays.

19.3.4 Cellular-Based Assays on CD Platform

Cell-based assays are often used in drug screening [19.27] and rely on labor-intensive microtiter plate technologies. Microtiter plate methods may be difficult to automate without the use of large and expensive liquid-handling systems and they present problems with evaporation when scaled down to small volumes. *Thomas* et al. [19.23] have reported on a CD platform-based automated adherent cell system. This adherent cell assay involved introducing the compounds to be screened to a cell culture, then determining if the cells were killed (a cell viability assay).

Reagents for cell growth, rinsing and viability staining were serially loaded into an annular, common distribution chamber and centripetal force was used for reagent loading, exchange, and rinsing of the cell growth chamber (Fig. 19.5). Individual inlets were used for the addition of compounds to be screened. The plastic channels (Fig. 19.5b) were capped with a polydimethylsiloxane (PDMS) sheet capable of fast gas transport in and out of the culture reservoirs.

HeLa, L929, CHO-M1, and MRC-5 cell lines were cultivated on the CD device. Cell viability assays were performed, on the CD, by removing the growth medium from the cells, washing the cells with PBS, and introducing a solution of the fluorescence assay reagents into the growth chamber. The LIVE/DEAD Viability Assay (Molecular Probes, Inc., Eugene) uses a mixture of calcein green-fluorescent nucleic-acid stain and the red-fluorescent nucleic-acid stain, ethidium. The assay performance is based on the differing abilities of the stains to penetrate healthy bacterial cells. The calcein green-fluorescent dye will label all cells, live or dead. The red-fluorescent ethidium stain will only label cells with damaged membranes. The red stain causes a reduction in the green stain fluorescence when both dyes are present. When the appropriate mixture of green and red stains is used, cells with intact membranes will have a green fluorescence and cells with damaged membranes will have a red fluorescence. The background remains almost completely nonfluorescent (Fig. 19.6). All liquid transfers were carried out using centripetal force from CD rotation with angular frequencies of 200–600 rpm. Quantitative detection of multiple cell viability assays, within 30 s, was carried out by measurement of calcein fluorescence with a charge-coupled device (CCD)-based fluorescence imaging system. These experimental results show linear fluorescence intensity across the range

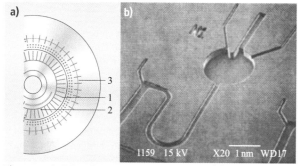

Fig. 19.5a,b Microfabricated cell-culture CD. (**a**) The CD caries a number of cell growth chambers (**1**) radially arranged around a common distribution channel (**2**) and is sealed with a silicone cover (**3**). (**b**) SEM close-up of an individual cell growth chamber and microfluidic connections (after [19.23])

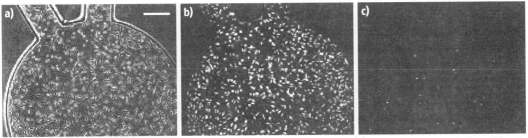

Fig. 19.6a–c L929 fibroblasts cultured for 48 hours in CD growth chambers. (**a**) Phase contrast (*scale bar* 100 μm), (**b**) epifluorescence image of calcein-stained viable cells, (**c**) epifluorescence image of ethidium-stained nonviable cells (after [19.23])

of 200–4000 cells and give an indication of the potential of this platform for miniaturized quantitative cell-based assays.

In the same work, the authors reported the results of experiments designed to investigate the effect on cells of using centripetal force to move liquids. The cells tested were shown to be compatible with centripetal forges of at least $600 \times g$, much larger than the $50–100 \times g$ needed for filling and emptying cell chambers. Furthermore, it was reported that cells grown in such devices appear to show the same cell morphology as cells grown under standard conditions.

In separate work done by our group in collaboration with NASA Ames [19.28] the LIVE/DEAD BacLight Bacterial Viability Kit (Molecular Probes, Inc., Eugene) has been integrated to a completely automated process on CD. Disposable and reusable CD structures, hardware, and software were developed for the LIVE/DEAD assay.

The CD design for assay automation must have the following functions or properties: contain separate reservoirs for each dye and the sample, retain those solutions in the reservoir until the disc is rotated at a certain velocity, evenly and completely mix the two dyes, evenly and completely mix the dye mixture with the sample containing the cells, collect this final mixture in a reservoir with good optical properties. Two methods for quick fabrication of prototype CDs were used. One method used molded PDMS structures. In a second method, a dry film photoresist (DF 8130, Think & Tinker, Palmer Lake) was laminated onto a 1 mm thick polycarbonate disc with predrilled holes for sample introduction. The microfluidic pattern was made using a photolithographic pattern on the negative photoresist. The fluidic system was capped with a polycarbonate disc that had been laminated with an optical-quality pressure-sensitive adhesive (3M 8142, 3M, Minneapo-

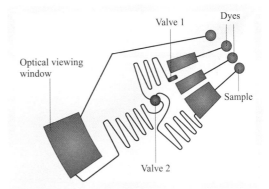

Fig. 19.7 Microfluidic pattern for LIVE/DEAD BacLight bacterial viability assay. The dyes and sample are introduced into the reservoir chambers using a pipette. The dyes fill the chamber stopping at a capillary valve (valve 1). Similarly, the sample containing cells is introduced into the sample reservoir. The disc is rotated to a velocity of 800 rpm, the dyes are forced through the capillary valves and they are mixed as they flow through the switchback turns of the microfluidic channels. Simultaneously, the sample passes from the reservoir into a fluid channel where it meets the dye mixture at valve 2. The velocity of the disc is increased to 1600 rpm and the dye mixture and sample combine and mix in the switchback microfluidic path leading to the optical viewing window

lis). Figure 19.7 shows the fluidic pattern for this assay. This pattern is based on the structure developed in a similar approach used to demonstrate multiple enzymatic assays on CD [19.10].

The dyes and sample were introduced into reservoir chambers using a pipette. The dyes fill the chamber stopping at a capillary valve (valve 1 in Fig. 19.7). Similarly, the sample containing cells was introduced

Fig. 19.8 Fluorescent microscopy overlaid images of *red*- and *green*-stained *E. coli* on CD from LIVE/DEAD BacLight bacterial viability assay

into the sample reservoir. Upon rotation, the dyes were forced through the capillary valves and were mixed as they flowed through the switchback turns of the microfluidic channels. Simultaneously, the sample passed from its reservoir into a fluid channel where it met the dye mixture at valve 2 of Fig. 19.7. The velocity of the disc was increased and the dye mixture and sample combine and mix in the switchback microfluidic path leading to the optical viewing window. The dye–sample mixture is allowed to incubate in the dark at room temperature for 5 min. The optical viewing chamber was imaged twice, once with optics for the green signal and then with optics for the red signal. A typical fluorescence microscopy image of an overlay of the red and green images of stained *E. coli* is shown in Fig. 19.8.

The instrument for disc rotation and fluorescence imaging (Fig. 19.9) used a programmable rotational motor for various velocities and acceleration/deceleration rates. The use of standard microscope objectives enabled magnification selection. An automatic focusing system was used. The light source was a mercury lamp, which used standard low-pass excitation filters for fluorescent excitation. A CCD camera was combined with standard emission filter cubes for imaging.

19.3.5 Integrated Nucleic-Acid Sample Preparation and PCR Amplification

Nucleic-acid analysis is often facilitated by the polymerase chain reaction (PCR) and requires substantial sample preparation that, unless automated, is labor extensive. After the initial sample preparation step of cell lyses to release the deoxyribonucleic acid (DNA)/ribonucleic acid (RNA), a step must be taken to prevent PCR inhibitors, usually proteins such as hemoglobin, from entering into the PCR thermocycle reaction. This can be done by further purification methods such as precipitation and centrifugation, solid-phase extraction, or by denaturing the inhibitory proteins. Finally, the sample must be mixed with the PCR reagents followed by thermocycling, a process that presents difficulty in a microfluidic environment because of the relatively high temperatures (up to 95 °C) required. In a small-volume microfluidic reaction chamber, the liquid will easily evaporate unless care is taken to prevent vapor from escaping.

Kellogg et al. [19.22] combine sample preparation with PCR on the CD. The protocol involves the following steps: (1) mixing raw sample (5 μL of dilute whole bovine blood or *E. coli* suspension) with 5 μL of 10 mM NaOH; (2) heating to 95 °C for 1–2 min (cell lyses and inhibitory protein denaturization); (3) neutralization of basic lysate by mixing with 5 μL of 16 mM tris-HCl (pH = 7.5); (4) neutralized lysate is mixed with 8–10 μL of liquid PCR reagents and primers of interest; and (5) thermal cycling. The CD fluidic design is shown schematically in Fig. 19.10. Three mixing channels are used in series to mix small volumes. A spinning platen allows control of the temperature by positioning thermoelectric devices against the appropriate fluidic chambers. The CD contacts the PC board platen on the

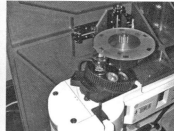

Fig. 19.9 *Left*: optical disc drive/imager with cover removed. Size of unit is made to fit in specific cargo bay of Space-Lab. *Right*: zoom of microscope objectives and a disc loaded in the drive

Fig. 19.10 Schematic illustration of the CD microfluidic PCR structure. The center of the disc is above the figure. The elements are (a) sample, (b) NaOH, (c) tris-HCl, (d) capillary valves, (e) mixing channels, (f) lysis chamber, (g) tris-HCl holding chamber, (h) neutralization lysate holding chamber, (i) PCR reagents, (j) thermal cycling chamber, (k) air gap. Fluids loaded in (a), (b), and (c) are driven at a first revolutions per minute (RPM) into reservoirs (g) and (f), at which time (g) is heated to 95 °C. The RPM is increased and the fluids are driven into (h). The RPM is increased and fluids in (h) and (i) flow into (j). On the *right*, the cross section shows the disc body (m), air gap (k), sealing layers (n), heat sink (l), thermoelectric (p), PC-board (q) and thermistor (o) (after [19.13]) ▶

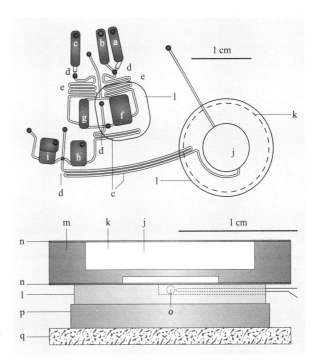

spindle of a rotary motor, with the correct angular alignment, which is connected by a slip ring to stationary power supplies and a temperature controller. Thermocouples are used for closed-loop temperature control and air sockets are used as insulators to isolate heating to reservoirs of interest. The thermoelectric at the PCR chamber both heats and cools and since the PCR reaction chamber is thin, 0.5 mm, fast thermocycling is achieved. Slew rates of $\pm 2\,°C/s$ with fluid volumes of 25 μL and thermal gradients across the liquid of 0.5 °C are reported. It is important to note here that the PCR chambers were not sealed; vapor generated inside the PCR chamber condensed on the cooler surfaces of the connecting microfluidic chamber and, since the CD is rotating, the condensed drops are centrifuged back into the hot PCR chamber. This microcondensation apparatus is unique for the centrifugal CD platform. Details of the experimental parameters used can be found in the original reference [19.22], but to summarize, sample preparation and PCR amplification for two types of samples, whole blood and *E. coli*, were demonstrated on the CD platform and shown to be comparable to conventional methods.

19.3.6 Sample Preparation for MALDI MS Analysis

MALDI MS peptide mapping is a commonly used method for protein identification. Correct identification and highly sensitive MS analysis require careful sample preparation. Manual sample preparation is quite tedious, time-consuming, and can introduce errors common to multistep pipetting. MALDI MS sample preparation protocols employ a protein digest followed by sample concentration, purification, and recrystallization with minimal loss of protein. Automation of the sample-preparation process, without sample loss or contamination, has been enabled on the CD platform by the Gyrolab MALDI SP1 CD and the Gyrolab Workstation (Gyros AB, Sweden) [19.20].

The Gyrolab MALDI SP1 sample-preparation CD will process up to 96 samples simultaneously using separate microfluidic structures. Protein digest from gels or solutions are concentrated, desalted, and eluted with matrix onto a MALDI target area. The CD is then transferred to a MALDI instrument for analysis without the need for further transfer to a separate target plate. The CD fluidic structure contains functions for common reagent distribution, volume definition (metering), valving, reverse phase column (RPC) for concentration and desalting, washing, and target areas for external calibrants. Figure 19.11 shows the Gyrolab MALDI SP1 sample preparation CD. The CDs are loaded with reagents and processed in a completely automated, custom workstation capable of holding up to five microtiter plates containing samples and reagents and up to five CD microlaboratories. The reagents are taken from the microplates to the CD inlets using a precision robotic arm fitted with multiple needles, the liquid is drawn into specific inlets by capillary forces, and then the needles are cleaned by rinsing at a wash sta-

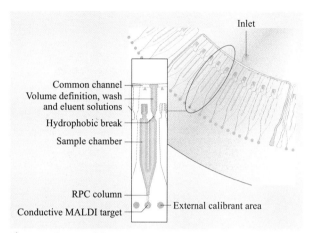

Fig. 19.11 Image of Gyrolab MALDI SP1 sample-preparation CD. The protein digest samples are loaded into the sample reservoir (*inset*) by capillary action. Upon rotation, the sample passes through the RPC column. The peptides are bound to the column and the liquid goes out of the system into the waste. A wash buffer is loaded into the common distribution channel and volume-definition chamber. The disc is rotated at a RPM that will empty the common distribution channel but not allow the wash solution to pass through the hydrophobic zone. A further increase of the RPM allows the well-defined volume of wash solution to pass the hydrophobic break and wash the RPC column then be discarded as waste. Next, a well-defined volume of the elution/matrix solution is loaded and passed through the column, taking the peptides to the MALDI target zone. The flow rate is controlled to optimize the evaporation of the solvent crystallization of the protein and matrix at the target zone (after [19.20])

tion. Samples are applied in aliquots from 200 nl up to 5 μl sequentially to each channel where it is contained using hydrophobic surface valves. The CD is then rotated, at an optimized rate, causing the sample to flow through an imbedded reverse phase chromatography column and liquid that passes through the column is collected in a waste container. Controlling the angular-velocity-dependent liquid flow rate maximizes protein binding to the column. A wash solution is introduced by capillary action into common distribution channels connected to groups of microstructures. The wash solution fills a volume definition chamber (200 nl) until it reaches a hydrophobic valve and the CD is rotated to clear the excess liquid in the distribution channel. Not until the rotational velocity is further increased is the defined wash volume able to pass through the hydrophobic valve and into the RPC column (SOURCE15 RPC). The peptides are eluted from the column and directly onto the MALDI target area using a solution that contains α-cyano-4-hydroxycinnamic acid and acetonitrile using the same common distribution channel and defined volume as the previous wash step. Optimization of rotational velocity during elution enables maximum recovery and balances the rate of elution with the rate of solvent evaporation from the target surface. Areas in and around the targets are gold-plated to prevent charging of the surface that would cause spectral mass shift and ensures uniform field strength. Well-defined matrix/peptide crystals form in the CD MALDI target area. Gyros reports high reproducibility, high sensitivity, and improved performances when compared to conventional pipette tip technologies. Data was shown that includes: comparison of 23 identical samples, processed in parallel on the same CD, from a bovine serum albumin (BSA) tryptic digest and analysis of identical samples processed on different CDs, run on different days. Sensitivities were shown in the attomole to femtomole range, indicating the ability to identify low-abundance proteins. The report attributed the superior performance of this platform to the pretreatment of the CD surface to minimized nonspecific adsorption of peptides, reproducible wash volume and flow, and reproducible elution (volume, flow, and evaporation) and crystallization.

19.3.7 Modified Commercial CD/DVD Drives in Analytical Measurements

The commercial CD/digital versatile disc (DVD) drive, commonly used for data storage and retrieval, can be thought of as a laser scanning imager. The CD drive retrieves optically generated electrical signals from the reflection of a highly focused laser light (spot size: full width at half maximum $\approx 1\,\mu\text{m}$), from a 1.2 mm thick polycarbonate disc that contains a spiral optical track feature. The track is fabricated by injection molding and is composed of a series of pits that are 1–4 μm long, 0.15 μm deep, and about 0.5 μm wide. The upper surface of a CD is made reflective by gold or aluminium metallization and protected with a thin plastic coating. Information is generated as the focused laser follows the spiral track by converting the reflected light signal into digital information. A flat surface gives a value of zero, an edge of a pit gives a value of one. The data is retrieved at a constant acquisition rate and the serial values (0/1) are converted to data of different kinds for various applications (music, data, etc.). In addition to the code generated by the spacing of the pits, optical signals necessary for focusing, laser tracking of the spiral track, and radial position determination of the read head are monitored and used in feedback loops for proper CD operation. The laser is scanned in a radial direction toward the outer diameter of the disc with an elaborate servo that maintains both lateral tracking and vertical focusing.

Researchers [19.29, 30] have taken advantage of this low-cost high-resolution optical platform in analytical DNA array applications. *Barathur* et al. [19.29] from Burstein Technologies (Irvine), for example, have modified the normal CD drive for use as a sophisticated laser-scanning microscope for analysis of a Bio Compact Disk assay, where all analysis is carried out in microfluidic chambers on the CD. The assay is carried out concurrently with the normal optical scanning capabilities of a regular CD drive. The authors report on the application of this device for DNA microspot-array hybridization assays and comment on its use in other diagnostic and clinical research applications. For the DNA spot-array application, arrays of captures probes for specific DNA sequences are immobilized on the surface of the CD in microfluidic chambers. Sample preparation and multiplexed PCR, using biotinylated primers, are carried out off-disc, then the biotinylated amplicons are introduced into the array chamber and hybridization occurs if amplicons with the correct sequence are present. Hybridization detection is achieved by monitoring the optical signal from the CD photodetector, while the CD is rotating. To generate an optical signal when hybridization has occurred a reporter is used, for the Bio Compact Disk assay the reporter is a streptavidin-labeled microsphere that will bind only to the array spots which have successfully captured biotinylated amplicons. The unbound microsphere reporters are removed from the array using simple centrifugation and no further rinsing is needed. As the laser is scanned across the CD surface, the microparticle scatters light that would have normally been reflected to the photodetector resulting in less light on the detector (bright-field microscopy) and a distinctive electronic signal is generated. The electronic signal-intensity data can be stored in memory then deconvoluted into an image. A $1\,\text{cm}^2$ microarray can be scanned in 20–30 s with a data-reduction time of 5 min and custom algorithms that perform the interpretations in real time. Data was shown for identification of three different species of the *Brucella* coccobacilli on the CD platform. Human infection occurs by transmission from animals by ingestion of infected food products, contact with an infected animal, or inhalation of aerosols. Multiplex PCR-amplified DNA from all three species (common forward primers and specific reverse primers resulting in amplicons of different length for various species, were used for verification of PCR on external gels) were incubated on arrays with species-specific capture probes. Removal of one of the species in the sample resulted in no probes present on that specific array spot, verifying the specificity of the assay.

Alexandre et al. [19.30] at Advanced Array Technology (Namur, Belgium), utilize the inner diameter area of a CD and standard servo optics for numerical information and operational control and employ a second scanning laser system to image DNA arrays on transparent surfaces at the outer perimeter of a CD. The second laser system, consisting of a laser-diode module that illuminates a $50\,\mu\text{m}$ spot on the CD surface, is scanned radially at a constant linear velocity of 20 mm/min while the CD is rotating. Each CD contains 15 arrays arranged in a single ring on the CD perimeter that extends in the radial direction for 15 mm. The arrays are rectangular and consist of four rows and 11 columns of $300\,\mu\text{m}$ spots. The normal CD servo optics are located below the disc and the added imaging optics are above the disc. A photodiode head follows the imaging laser and the refracted light intensity is stored digitally at a high sampling rate. An image of each array on the disc is reconstructed by deconvolution of the light-intensity data. The entire CD can be scanned in less than one minute, producing a total of 6 MB of information. Sample preparation and PCR amplification was carried out off-disc. Specific DNA capture probes were spotted on the surface of the CD using a custom arrayer that transfers the probes from a multiwell plate onto the surface of up to 12 discs using a robotic arm. Biotinylated amplicons are introduced onto the array chambers (one chamber for each array) and hybridization occurs if amplicons with the correct sequence are present. In order to get an optical signal that can be detected, after a rinse step, a solution of streptavidin-labeled colloidal gold particles is applied to the array followed by a Silver Blue solution (AAT, Namur, Belgium). The silver solution causes silver metal to grow on the gold particles, thereby making the hybridization-positive microarray spots refractive to the incident laser light. Results were shown for the detection of the five most common species of *Staphylococci* and an antibiotic-resistant strain. The *fem A* and *mec A* genes of the various species of *Staphylococci* were amplified by primers common to all *Staphylococci* species then hybridized to a microarray containing spots with probes specific for the different *Staphylococci* species. The array also included a capture probe for the genus *Staphylococci* and a probe for the *mec A* gene that is associated with methicillin resistance of the *Staphylococci* species. The results were digitized and quantified with software that is part of the custom Bio-CD workstation. Signal-to-noise ratios were above 50 for all positive signals.

19.3.8 Microarray Hybridization for Molecular Diagnosis of Infectious Diseases

In recent years, microarrays have become important tools for nucleic-acid analysis and gene-expression profiling. The expression of thousands of genes can be monitored in a single experiment using this technology.

A number of investigators have attempted to adapt this technology to rapidly detect infectious agents in clinical specimens for diagnostic purposes [19.31–35]. However, such systems are still in their infancy and most of them require technologically complex biochips with integrated heating/cooling systems [19.31, 32, 36]. The Madou group at UCI together with the Bergeron group at Laval University have reported [19.37] a CD-based microfluidic platform for DNA microarray analysis of infectious disease, presenting an elegant solution to automate and speed up microarray hybridization. Staphiloccocal-specific oligonucleotides were used as capture probes immobilized in 4×5 arrays of 125 μm spots on a standard 3×1 in glass slide. The layout of the array is shown in Fig. 19.12a. A flow cell is designed to realize the self-contained hybridization process in the CD platform. As shown in Fig. 19.12b, the flow cell consists of a hybridization column 1, aligned with the DNA microarray on the glass slide, a sample chamber 2, and a rinsing chambers 3 and 4. The reagent chambers are connected to the hybridization column with a microchannel which is 50 μm in width and 25 μm in depth. The flow cell is aligned with and adhered to the glass slide to form a DNA hybridization detection unit, up to five of which can be mounted into the CD platform fabricated from acrylic plastic using computer numerical control (CNC) machining (Fig. 19.12c). The reagents are positioned to be pumped through the hybridization column by centrifugal force in a sequence beginning with chamber 2 up to chamber 4 and this flow sequence is achieved by manipulating the balance between the capillary force and centrifugal pressure. The sample (chamber 2) is released first and flows over the 140 nl hybridization chamber (chamber 1) where the oligonucleotide capture

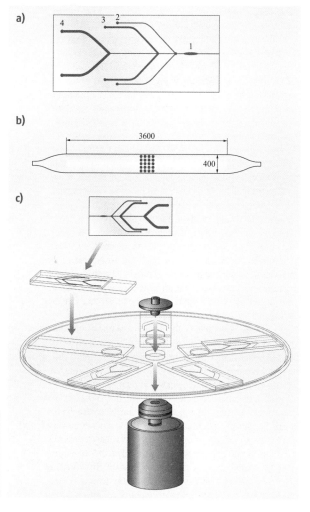

Fig. 19.12a–c Schematic representation of the microfluidic system. (**a**) PDMS microfluidic unit: The test sample (chamber 2) is released first and flows over the hybridization chamber (chamber 1) where the oligonucleotide capture probes are spotted onto the glass support. The wash buffer in chamber 3 and the rinsing buffer in chamber 4 then start to flow at a higher angular velocity. (**b**) Schematic view of the hybridization chamber showing the dimension in μm and the area of the chamber (*shaded section*) that can accommodate up to 150 microarray spots. Layout of the staphylococcal microarray used in the present study is also showed (five capture probes for each species). (**c**) Engraved PDMS is applied to a glass slide on which are arrayed nucleic-acid capture probes. The glass slide is placed on a compact disc support that can hold up to five slides ◄

probes are spotted onto the glass support. The rinsing buffers (chamber 3 and 4) are then released sequentially at a higher angular velocity and are used to wash the nonspecifically bound targets following the hybridization process.

This custom microarray hybridization microfluidic platform is easy to use, automated, and rapid. It uses standard glass slides which are compatible with commercial arrayers and standard commercial scanners found in most academic departments. In this removable microfluidic system, the hybridization chamber is composed of a low-cost elastomeric material, PDMS using standard moulding methods [19.26], engrafted with a microfluidic network. This elastomeric material reversibly sticks to the glass slide without any adhesives or chemical reactions, forming the microfluidic unit. Placed onto a plastic compact disc-like support, the microfluidic units are spun at different speeds to control fluid movements. To simplify hybridization experiments using this device, buffer compositions and capture probe sequences were optimized to be compatible with room-temperature hybridizations to prevent the need for a heating device. Furthermore, this microfluidic system allows one to drastically reduce the volume of reagents needed for microarray hybridizations and does not require a PCR amplicon purification step, which may be time-consuming.

In a passive hybridization system, a hybridization event requiring collision between a capture probe and the analyte relies solely on diffusion. In such systems, sensitivity is increased by using longer hybridization periods [19.38, 39]. One advantage of flow through hybridization is that the probability of collision between the probe and the analyte is increased by the much shorter diffusion distance allowed by the shallow hybridization chamber, thereby accelerating the hybridization kinetics [19.39–41]. In the study it was shown that for the same concentration of 15-mer oligonucleotides or 368 bp amplicons, a five-minute flow through hybridization increased the kinetics of hybridization respectively by a factor of 2.5 and 7.5, respectively, in comparison with the passive hybridization. These results are in line with a previous study. Using a microfluidic system, Chung et al. have shown a sixfold rate increase between flow-through hybridization versus passive hybridization. However, this system required a 30 min hybridization step [19.42]. Interestingly, the difference between passive and flow-through hybridization was about three times more important for the amplicons compared to the shorter 15 mer oligonucleotides [19.38]. This could be explained by the higher diffusion coefficient of the smaller oligonucleotide molecules.

To be used for clinical applications, in addition to being rapid and inexpensive, a molecular test should be sensitive and specific. In 5 min of hybridization, the CD system showed a detection limit of 500 amol of amplified target. This result is comparable with results obtained with more complex microfluidic devices [19.31, 43]. One system using chemiluminescence shows a detection limit of 250 amol, but requires a 3 h hybridization time [19.44]. In order to detect a significant fluorescent signal, an amplification step is required with microarray technology. The CD microfluidic system reported allows detection of amplicons amplified from 10 bacterial genome copies, which is at least 1000 times more sensitive than results obtained by other groups showing microarray hybridization using microfluidic devices [19.45].

In terms of specificity, the CD system was able to discriminate four different *Staphylococcus* species using a post-PCR hybridization protocol of only 15 min. The *S. aureus* probe designed with only one mismatch in the *S. epidermidis* amplicon sequence, did not show any significant cross-hybridization. This clearly demonstrates the possibility to discriminate one SNP using the CD system at room temperature and with only 10 µl of washing and rinsing buffer. This SNP discrimination capacity will allow rapid identification of bacteria and their antibiotic-resistance genes.

19.3.9 Cell Lysis on CD

There are many types of cell-lysis methods used today that are based on mechanical [19.46], physicochemical [19.47], chemical [19.48] and enzymatic [19.48] principles. The most commonly used methods in biology research labs rely on chemical and enzymatic principles. The main drawbacks of those procedures include intensive labor, adulteration of cell lysate, and the need for additional purification steps. In order to minimize the required steps for cell lysis, a rapid and reagent-less cell-lysis method would be very useful. Recently, cell lysis has been demonstrated by *Kim* et al. [19.49] on a microfluidic CD platform. In this purely mechanical lysis method, spherical particles (beads) in a lysis chamber microfabricated in a CD caused disruption of mammalian (CHO-K1), bacterial *Escherichia coli*, and yeast (*Saccharomyces cerevisiae*) cells. Investigators took advantage of interactions between beads and cells generated in rimming flow [19.50, 51] established inside a partially filled annular chamber

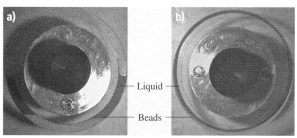

Fig. 19.13a,b Flow patterns for two rotational states of the CD. (**a**) At rest: beads sediment at the bottom of the annular chamber. (**b**) While spinning: two circumferential bands of beads (*lighter*) and liquid (*darker*) are observed

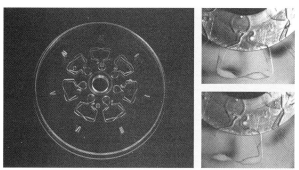

Fig. 19.14 *Left:* photograph of CD. *Right:* still images of a rotating CD (zirconia-silica beads and water-loaded). *Upper right*: more beads are observed on the left because of a rapid stop from a clockwise rotation. *Lower right*: more beads are on the right because of a rapid stop from a counterclockwise rotation

in the CD rotating around a horizontal axis (Fig. 19.13). To maximize bead–cell interactions in the lysis chamber, the CD was spun forward and backward around this axis, using high accelerations for 5–7 min. Cell disruption efficiency was verified either through direct microscopic viewing or measurement of the DNA concentration after cell lysing. Lysis efficiency relative to a conventional lysis protocol was $\approx 65\%$. Experiments identified the relative contribution of control parameters such as bead density, angular velocity, acceleration rate, and solid-volume fraction.

More recent work [19.52] by the same investigators used the multiplexed lysis design shown in Fig. 19.14. Bead–cell interactions for lysing arise while the beads and cells are pushed back and forth (by switching the CD rotational direction) through a continuously narrowing chamber wall. This phenomenon is called the *keystone effect*. There are two interaction forces associated with the keystone effect: collision induced by the geometry and friction due to a velocity gradient set up along the chamber (i.e., fast in the core and slow around the wall). The investigators used real-time PCR to characterize the performance of this CD design and achieved 95% lysis efficiency of *B. globigii* spores.

All prototype CDs in this work were fabricated using photolithography and PDMS molding. For the purpose of mechanical cell disruption, an ultra-thick SU-8 process was developed to fabricate a mold featuring high structures (≈ 1 mm) so that sufficiently high lysing chambers could be formed in the PDMS.

In the long term, this work is geared toward CD-based sample-to-answer nucleic-acid analysis which will include cell lysis, DNA purification, DNA amplification, and DNA hybridization detection.

19.3.10 CD Automated Culture of *C. Elegans* for Gene Expression Studies

Kim et al. [19.53] are developing a CD platform for automated cultivation and gene-expression studies of *C. elegans* nematodes. In this research, funded by NASA, the ultimate goal was to understand how a space environment, such as microgravity, hypergravity and radiation, affect various living creatures. The space environment can cause various physiological changes in organisms that have evolved in unit gravity ($1 \times g$) [19.54]. The CD platform is of particular interest in space studies because of its ability to provide a $1 \times g$ control using centripetal force, however its use is not particularly limited to these space study applications. A CD capable of the automated culture of *C. elegans* has been developed and is discussed in this section.

The culture system for *C. elegans* contains cultivation chambers, waste chambers, microchannels, and venting holes. Feeding and waste-removal processes are achieved automatically using centrifugal-force-driven fluidics. In this microfluidic system, the nutrient, Escherichia coli (*E. coli*) and the liquid media are automatically managed for the feeding and waste-removal processes. *C. elegans* was selected as a model organism for the gene-expression experiment in space due to its short lifespan (2–3 weeks), availability of green fluorescent protein (GFP) mutants, ease of laboratory cultivation and completely sequenced genome. Moreover, one can observe its transparent body with a microscope.

The main fabrication material of microfluidic platform is polydimethylsiloxane (PDMS), which is highly permeable to gases (a requirement for any aerobic culture), a chemically inert surface and optically trans-

Fig. 19.15 (a) Schematic illustration of the microfluidic structure employed for the CD cultivation system. The fluidic structure contains a nutrient reservoir (1), a cultivation chamber (2), and a waste reservoir (3). A liquid nutrient is loaded in a nutrient reservoir (1). Upon increasing the rotation rate of the system, the nutrient solution is gated into the cultivation chamber and some of the waste from the cultivation chamber can drain through the microchannels ($50 \times 40\,\mu m^2$). (b) A cross section of the waste chamber (4), waste-removal channels (5) in (b). Note that (a) 3 is the same as (b) 4 ▶

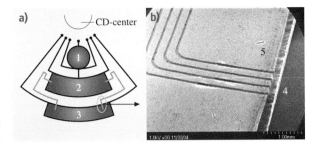

Fig. 19.16 The number of *C. elegans* nematodes cultivated on *E. coli* in S-medium in a CD-based culture under unit gravity over a 14 d period. Values are mean ± standard error of the mean (SEM); $n = 9$ (cultivation discs) ▶

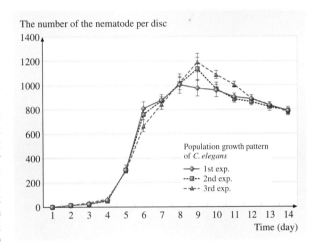

parent down to 300 nm such that it can be used to observe the behavior of *C. elegans*.

The CD assembly has a two-layer PDMS structure (Fig. 19.15) [19.55]. One layer contains the low-height channels (40 μm, Figs. 19.2 and 19.5) for draining waste from the cultivation chamber, and the other layer contains a cultivation chamber, a loading chamber and microfluidic connections all with a height of 1 mm. This design allows only wastes such as ammonia, not adult worms (80 μm diameter), to be moved from the cultivation chamber to the waste chamber. Cultivation of *C. elegans* was successfully carried out in the CD cultivation system for a period of up to two weeks. *C. elegans* show a specific population growth pattern (Fig. 19.15). Based on these results, the Madou Group and NASA have begun further development of the CD platform for gene-expression experiments to evaluate gene-expression changes in *C. elegans* upon exposure to altered gravity conditions and other factors.

19.4 Conclusion

In comparing miniaturized centrifugal fluidic platforms to other available microfluidic propulsion methods we have demonstrated how CD-based centrifugal methods are advantageous in many analytical situations because of their versatility in handling a wide variety of sample types, ability to gate the flow of liquids (valving), simple rotational motor requirements, ease and economic fabrication methods, and the large range of flow rates attainable. Most analytical functions required for a lab-on-a-disc, including metering, dilution, mixing, calibration, separation, etc., have all been successfully demonstrated in the laboratory. Moreover, the possibility of maintaining simultaneous and identical flow rates, to perform identical volume additions, to establish identical incubation times, mixing dynamics, and detection in a multitude of parallel CD assay elements makes the CD an attractive platform for multiple parallel assays. The platform has been commercialized by Tecan Boston for high-throughput screening (HTS) [19.4], by Gyros AB for sample-preparation techniques for MALDI [19.20] and by *Abaxis* (in a somewhat larger and less-integrated rotor format compared to the CD format) for human and veterinary diagnostic blood analysis [19.5]. The Abaxis system for human and veterinary medicine uses only dry reagents, but for many diagnostic assays, requiring more fluidic steps, there are severe limitations in progressing toward the lab-on-a-disc goal, as liquid storage on the disc becomes necessary. In high-throughput screening (HTS) applications, the CD platform is being coupled to automated liquid-reagent loading systems and no liquids/reagents need to be stored on the disc. The latter makes the commercial introduction of the CD platform for HTS somewhat simpler [19.4, 20, 30]. There is an urgent need though for the development of methods for long-term reagent storage that incorporates both liquid and vapor barriers to enable the introduction of lab-on-a-disc platforms for a wide variety of fast diagnostic tests. One possible solution to this problem involves the use of lyophilized reagents with common hydration reservoir feeds, but the issue in this situation becomes the speed of the test as the time required to redissolve the lyophilized reagents is often substantial.

The CD platform is easily adapted to optical detection methods because it is manufactured with high-optical-quality plastics, enabling absorption, fluorescence, and microscopy techniques. Additionally, the technology developed by the optical disc industry is being used to image the CD at micrometer resolution and move to DVD and high-density (HD) DVD will allow submicrometer resolution. The latter evolution will continue to open up new applications for the CD-based fluid platform. Whereas today the CD fluidic platform may be considered a smart microcentrifuge, we believe that in the future the integration of fluidics and informatics on DVDs and HD DVDs may lead to a merging of informatics and fluidics on the same disc. One can then envision making very sharp images of the bacteria under test and correlate both test and images with library data on the disc.

References

19.1 Alpha-MOS: http://www.alpha-mos.com (Alpha-MOS, Hillsborough 2008) Example for commercial electronic noses and tongues

19.2 A. Manz, E. Verpoorte, C.S. Effenhauser, N. Burggraf, D.E. Raymond, D.J. Harrison, H.M. Widmer: Miniaturization of separation techniques using planar chip technology, HRC J. High Resolut. Chromatogr. **16**, 433–436 (1993)

19.3 Caliper Life Science: http://www.caliperLS.com (Caliper Life Science, Hopkinton 2008)

19.4 Tecan: *Look for LabCD-ADMET System* (Tecan, Boston 2008), http://www.tecan-us.com/us-index.htm

19.5 Abaxis: http://www.abaxis.com (Abaxis, Union City 2008)

19.6 M.J. Madou: *Fundamentals of Microfabrication*, 2nd edn. (CRC, Boca Raton 2002)

19.7 S. Miyazaki, T. Kawai, M. Araragi: A piezo-electric pump driven by a flexural progressive wave, Proc. IEEE Micro Electro Mech. Syst. (MEMS '91) (Nara 1991) pp. 283–288

19.8 J.W. Jorgenson, E.J. Guthrie: Liquid chromatography in open-tubular columns, J. Chromatogr. **255**, 335–348 (1983)

19.9 D.J. Harrison, Z. Fan, K. Fluri, K. Seiler: Integrated electrophoresis systems for biochemical analyses, Solid State Sens. Actuator Workshop, Tech. Dig. (Hilton Head Island 1994) pp. 21–24

19.10 D.C. Duffy, H.L. Gills, J. Lin, N.F. Sheppard, G.J. Kellogg: Microfabricated centrifugal microfluidic systems: Characterization and multiple enzymatic assays, Anal. Chem. **71**(20), 4669–4678 (1999)

19.11 M.J. Madou, G.J. Kellogg: A centrifuge-based microfluidic platform for diagnostics, LabCD **3259**, 80–93 (1998)

19.12 G.T.A. Kovacs: *Micromachined Transducers Sourcebook* (Dordrecht/WCB/McGraw-Hill, Boston 1998) pp. 787–793

19.13 G. Ekstrand, C. Holmquist, A. Edman Örlefors, B. Hellman, A. Larsson, P. Anderson: Microfluidics in a rotating CD. In: *Micro Total Analysis Systems 2000*, ed. by A. van den Berg, W. Olthuis, P. Bergveld (Kluwer, Dordrecht 2000) pp. 311–314

19.14 A.-L. Tiensuu, O. Öhman, L. Lundbladh, O. Larsson: Hydrophobic valves by ink-jet printing on plastic CDs with integrated microfluidics. In: *Micro Total Analysis Systems 2000*, ed. by A. van den Berg, W. Olthuis, P. Bergveld (Kluwer, Dordrecht 2000) pp. 575–578

19.15 M.J. Madou, Y. Lu, S. Lai, J. Lee, S. Daunert: A centrifugal microfluidic platform – A comparison. In: *Micro Total Analysis Systems 2000*, ed. by A. van den Berg, W. Olthuis, P. Bergveld (Kluwer, Dordrecht 2000) pp. 565–570

19.16 J. Zeng, D. Banerjee, M. Deshpande, J.R. Gilbert, D.C. Duffy, G.J. Kellogg: Design analysis of capillary burst valves in centrifugal microfluidics. In: *Micro Total Analysis Systems 2000*, ed. by A. van den Berg, W. Olthuis, P. Bergveld (Kluwer, Dordrecht 2000) pp. 579–582

19.17 I.H.A. Badr, R.D. Johnson, M.J. Madou, L.G. Bachas: Fluorescent ion-selective optode membranes incorporated onto a centrifugal microfluidics platform, Anal. Chem. **74**(21), 5569–5575 (2002)

19.18 R.D. Johnson, I.H.A. Badr, G. Barrett, S. Lai, Y. Lu, M.J. Madou, L.G. Bachas: Development of a fully integrated analysis system for ions based on ion-selective optodes and centrifugal microfluidics, Anal. Chem. **73**(16), 3940–3946 (2001)

19.19 M. McNeely, M. Spute, N. Tusneem, A. Oliphant: Hydrophobic microfluidics, Proc. Microfluid. Dev. Syst. **3877**, 210–220 (1999)

19.20 Gyros AB: *Gyrolab MALDA SP1* (Gyros AB, Uppsala 2008), Application Report 101

19.21 S. Lai, S. Wang, J. Luo, J. Lee, S. Yang, M.J. Madou: Design of a compact disk-like microfluidic platform for enzyme-linked immunosorbent assay, Anal. Chem. **76**(7), 1832–1837 (2004)

19.22 G.J. Kellogg, T.E. Arnold, B.L. Carvalho, D.C. Duffy, N.F. Sheppard: Centrifugal microfluidics: Applications. In: *Micro Total Analysis Systems 2000*, ed. by A. van den Berg, W. Olthuis, P. Bergveld (Kluwer, Dordrecht 2000) pp. 239–242

19.23 N. Thomas, A. Ocklind, I. Blikstad, S. Griffiths, M. Kenrick, H. Derand, G. Ekstrand, C. Ellström, A. Larsson, P. Anderson: Integrated cell based assays in microfabricated disposable CD devices. In: *Micro Total Analysis Systems 2000*, ed. by A. van den Berg, W. Olthuis, P. Bergveld (Kluwer, Dordrecht 2000) pp. 249–252

19.24 A.W. Anderson: *Physical Chemistry of Surfaces* (Wiley, New York 1960) pp. 5–6

19.25 Private communication with G.J. Kellogg of Tecan, Boston (2008)

19.26 D.C. Duffy, J.C. McDonald, O.J.A. Schueller, G.M. Whitesides: Rapid prototyping of microfluidic systems in poly(dimethylsiloxane), Anal. Chem. **70**, 4974–4984 (1998)

19.27 J. Burbaum: Miniaturization technologies in HTS: How fast, how small, how soon?, Drug Discov. Today **3**(7), 313–322 (1998)

19.28 J.V. Zoval, R. Boulanger, C. Blackwell, B. Borchers, M. Flynn, D. Smernoff, R. Landheim, R. Mancinelli, M.J. Madou: Work done by this group, SETI Institute, Orbital Sciences, Dysyscon Inc., Stanford University and NASA Ames (2008)

19.29 R. Barathur, J. Bookout, S. Sreevatsan, J. Gordon, M. Werner, G. Thor, M. Worthington: New disc-based technologies for diagnostic and research applications, Psychiatr. Genet. **12**(4), 193–206 (2002)

19.30 I. Alexandre, Y. Houbion, J. Collet, S. Hamels, J. Demarteau, J.-L. Gala, J. Remacle: Compact disc with both numeric and genomic information as DNA microarray platform, BioTechniques **33**(2), 435–439 (2002)

19.31 Y. Wang, B. Vaidya, H.D. Farquar, W. Stryjewski, R.P. Hammer, R.L. McCarley, S.A. Soper, Y.W. Cheng, F. Barany: Microarrays assembled in microfluidic chips fabricated from poly(methyl methacrylate) for the detection of low-abundant DNA mutations, Anal. Chem. **75**, 1130–1140 (2003)

19.32 V. Mikhailovich, S. Lapa, D. Gryadunov, A. Sobolev, B. Strizhkov, N. Chernyh, O. Skotnikova, O. Irtuganova, A. Moroz, V. Litvinov, M. Vladimirskii, M. Perelman, L. Chernousova, V. Erokhin, A. Zasedatelev, A. Mirzabekov: Identification of rifampin-resistant mycobacterium tuberculosis strains by hybridization, PCR, and ligase detection reaction on oligonucleotide microchips, J. Clin. Microbiol. **39**, 2531–2540 (2001)

19.33 S. Bekal, R. Brousseau, L. Masson, G. Prefontaine, J. Fairbrother, J. Harel: Rapid identification of Escherichia coli pathotypes by virulence gene detection with DNA microarrays, J. Clin. Microbiol. **41**, 2113–2125 (2003)

19.34 S.G. Bavykin, J.P. Akowski, V.M. Zakhariev, V.E. Barsky, A.N. Perov, A.D. Mirzabekov: Portable system for microbial sample preparation and oligonucleotide microarray analysis, Appl. Environ. Microbiol. **67**, 922–928 (2001)

19.35 L. Westin, C. Miller, D. Vollmer, D. Canter, R. Radtkey, M. Nerenberg, J.P. O'Connell: Antimicrobial resistance and bacterial identification utilizing a microelectronic chip array, J. Clin. Microbiol. **39**, 1097–1104 (2001)

19.36 H.Z. Fan, S. Mangru, R. Granzow, P. Heaney, W. Ho, Q. Dong, R. Kumar: Dynamic DNA hybridization on a chip using paramagnetic beads, Anal. Chem. **71**, 4851–4859 (1999)

19.37 R. Peytavi, F. Raymond, D. Gagné, K. Boissinot, F. Picard, M. Boissinot, L. Bissonnette, M. Ouellette, M. Bergeron: Microfluidic device for rapid (<15 min) automated microarray hybridization, Clin. Chem. **51**, 10 (2005)

19.38 V. Chan, D.J. Graves, S.E. McKenzie: The biophysics of DNA hybridization with immobilized oligonucleotide probes, Biophys. J. **69**, 2243–2255 (1995)

19.39 M.K. McQuain, K. Seale, J. Peek, T.S. Fisher, S. Levy, M.A. Stremler, F.R. Haselton: Chaotic mixer improves microarray hybridization, Anal. Biochem. **325**, 215–226 (2004)

19.40 E. Bringuier, A. Bourdon: Colloid transport in nonuniform temperature, Phys. Rev. E **67**, 011404 (2003)

19.41 D. Axelrod, M.D. Wang: Reduction-of-dimensionality kinetics at reaction-limited cell-surface receptors, Biophys. J. **66**, 588–600 (1994)

19.42 Y.C. Chung, W.N. Chang, Y.C. Lin, M.Z. Shiu: Microfluidic chip for fast nucleic acid hybridization, Lab Chip **3**, 228–233 (2003)

19.43 R.H. Liu, R. Lenigk, R.L. Druyor-Sanchez, J. Yang, P. Grodzinski: Hybridization enhancement using cavitation microstreaming, Anal. Chem. **75**, 1911–1917 (2003)

19.44 B.J. Cheek, A.B. Steel, M.P. Torres, Y.Y. Yu, H. Yang: Chemiluminescence detection for hybridization assays on the flow-thru chip, a three-dimensional microchannel biochip, Anal. Chem. **73**, 5777–5783 (2001)

19.45 R. Lenigk, R.H. Liu, M. Athavale, Z. Chen, D. Ganser, J. Yang, C. Rauch, Y. Liu, B. Chan, H. Yu, M. Ray, R. Marrero, P. Grodzinski: Plastic biochannel hybridization devices: A new concept for microfluidic DNA arrays, Anal. Biochem. **311**, 40–49 (2002)

19.46 D. Di Carlo, K.-H. Jeong, L.P. Lee: Reagentless mechanical cell lysis by nanoscale barbs in microchannels for sample preparation, Lab Chip **3**, 287–291 (2003)

19.47 S.W. Lee, Y.-C. Tai: A micro cell lysis device, Sens. Actuators A **73**, 74–79 (1999)

19.48 J. Sambrook, D.W. Russell: *Molecular Cloning* (CSHL, Cold Spring Harbor 2001)

19.49 J. Kim, S.H. Jang, G. Jia, J.V. Zoval, N.A. Da Silva, M.J. Madou: Cell lysis on a microfluidic CD (compact disc), Lab Chip **4**, 516–522 (2004)

19.50 K.K.J. Ruschak, L.E. Scriven: Rimming flow of liquid in a rotating horizontal cylinder, J. Fluid Mech. **76**, 113–125 (1976)

19.51 S.T. Thoroddsen, L. Mahadevan: Experimental study of coating flows in a partially-filled horizontally rotating cylinder, Exp. Fluids **23**, 1–13 (1997)

19.52 J. Kim, H. Kido, J.V. Zoval, D. Gagné, R. Peytavi, F.J. Picard, M. Bastien, M. Boissinot, M.G. Bergeron, M.J. Madou: Rapid and automated sample preparation for nucleic acid extraction on a microfluidic CD, Int. Symp. LifeChips (2006), Poster Abstracts

19.53 N. Kim, C.M. Dempsey, J.V. Zoval, J. Sze, M.J. Madou: Automated microfluidic compact disc (CD) cultivation system of C. elegans, Sens. Actuators B **122**(2), 511–518 (2007)

19.54 E. Le Bourg: A review of the effects of microgravity and of hypergravity on aging and longevity, Exp. Gerontol. **34**, 319–336 (1999)

19.55 H. Wu, T.W. Odom, D.T. Chiu, G.M. Whitesides: Fabrication of complex three-dimensional microchannel systems in PDMS, J. Am. Chem. Soc. **125**, 554–559 (2003)

20. Micro-/Nanodroplets in Microfluidic Devices

Yung-Chieh Tan, Shia-Yen Teh, Abraham P. Lee

Fluid is often transported in the form of droplets in nature. From the formation of clouds to the condensation of dew on leaves, droplets are formed spontaneously in the air, on solids, and in immiscible fluids. In biological systems, droplets with lipid bilayer membranes are used to transport subnanoliter amounts of reagents between organelles, between cells, and between organs, in processes that control our day-to-day metabolic activities. The precision of such systems is self-evident and proves that droplet-based systems provide intrinsically efficient ways to perform controlled transport, reactions, and signaling.

This precision and efficiency can be utilized in many lab-on-chip applications by manipulating individual droplets using microfabricated force gradients. Complex segmented flow processes involving generating, fusing, splitting, and sorting droplets have been developed to digitally control fluid volumes and concentrations to nanoliter levels. In this chapter, microfluidic techniques for manipulating droplets are reviewed and analyzed.

20.1	Active or Programmable Droplet Systems	554
	20.1.1 Electrowetting on Dielectric-Based Droplet Microfluidic Devices	554
	20.1.2 Operational Principle of EWOD	554
	20.1.3 Reagent Mixing in EWOD	556
	20.1.4 Improvements in EWOD	556
	20.1.5 Droplet Manipulation via Dielectrophoresis	556
20.2	Passive Droplet Control Techniques	557
	20.2.1 Generation of Monodispersed Droplets	558
	20.2.2 Devices Based on Microcapillary Arrays	559
	20.2.3 Double Emulsions	559
20.3	Applications	564
	20.3.1 Droplet as Microtemplate and Encapsulation Agents	564
	20.3.2 Droplets as Real-Time Chemical Processors and Combinatorial Synthesizers	565
	20.3.3 Droplets as Micromechanical Components	566
20.4	Conclusions	566
	References	566

Droplet microfluidics, also termed *digital microfluidics*, has been the focus of much interest due to its potential for manipulating small quantities of reagent volumes and controlling complex reaction processes. Conventional microfluidics, or analog microfluidics, is governed by diffusion and low-Reynolds-number laminar flow that prevent rapid mixing of miscible liquids. When a liquid is dispersed into another immiscible liquid as droplets, mixing is on the millisecond scale due to the decreased striation area between miscible fluid interfaces. The ability to split and fuse individual droplets further simplifies the control of reagent volume and concentrations. Multiphase interactions, such as polymerization at the monomer–initiator fluid interface, lipid self-assembly at the oil–water interface, and ion exchange at pH-sensitive interfaces, all add to the benefits of droplet-based systems. These advantages have transformed the application of droplet technology, making it useful not just for making bulk quantities of particles, but also as an integrated multidisciplinary tool that can be applied to fields ranging from material synthesis to molecular biology [20.1].

The versatility of droplet microfluidic systems lies essentially in the ability to selectively *dial-in* the fluidic volume and concentration of a single droplet. This can be achieved either actively or passively. Active devices use dynamic modulation of forces around the droplet to achieve reconfigurable droplet flows. Passive devices use passive gradients to manipulate droplets through a set of operations governed by channel geometries and predefined wetting properties. In the next two sections, mechanisms for the actuation and control of droplets will be introduced, and important parameters and design criteria will be reviewed.

20.1 Active or Programmable Droplet Systems

Active droplet microfluidic systems dynamically modulate droplet motion and offer real-time control of changes in droplet chemistry. Individual droplets can be transported on demand in opposite directions and then fused to mix reagents in order to induce a chemical reaction. The system can also be programmed to split droplets in order to break up reaction products into desired volumes.

While a vast literature exists on different droplet actuation mechanisms, including the use of Marangoni flows [20.3], active chemical gradient [20.4], acoustic energy, pneumatic pressure [20.5], electrowetting [20.2, 6–19], and thermocapillary flows [20.20–24], given the scope of this chapter it is not possible to review all of these methods of droplet actuation. Instead, this section will focus on actuation mechanisms that have demonstrated the ability to accurately dispense liquid volumes and consistently split and mix droplet volumes, as well as rapid actuation schemes.

20.1.1 Electrowetting on Dielectric-Based Droplet Microfluidic Devices

Electrowetting is a method that modulates the surface tension of material by applying an electric field. Since most other forces diminish at the microscale, surface tension becomes a dominant force, and thus controlling the interfacial tension is an attractive way to actuate fluids at the microlevel. While there are several different types of electrowetting (EW) configurations, electrowetting on dielectric (EWOD) is the method most widely used to control droplets. EWOD can be applied to virtually any aqueous liquid [20.11]. Furthermore, since the interfacial energy is varied on a dielectric layer in EWOD, the problem of electrode electrolysis is avoided [20.12], and it can be used to drive droplets in both oil and air. While applications of EW and continuous electrowetting (CEW) have been mostly limited to droplet dispensing and actuation of metallic liquids, EWOD has found applications in the generation, fusion, and fission of droplets of a wide range of chemicals and biological fluids.

20.1.2 Operational Principle of EWOD

EW, CEW, and EWOD are all based upon Lippman's principle [20.25]

$$\gamma_{\text{SL}} = \gamma_{\text{SL}}^0 - \frac{\varepsilon V^2}{2d}, \tag{20.1}$$

where γ_{SL}^0 is the interfacial tension in the absence of the applied potential V, ε is the dielectric constant, and d is the thickness of the insulating layer [20.12]. When the droplet is on a dielectric surface covering an electrode source, the contact angle of the droplet may be lowered by applying an electric field. The change in contact angle is predicted by Lippman and Young's equation [20.6, 10]

$$\cos\theta = \cos\theta_0 + \frac{\varepsilon V^2}{2d\gamma_{\text{LG}}}, \tag{20.2}$$

where θ_0 is the contact angle when the electric field across the interfacial layer is zero [20.6]. While many

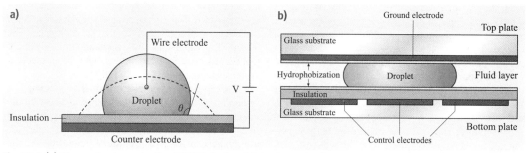

Fig. 20.1 (a) The conductive path of an EWOD device. (b) A schematic of an EWOD device (after [20.2])

have suggested that this change in contact angle is required for droplet transport, it has been suggested by *Zeng* and *Korsmeyer* [20.26] that the change in contact angle merely reflects the difference in interfacial energy and is not required for droplet transport.

For droplet-based EWOD, the device consists of a top ground electrode layered on glass and a bottom layer of control electrodes underneath an insulating layer of dielectric material. A thin hydrophobic layer is coated on the surface of the electrodes and the insulating material. The hydrophobic coating acts to prevent droplets from spreading into the channel and is not considered to be insulative. A typical EWOD setup is shown in Fig. 20.1 [20.2].

During operation, the activation of one electrode induces a local interfacial energy difference between adjacent electrodes. When the droplet experiences the energy difference, it moves toward surface of lower energy, and so through sequential activation and deactivation of electrodes the droplet can be transported as shown in Fig. 20.2 [20.2]. Due to the localized activated interfacial energy difference, *Cho* et al. [20.18] have shown that droplet volume large enough to cover the edge between two electrode surfaces is required to actuate the droplet. In addition, *Pollack* et al. [20.13] reported that a threshold voltage is required to actuate the droplet, and it was found that the threshold voltage for a water droplet dispersed in silicon oil is much lower than for a water droplet in air. It was also shown by *Moon* et al. [20.7] that the threshold voltage decreases with the thickness of the dielectric layer. Using barium strontium titanate (BST) as the dielectric material, an actuation voltage as low as 15 V can be used to transport droplets [20.6]. Once the applied voltage exceeds the threshold voltage, the droplet transport speed increases with the driving voltage. A transport speed of 250 mm/s with an alternating-current (AC) voltage of > 150 V has been reported by *Cho* et al. [20.18], and the transport of droplet volume as small as 5 nl has been reported by *Lee* et al. [20.11].

Similarly, more complex droplet manipulations can be achieved through the simultaneous operation of multiple droplet control sequences. Droplets can be fused by transporting two droplets toward the same electrode, or a single droplet can be split into smaller droplets by simultaneously transporting the two halves of a droplet in different directions. Droplet operations can also be combined to form sequential fuse and split operations, as shown in Fig. 20.3. The efficiency of the droplet splitting process, due to the energy required to overcome capillary pressure, is dependent on the geometry of the electrodes. *Cho* et al. [20.18] have shown that smaller channel gaps and larger electrode sizes favor droplet fission. The geometric constraint for a square electrode is $d/R_2 < 0.22$, where d is

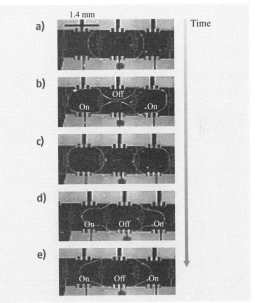

Fig. 20.3a–e Split and fuse operation for a droplet in EWOD. (**a**) A single DI water droplet is formed on an electrode. (**b,c**) The droplet is split by activating the two adjacent electrodes. (**d,e**) The droplet is then merged through the simultaneous activation of the middle electrode and the subsequent deactivation of the side electrodes (after [20.18], © IEEE 2003)

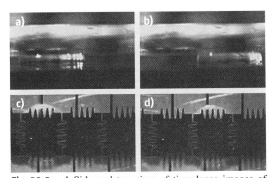

Fig. 20.2a–d Side and top-view of time-lapse images of the droplet transport process. The electrodes are sequentially activated to transfer the droplet from *left* (**a,c**) to *right* (**b,d**) (after [20.2], © R. Soc. Chem.)

the gap size and R_2 is roughly half the electrode width [20.18].

20.1.3 Reagent Mixing in EWOD

Reagent mixing in droplets is efficient in EWOD; *Paik* et al. demonstrated that complete mixing of two 800 nl droplets can be achieved as rapidly as 1.7 s [20.27]. It was shown that the droplet mixing rate is dependent on the subsequent motion of the coalesced droplet. In a linear array such as that shown in Fig. 20.4, movement of the droplet in one direction promotes mixing, while movement in the reverse direction undoes the mixing, due to flow reversibility at low Reynolds number [20.14]. The mixing rate can be further improved by increasing the rate of oscillation of the droplet between electrodes, increasing the number of electrodes for a larger transport area, and increasing the complexity of movement of the droplet through the use of multidimensional arrays. When the aspect ratio, defined by *Paik* et al. [20.27] as the ratio of the gap size to the width of the electrodes, is 0.4, the mixing efficiency is optimized; lower aspect ratios inhibit vertical flow and result in longer mixing times [20.27].

20.1.4 Improvements in EWOD

Despite the ability to rapidly transport, mix, and split droplets, there are problems that limit the use of EWOD.

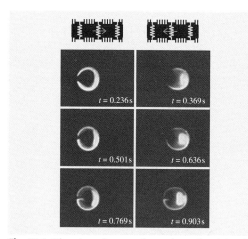

Fig. 20.4 Time-lapse images of two-electrode mixing. The pattern created by the dyes show that the droplet can be transported in the opposite direction. The *arrows* indicate the direction of droplet movement

First, the surface contact required for droplet actuation limits its use for complex biological fluids; absorbance of biomolecules onto the electrodes through electrostatic interactions and passive hydrophobic absorption eventually renders the electrode unusable after repeated use [20.12]. Second, while it is simple to control the actuation of a small number of electrodes, complex logarithms would be required to control large arrays of electrodes [20.15]. Third, only aqueous droplets can be actuated, due to the need for conductivity between the droplet and the ground electrode. Lastly, a droplet volume larger than the size of the electrode is required for EWOD to work, which limits the fluidic volume that can be used. However, improvements in EWOD have addressed the first two limitations. *Yoon* et al. [20.12] reported that reducing the duration of the square-wave electric field applied improved the durability of the electrode, and *Chiou* et al. [20.15] demonstrated that the addition of a photoconductive layer permits the use of light signals to control up to 20 000 electrodes.

20.1.5 Droplet Manipulation via Dielectrophoresis

Droplet actuation by dielectrophoresis (DEP) originates from the polarization of droplets in a nonuniform electric field. The electric field induced on the droplet interacts with the imposed spatially varying electric field to produce controlled droplet motion. The DEP force F_{DEP} for a droplet of volume V suspended in a medium of dielectric constant ε_S under the effect of an inhomogeneous electric field E can be mathematically described as

$$F_{DEP} = \tfrac{3}{2} V \varepsilon_S f_{CM} \nabla E^2 \, , \quad (20.3)$$

where f_{CM} is the real part of the Clausius–Mossotti factor [20.28]. For typical droplet processing applications, the difference in permittivity between the droplet and the medium is large, which allows f_{CM} to be expressed as

$$f_{CM} = \mathrm{Re} \left(\frac{\varepsilon_d - \varepsilon_s}{\varepsilon_d + 2\varepsilon_s} \right) \, , \quad (20.4)$$

where ε_d is the dielectric constant of the droplet [20.28].

When $\varepsilon_d > \varepsilon_s$, positive dielectrophoresis causes the droplet to move toward the region of high field strength. In contrast, when $\varepsilon_d < \varepsilon_s$, the droplet is displaced away from the high field. Since DEP is conducted through bulk liquid, no physical contact between the droplet

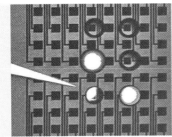

Fig. 20.5 Device used for DEP droplet actuation (after [20.28], © R. Soc. Chem.)

and the substrate is required to actuate the droplet. This allows droplet DEP to work with polar, nonpolar, aqueous, and organic droplets [20.26]. *Gascoyne* et al. [20.28] demonstrated a complete DEP-based platform that is capable of transporting, dispensing, fusing, and splitting droplets. Shown in Fig. 20.5, the device consists of microfabricated electrodes, all of which are independently addressable through a user interface. The surface of the electrode is coated with an electrically insulating material to prevent current leakage, and a droplet-repelling layer to minimize droplet–surface contact. When $\varepsilon_d > \varepsilon_s$ the droplets move towards the highest field region, and when $\varepsilon_d < \varepsilon_s$ droplets can be trapped inside energy *cages*, as indicated in Fig. 20.6 [20.28]. Thus, through controlled excitation of the electrodes, droplets may be shifted from position to position.

To dispense a droplet, the liquid is initially pushed out from a small orifice; Laplace pressure on the dispensing medium causes this liquid to bulge. Then an electric field is applied in order to induce a DEP force on the liquid droplet so that the injection can be modulated. For a liquid with positive DEP characteristics, the DEP force acts to pull the liquid volume from the tube, and the size of the droplet is controlled by the duration that the electric field is applied [20.28]. For liquids with

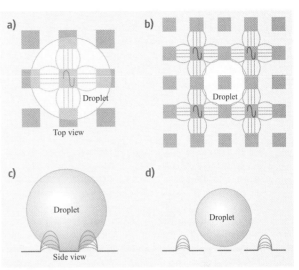

Fig. 20.6a–d Two types of droplet DEP control are possible. Panels (**a**) and (**c**) show top and side views of positive droplet DEP. Panels (**b**) and (**d**) show top and side views of negative droplet DEP, where the droplet is *caged* by the electrodes (after [20.28], © R. Soc. Chem.)

negative DEP, the DEP force pulls liquid into the tube and an increase in the threshold pressure is required to maintain the curvature of the droplet. In this case, the net threshold pressure difference causes fluid injection once the electric field is removed.

Droplet fusion is achieved by moving two droplets into close proximity in order to cause spontaneous coalescence due to the reduction of surface energy. Droplet fission is achieved by using small dielectric beads to remove a portion of the droplet under DEP [20.28].

Using this system, sample volumes as small as 4 pl can be dispensed, and a droplet 0.29 nL in volume can be moved at up to 670 μm/s across a 60 μm-pitch electrode array actuated by 130 V [20.28].

20.2 Passive Droplet Control Techniques

Unlike in active microfluidic systems, where one actuation mechanism and thus a single-component setup is responsible for multipurpose processing, passive microfluidic devices are controlled by immiscible flows. Likewise, almost all passive microfluidic devices are driven by fluidic pumps that provide constant output flow rates or deliver fluids at constant pressure. By manipulating the properties of the immiscible fluid flow within different channel geometries, the operations of droplet generation, fusing, splitting, and sorting can be achieved. Droplet transport is naturally achieved within any fluid flow carrying droplets. Since these different operations are driven by the same fluid actuation mechanism, they can be combined to form complex processors for sequential droplet operations in one integrated device. In addition, the device does not have any moving components, which makes its fabrication simple and more reliable.

20.2.1 Generation of Monodispersed Droplets

The key features of passive droplet-generation systems are that the generated droplet sizes can be much smaller than the features of the device, the generated droplets have narrow size distributions, and the generation of droplet is a continuous process. Droplet size distributions of < 2% coefficient of variation have been reported [20.30], and droplet sizes as small as 100 nm have been demonstrated [20.30, 31]. Most passive microfluidic droplet-generating devices utilize either the shear stress created at the immiscible flow interface or the pressure gradient created at a junction of narrow pores to initiate the continuous break-up of droplets. The shear break-up device uses either an asymmetric or a symmetric channel junction to introduce the immiscible fluids, and the pressure gradient break-up device uses either straight-through holes or microcapillary channels with flat terraces to pressurize the dispersed phase. In the microfluidic flow, the Reynolds number Re (inertial force/viscous force) is much less than 1, so viscous forces dominate over inertial forces. The viscous force is also weaker in magnitude when compared with the surface tension force, and thus, regardless of the channel geometries, in order to achieve steady generation of droplets, it is critical that the dispersed phase does not wet the droplet-generating surface [20.32, 33].

Shear-Induced Droplet Generation

The generation of droplets in a microfluidic device is governed by the interaction of shear stress with surface tension. The shear stress exerted by the continuous phase acts to deform the liquid surface, while the interfacial force at the immiscible fluid interface acts to restore the deformation. This can be described by the dimensionless capillary number $Ca = \eta\varepsilon r/\sigma$, where r is the radius of the droplet, η is the viscosity of the continuous phase, ε is the shear rate in the channel, and σ is the interfacial tension at the immiscible fluid interface [20.34]. Assuming that $Ca = 1$ is the critical condition when the shear stress is large enough to break up the liquid thread, $r \sim \sigma/\eta\varepsilon$, which is the scaling of droplet sizes observed by *Thorsen* et al. [20.35], corroborating the observations of *Nisisako* et al. [20.29]. As the diameter of the droplet increases to beyond the width of the channel, the effect of the wall becomes dominant over the effect of shear stress [20.36], and hence the size is weakly dependent on the flow rate, as reported by *Tan* et al. [20.32] and *Tice* et al. [20.36, 37]. This is especially true for Ca < 1, where Ca can also be defined by $Ca = \eta U/\sigma$, in which $U = \varepsilon r$, and the observed droplets become confined to elongated plugs [20.37, 38].

Where Ca is less useful for predicting droplet size, the conservation of mass flow can be used to determine droplet volume. *Tice* et al. demonstrated that the length of the plug can be predicted from $l = p[V_d/(V_d + V_c)]$, where l is the length of the plug, p is the period of break-up, V_d is the volume flow of the dispersed phase, and V_c is the volume of the continuous phase [20.37]. In addition, *Tice* et al. [20.37] observed different regimes of droplet formation. Stable droplet formation occurs within a limited regime. This change in behavior for different regimes was also observed by *Anna* et al. [20.31] and *Dreyfus* et al. [20.39]. The flow regime behavior observed by Dreyfus et al. indicates that, when the flow rate of the dispersed phase is much lower than the flow rate of the continuous phase, isolated drop formation is observed, but as the flow rate of the dispersed phase increases the stratified regime is observed.

Asymmetric and Symmetric Shearing Design

The asymmetric shearing of immiscible fluids is achieved using a T-type intersection such as that shown in Fig. 20.7 [20.29]. The dispersed phase is sheared at the junction by the continuous phase, and the droplet size and frequency of generation are controlled by the flow rates of the continuous and the dispersed phases. Due to the asymmetric nature of the droplet break-up process, the reagents injected for mixing are partially exchanged when the liquid droplets break off. This effect is known as *twirling* and will be discussed in more detail later [20.37]. Twirling can be an advan-

Fig. 20.7 T-junction device from *Nisisako* et al. [20.29], © R. Soc. Chem.

tage if quick mixing is desired, but a disadvantage if the reagents need to be aligned to create, e.g., bicolor polymeric beads. Furthermore, the simplicity of the design allows two or more generator units to be aligned adjacent or opposite to each other in the same channel for synchronized generation of droplets. This geometry has been exploited in order to index mixing conditions [20.30, 40].

Droplet generation from symmetric fluidic junctions results in monodispersed droplets at controlled frequency [20.31, 32, 41]. Figure 20.8 shows an example of the droplet-generation process [20.32]. As the flow passes through the junction, the narrow width creates a maximum velocity and, after the flow has passed the junction, the fluid velocity decreases due to the increase in channel width. This creates a shear gradient that is maximized at the orifice, producing *shear-focusing* break-up. Droplets are generated from the shear-focusing break-up precisely at the orifice, and the droplet size and generation frequency are controlled by the relative flow rates.

20.2.2 Devices Based on Microcapillary Arrays

Straight silicon through-holes and microcapillary arrays (MC) are attractive designs for forming emulsions. Both of these methods can be applied under conditions of no external flow to produce narrow size distributions. Variations of $< 2\%$ have been reported for straight silicon through-holes [20.42] and $\approx 5\%$ for MC designs [20.43].

In these devices, the production rate is rather low; *Kobayashi* et al. [20.41] reported a production rate of 3–11 drops/s for a device based on silicon through-holes. Furthermore, the size of the generated droplet is limited by the size of the pores [20.44].

In the silicon through-hole method, the dispersed phase is pushed through small holes etched in a silicon wafer. The droplet detaches when the interfacial tension pinches off the dispersed phase. The generated droplets are monodispersed in terms of droplet size [20.42].

In microcapillary array designs, a glass plate is bonded to an etched silicon channel. The channel connects to a flat terrace that leads to an indented well. The droplet-generation process is characterized by a two-step process of filling the terrace with the dispersed phase and the detachment of the liquid as it deforms at the terrace. Upon increasing the pressure of the dispersed phase, the dispersed liquid is entrained on the terrace surface. When liquid reaches the indented well,

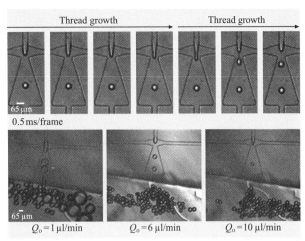

Fig. 20.8 The shear-focusing design demonstrated by *Tan* et al. [20.32], © R. Soc. Chem.

the edge of the liquid deforms, causing the surface tension to pinch off liquid near the edge of the terrace.

When designing the channel terrace, *Sugiura* et al. [20.43] reported that the diameters of the droplets generated from MC can be predicted using the terrace length L and depth H, according to the geometric parameters presented in Fig. 20.9 by

$$D = \left(\frac{6(H + 0.626)}{\pi} \right.$$
$$\times \left\{ \frac{L^2}{4} \cos^{-1}\left(\frac{L - 13.76H - 8.61}{L} \right) \right.$$
$$- \frac{L(L - 13.76H - 8.61)}{4}$$
$$\left. \left. \times \sin\left[\cos^{-1}\left(\frac{L - 13.76H - 8.61}{L} \right) \right] \right\} \right)^{1/3}.$$
(20.5)

The effect of external flow can also reduce droplet size, as reported by *Kawakatsu* et al. [20.44]. The droplet size decreases from $47.6 – 37.3\,\mu\text{m}$ as the flow is increased from 1.4×10^{-2} to $2.4\,\text{ml/min}$. However, for 16 and $20\,\mu\text{m}$ droplets generated with a smaller MC channel, varying the flow has no effect on droplet size.

20.2.3 Double Emulsions

Double emulsions are formed when a liquid is dispersed in an immiscible fluid that is further dispersed in another immiscible fluid. Since the generated double emulsion contains both organic and aqueous phases, it is

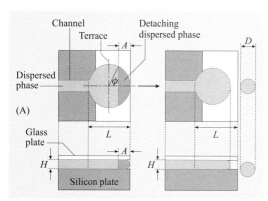

Fig. 20.9 Model for predicting droplet diameter (after [20.43])

a versatile way of encapsulating and delivering polar or nonpolar substances. *Kawakatsu* et al. [20.45] was able to generate monodispersed double emulsions by using MC devices where the core droplets were generated by the homogenization of water in oil. Controlled forma-

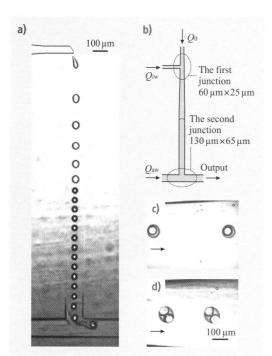

Fig. 20.10 Generation of compound drops through a sequential droplet break-up process (after [20.46], © Am. Soc. Chem.)

tion of both the inner phase and the middle phase has been demonstrated by *Okushima* et al. [20.46] using sequential break-up through two T-junctions, as shown in Fig. 20.10. *Utada* et al. [20.47] also demonstrated precise generation of double emulsions using coaxial flow to simultaneously break up the immiscible fluid interfaces between the inner and outer droplet, and the outer droplet and continuous phase, as shown in Fig. 20.11. Both methods provide the ability to control the size of the inner droplet, the thickness of the middle or shell layer, and the number of encapsulated inner droplets. Ratios of shell thickness to outer drop radius as low as 3% and as high as 40% have been reported [20.47].

Reagent Mixing

There are two ways to mix reagents in a passive microfluidic system. In the first technique, the reagents to be mixed are introduced as adjacent laminar streams that are broken into single droplets using either an asymmetric or a symmetric shearing system. The reagents are then mixed by diffusion or convection induced either by the surrounding flow or the walls of the channel. The other mixing mechanism is based on fusing two droplets in the microfluidic channel, as detailed in the next section.

Mixing in moving plugs is facilitated by recirculation flow, which distributes reagents from the center to the edge of the droplet [20.48]. When the reagent gradient is perpendicular to the direction of transport, recirculation is not as effective at accelerating the mixing, as illustrated in Fig. 20.12 [20.37]. As shown in Fig. 20.13, for droplets generated by a symmetric shearing system, the reagent gradient in the laminar stream is directly transferred into the droplet [20.49]. However, for droplets generated from an asymmetric shearing system, mixing is facilitated by an effect called *twirling*, which is an eddy that transports reagents to different parts of the droplet. The effect of twirling is finite and so it increases the mixing rate for short plugs, but does not significantly increase the mixing rate for long plugs, as shown in Fig. 20.14 [20.37].

Mixing can be further improved through the use of winding channels. As the droplet passes through these winding channels, it is stretched, folded, and reoriented to induce chaotic mixing inside the droplets [20.50, 51]. The time of mixing is verified experimentally to be

$$t_{\mathrm{mix,ca}} \sim \left(\frac{aw}{U}\right) \log(\mathrm{Pe}) \,, \tag{20.6}$$

$$\mathrm{Pe} = \frac{wU}{D} \,, \tag{20.7}$$

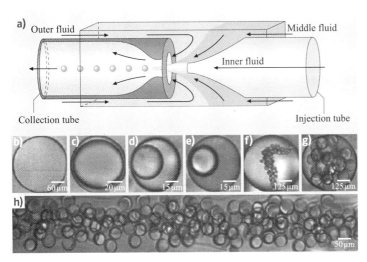

Fig. 20.11a–h Generation of compound drops through coaxial flow in a microfluidic device. Parameters such as shell thickness, the internal droplet number, and the sizes of the internal droplets could be individually controlled. (a) Schematic of the glass microcapillary device. (b–e) The diameter of the inner and outer microcapillary tubes were varied from 10 to 50 μm and 50 to 500 μm, respectively. This allowed control over the thickness of the middle fluid phase. (f,g) Multiple droplets of controlled size can be contained within a single droplet. (h) A large number of double emulsions containing one droplet can be generated (after [20.47], © AAAS)

where w is the cross-section dimension of the microchannel, a is the dimensionless length of the plug measured relative to w, U is the flow velocity, and Pe is the Péclet number [20.51]. Submillisecond mixing times have been reported for mixing in winding channels [20.48].

Droplet Fusion

Similar to the principle of pipetting volumes in and out of a single mixing well, the fission (splitting) and fusion of droplets in a microfluidic channel control both the concentration of reagents and the volume of the mixed samples. While in a channel, droplets are spatially confined, such that droplets must travel through fixed channel geometries to be split or fused, the rate of operation is controlled by the velocity of the continuous phase, and millisecond-scale operations are possible.

Droplet fusion or coalescence is due to film drainage, which has been reviewed elsewhere [20.36]. Film drainage occurs when drops are close to each other. In a microfluidic channel, however, droplets are separated by plugs of immiscible fluids, meaning that film drainage is unlikely to occur between droplets. The challenge is to control the flow of the liquid separating the droplets. There are several ways to achieve this. *Song* et al. [20.51] utilized the difference in traveling velocity between droplets of different sizes in straight channels to fuse large and small drops. *Köhler* et al. [20.52] and *Tan* et al. [20.30] used passive channel geometries to temporarily trap and fuse droplets

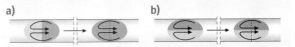

Fig. 20.12 (a) When the concentration gradient is parallel to the direction of transport, recirculation flow mixes the reagents efficiently. (b) When the concentration gradient is perpendicular to the direction of transport, mixing by recirculation flow is not efficient, and the reagents remain primarily in their own halves throughout the channel (after [20.37, 47], © AAAS)

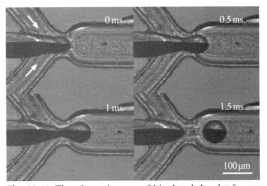

Fig. 20.13 Time-lapse images of bicolored droplet formation. Laminar flow preserves the separate dye flow stream even during formation of the droplet (after [20.49], © Elsevier)

at fluidic junctions. Alternatively, by designing the channel geometry appropriately, the fluid separating the droplets can be continuously drained at bifurcat-

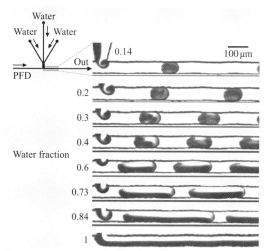

Fig. 20.14 The *twirling effect* transports small amounts of the reagents across the interface immediately after break-up (after [20.37], © Am. Chem. Soc.)

ing junctions to achieve coalescence of a series of droplets, as demonstrated by *Tan* et al. [20.30] with the *flow-rectifying junction*. This design, shown in Fig. 20.15, allows various numbers of droplets to coalesce at various generation frequencies and traveling velocities.

Droplet fusion mixing is analogous to the digital mixing of droplets in active devices. Since the mixing process can be made to be weakly dependent on the generation process, it allows reagent volumes and concentrations to be controlled independently. Controlled mixing of two different reagents by fusion is also shown in Fig. 20.16 [20.30, 53].

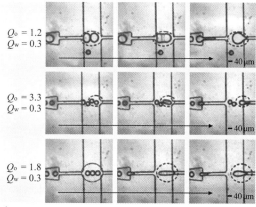

Fig. 20.15 Controlled fusion of droplets in a microfluidic channel using the flow-rectifying design (after [20.30], © R. Soc. Chem.)

Droplet Fission

Droplet fission occurs at a bifurcating junction of a channel. Similar to the splitting of a droplet from a thread, droplets at the bifurcating junction are continuously elongated by the extensional shear stress exerted by the flow, eventually reaching a critical length that can no longer be sustained by the interfacial tension of the droplet surface, which results in droplet break-up. Droplet break-up can be symmetric, where a single droplet is broken into two equal halves, or asymmetric, where a single droplet is broken into multiple unequal parts. Symmetric fission is achieved at a junction with equal bifurcating flows. For a channel with square cross-sectional geometry, the critical break-up conditions can be expressed according to the initial droplet length and the width of the channel, as indicated in

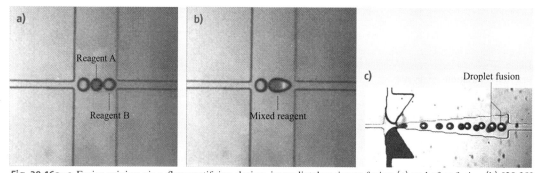

Fig. 20.16a–c Fusion mixing via a flow-rectifying design, immediately prior to fusion (**a**) and after fusion (**b**) [20.30] and fusion within a channel expansion design (**c**) (after [20.53])

Fig. 20.17. *Link* et al. [20.54] derived the critical capillary number as

$$C_{\text{cr}} = \alpha \varepsilon_0 \left(\frac{1}{\varepsilon_0^{2/3}} - 1 \right)^2, \qquad (20.8)$$

$$\varepsilon_0 = \frac{\ell_0}{\pi w_0}, \qquad (20.9)$$

which shows good agreement with experimental results, as shown in Fig. 20.17 [20.54]; ε_0 is the initial extension ratio of length ℓ_0 to the circumference (πw_0), and α is the fitting parameter, which is equal to 1 for square channels [20.54].

Droplets of various sizes can be created through asymmetric break-up, such that even submicrometer-sized droplets can be split from large droplets tens of microns in radius [20.30]. It was shown that the sizes of the split droplets depend on the size of the original droplet [20.30] and the bifurcating flow distribution at the junction [20.30, 50, 54]. Using droplet break-off from unmixed reagents, by controlling the location and time of fission, the reagent concentration inside the droplet could be redistributed according to the mixed gradient at the time of break-up to produce arrays of split droplets with different reagent concentrations [20.30].

Droplet Sorting

In an active droplet control system, the complex droplet transport process is modulated by algorithms that control the electrode switches. In a passive system, transport is guided by the flow distribution in the channel, and controlled multipath transport is difficult to achieve because individual droplets are simply distributed according to the flow rates. The ability to switch droplets between different continuous phases is useful because it makes it possible to filter contaminants from the droplet stream, it allows the concentration of reagents present in the continuous phase to be changed, it means that we can organize unknown particulates by size, and it allows us to set up a passive monitoring system for variations in droplet size.

During the droplet-generation process, the continuous break-up of the neck connecting the droplet to the liquid thread leads to the formation of small satellite droplets. The presence of satellite droplets increases the size distribution and decreases the mixing accuracy due to the fusion of satellite droplets with primary droplets. *Tan* et al. [20.30, 55] demonstrated droplet sorting in a microfluidic channel by controlling the shear stress gradient at bifurcating junctions. As shown

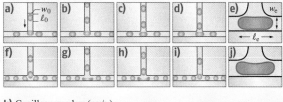

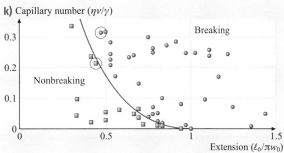

Fig. 20.17a–k The parameters used to predict droplet break-up conditions are shown in (**a**). Panels (**a–j**) show the time-lapse images of the break-up process. As shown in panels (**e**) and (**j**), the droplet reaches maximum extension with width w_e and length ℓ_e, obtaining values for ε of 0.95 and 1.15 respectively. (**k**) The predicted capillary number agrees with the experimental results (after [20.54])

in Fig. 20.18, a droplet at the junction will be transported toward a region of higher shear stress. Since the shear force experienced by the droplet depends on the surface area of the droplet, passive sorting by droplet size can also be achieved. To collect the individual satellite droplets efficiently, *Tan* et al. developed a dynamic flow technique that modulates the stress exerted on the liquid thread to control the location of break-up

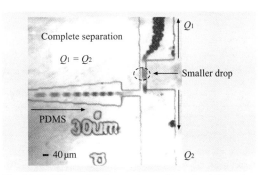

Fig. 20.18 When the flow rates exiting the bifurcating junctions are balanced, droplets are transported toward regions of higher shear stress created by the narrow inlet channel (after [20.30], © R. Soc. Chem.)

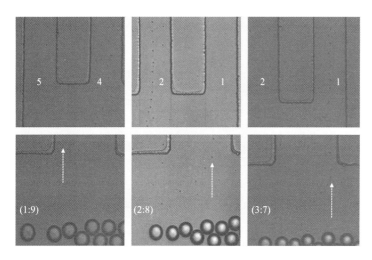

Fig. 20.19 Satellite droplets of the desired size can be selectively generated into specific collection zones, labeled 1–5 (zone 3 is not shown), by changing the oil/water flowrate ratios from 1 : 9 to 3 : 7 (after [20.55], © R. Soc. Chem.)

and the distribution of different satellite droplets in the channel, as shown in Fig. 20.19, where larger droplets and smaller droplets are separated into separate chambers [20.55].

20.3 Applications

Droplet-based microfluidic systems possess tremendous potential to improve current emulsion technologies that are widely used in industry to produce sol-gels, drugs, synthetic materials, and food products. In addition, a wide variety of new applications can be developed due to their precise metering capabilities, rapid and controllable mixing response, and automated combinatorial capabilities.

20.3.1 Droplet as Microtemplate and Encapsulation Agents

Polymer precursor droplets can be generated in microfluidic devices and subsequently polymerized either by ultraviolet (UV) [20.49] or chemical agents [20.57] to produce polymeric beads with narrow size distributions. Solvent evaporation and extraction methods have also been developed to form polymer particles on a droplet microfluidic platform [20.58–60]. This process involves the formation of oil-in-water single or double emulsions and polymerization of the emulsion by removal of solvent from the system. Alternatively, aqueous droplets can be used as a template for synthesizing uniform colloid structures. *Yi* et al. [20.56, 61, 62] developed a technique to generate monodispersed aqueous droplets containing latex beads. Evaporation of the droplets results in the formation of photonic balls, as shown in Fig. 20.20 [20.56], and other colloid structures that show unique responses to flow and magnetic fields [20.56, 61, 62].

By transferring laminar patterns of two reagents to droplets using a symmetric shearing device, *Nisisako*

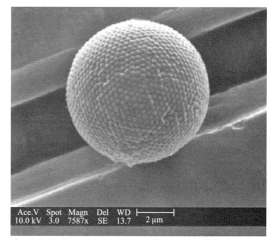

Fig. 20.20 Scanning electron microscopy (SEM) images of latex photonic ball (after [20.56], © Elsevier)

et al. [20.49] demonstrated the production of polymeric beads with bichromal and oriented charge polarities. Similarly, as demonstrated by *Millman* et al. [20.63] using dielectrophoretic-based digital fusion of different polymeric droplets, anisotropic particles with tailored properties can be synthesized.

For biological applications, single and multiple cells and organelles can be trapped inside droplets [20.66, 67] for analysis or to provide scaffolds for cell [20.52] and tissue growth. Biocompatible and biodegradable hydrogels such as alginate are commonly used for cell encapsulation due to their relatively simple preparation. Since gelation occurs immediately upon contact between alginate and polycations, droplet fusion techniques have been implemented to encapsulate cells and polymerize particles simultaneously on a chip [20.68, 69]. Control over particle size and monodispersity is important for the use of particles in the administration and controlled release of encapsulated substances such as drugs, dyes, and enzymes [20.70]. Current microfluidic platforms allow droplets to be filled with various hydrophilic or hydrophobic materials and the capsule shell thickness to be altered to control compound release rates [20.71, 72].

20.3.2 Droplets as Real-Time Chemical Processors and Combinatorial Synthesizers

Mixing assays that result in photodetectable changes can be rapidly carried out by mixing reagents in microfluidic droplets. Utilizing an EWOD-based droplet microfluidic system, *Srinivasan* et al. [20.64, 73] have demonstrated a rapid on-chip glucose assay involving three steps – dispensing, mixing, and detection – such that glucose concentrations in the range of 25–300 mg/dl could be detected in less than 60 s [20.73]. Subsequently, a similar programmable lab-on-a-chip device (shown in Fig. 20.21) was applied to the detection of glucose in human physiology fluids including human whole blood, serum, plasma, urine, saliva, sweat, and tears. This latter device has shown great reliability, lasting > 25 000 continuous cycles of reagent transport performed at a frequency of 20 Hz and actuated by less than 65 V [20.64].

Droplet-based systems have also been applied to millisecond-scale nanoparticle synthesis [20.74], polymerase chain reactions (PCR) [20.75], DNA analysis [20.76], and to screen protein crystallization conditions [20.40, 65, 77, 78]. In these systems, the variation in the flow rate automatically produces a time-dependent concentration gradient for the reagents inside the droplets. These properties can be used to screen protein crystallization conditions, as demonstrated by *Zheng* et al. [20.77]. Volumes of < 4 nl of reagent could be used for each trial inside a 7.5 nl aqueous droplet, and hundreds of trials performed at rates of several trials per second can be achieved with computer control [20.77]. Subsequent crystal growth by vapor diffusion from droplets generated x-ray-analyzable crystals, as shown in Fig. 20.22 [20.65]. To mitigate the complexity of resolving the time-dependent screening concentrations, an indexing stream of fluorescent droplets generated at the same time as the screening droplets can be used to indicate the concentration of the protein [20.77].

Recent work has been done with droplet-based platforms for biomolecule synthesis. Since droplets can be generated at micron sizes or smaller, encapsulation of a single template copy of DNA is achievable. The integration of heating elements, the ability to precisely manipulate droplet movement, and increased mixing rates enable droplets to serve as microreactors for in vitro protein expression, DNA amplification, and other biochemical reactions. *Dittrich* et al. demonstrated cell-free expression of green fluorescent protein (GFP) within monodispersed water-in-oil droplets generated in a microfluidic platform [20.79]. PCR [20.80] and DNA assays [20.81] in droplets recently showed

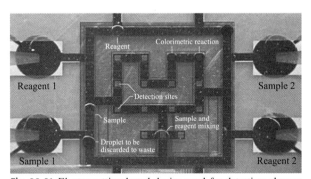

Fig. 20.21 Electrowetting-based device used for detecting glucose in human physiological fluids (after [20.64], © R. Soc. Chem.)

Fig. 20.22 Polarized-light micrograph of crystals generated in droplets (after [20.65], © Wiley)

improved sensitivity and decreased reaction times, enabling higher-throughput assays.

20.3.3 Droplets as Micromechanical Components

Since droplets can be deformed by the flow, they can be fixed into a variety of shapes and sizes that could be used as microcomponents; furthermore, these structures can be made into permanent microscale building blocks for device assembly if they are polymerized [20.82]. Microcomponents such as colloidal microspheres can be made into pumps and valves in microfluidic channels [20.83]. Droplets driven by electrowetting have also been used as the main driving components for liquid micromotors [20.10] and to regulate the frequencies of optical-fiber devices [20.9, 83, 84].

20.4 Conclusions

The field of droplet-based microfluidic technology is diverse in terms of the actuation mechanisms and methods of operation used. Essentially, all droplet-based systems are used to control liquid dispensing, mixing, splitting, and localization. Microfluidic methods allow droplets to be processed individually, provide accurate dispensing of fluid volume, and improve speed of reagent mixing. These factors have made droplet technology a valuable new tool for controlling micro- and nanointeractions.

References

20.1 S.-Y. Teh, R. Lin, L.-H. Hung, A.P. Lee: Droplet microfluidics, Lab Chip **8**(2), 198–220 (2008)

20.2 M.G. Pollack, A.D. Shenderov, R.B. Fair: Electrowetting-based actuation of droplets for integrated microfluidics, Lab Chip **2**, 96–101 (2002)

20.3 R.H. Farahi, A. Passian, T.L. Ferrell, T. Thundat: Microfluidic manipulation via Marangoni forces, Appl. Phys. Lett. **85**(18), 4237–4239 (2004)

20.4 B.S. Gallardo, V.K. Gupta, F.D. Eagerton, L.I. Jong, V.S. Craig, R.R. Shah, N.L. Abbott: Electrochemical principles for active control of liquids on submillimeter scales, Science **283**, 57–60 (1999)

20.5 K. Hosokawa, T. Fujii, I. Endo: Handling of picoliter liquid samples in a poly(dimethysiloxane)-based microfluidic device, Anal. Chem. **71**, 4781–4785 (1999)

20.6 C. Quillet, B. Berge: Electrowetting: A recent outbreak, Curr. Opin. Colloid Sci. **6**, 34–39 (2001)

20.7 H. Moon, S.K. Cho, R.L. Garrell, C. Kim: Low voltage electrowetting-on-dielectric, J. Appl. Phys. **92**(7), 4080–4087 (2002)

20.8 J. Ding, K. Chakrabarty, R.B. Fair: Scheduling of microfluidic operations for reconfigurable two-dimensional electrowetting arrays, IEEE Trans. Comput.-Aided Des. Integr. Circuits Syst. **20**(12), 1463–1468 (2001)

20.9 J. Hsieh, P. Mach, F. Cattaneo, S. Yang, T. Krupenkine, K. Baldwin, J.A. Rogers: Tunable microfluidic optical-fiber devices based on electrowetting pumps and plastic microchannels, IEEE Photon. Technol. Lett. **15**(1), 81–83 (2003)

20.10 J. Lee, C. Kim: Surface-tension-driven microactuation based on continuous electrowetting, J. Microelectromech. Syst. **9**(2), 171–180 (2000)

20.11 J. Lee, H. Moon, J. Fowler, T. Schoellhammer, C. Kim: Electrowetting and electrowetting-on-dielectric for microscale liquid handling, Sens. Actuators A **95**, 259–268 (2002)

20.12 J. Yoon, R.L. Garrell: Preventing biomolecular adsorption in electrowetting-based biofluidic chips, Anal. Chem. **75**(19), 5097–5102 (2003)

20.13 M.G. Pollack, R.B. Fair: Electrowetting-based actuation of liquid droplets for microfluidic applications, Appl. Phys. Lett. **77**(11), 1725–1726 (2000)

20.14 P. Paik, V.K. Pamula, M.G. Pollack, R.B. Fair: Electrowetting-based droplet mixers for microfluidic systems, Lab Chip **3**, 28–33 (2003)

20.15 P.Y. Chiou, H. Moon, H. Toshiyoshi, C. Kim, M.C. Wu: Light actuation of liquid by optoelectrowetting, Sens. Actuators A **104**, 222–228 (2003)

20.16 R.A. Hayes, B.J. Freenstra: Video-speed electronic paper based on electrowetting, Nature **425**, 383–385 (2003)

20.17 S.K. Cho, H. Moon, J. Fowler, C. Kim: Splitting a liquid droplet for electrowetting-based microfluidics, ASME Int. Mech. Eng. Congr. Expo. (ASME International, New York 2001)

20.18 S.K. Cho, H. Moon, C. Kim: Creating, transporting, cutting, and merging liquid droplets by electrowetting-based actuation for digital microfluidic circuits, J. Microelectromech. Syst. **12**(1), 70–80 (2003)

20.19 T.B. Jones, J.D. Fowler, Y.S. Chang, C. Kim: Frequency-based relationship of electrowetting and dielectrophoretic liquid microactuation, Langmuir **19**, 7646–7651 (2003)

20.20 A.A. Darhuber, J.M. Davis, S.M. Troian, W.W. Reisner: Thermocapillary actuation of liquid flow on chemically patterned surfaces, Phys. Fluids **15**(5), 1295–1304 (2003)

20.21 A.A. Darhuber, J.P. Valentino, S.M. Troian, S. Wagner: Thermocapillary actuation of droplets on chemically patterned surfaces by programmable microheater arrays, J. Microelectromech. Syst. **12**(6), 873–879 (2003)

20.22 A.A. Darhuber, J.P. Valentino, J.M. Davis, S.M. Troian: Microfluidic actuation by modulation of surface stresses, Appl. Phys. Lett. **82**(4), 657–659 (2003)

20.23 A.A. Darhuber, J.Z. Chen, J.M. Davis, S.M. Troian: A study of mixing in thermocapillary flows on micropatterned surfaces, Philos. Trans. R. Soc. Lond. Ser. A **362**, 1037–1058 (2003)

20.24 A.A. Darhuber, S.M. Trojan: Dynamics of capillary spreading along hydrophilic microstripes, Phys. Rev. E. **64**, 031603 (2001)

20.25 M.G. Lippmann: Relations entre les phénomènes électriques et capillaires, Anal. Chim. Phys. **5**(11), 494–549 (1875), in French

20.26 J. Zeng, T. Korsmeyer: Principles of droplet electrohydrodynamics for lab-on-a-chip, Lab Chip **4**(4), 265–277 (2004)

20.27 P. Paik, V.K. Pamula, R.B. Fair: Rapid droplet mixers for digital microfluidic systems, Lab Chip **3**, 253–259 (2003)

20.28 P.R.C. Gascoyne, J.V. Vykoukal, J.A. Schwartz, T.J. Anderson, D.M. Vykoukal, K.W. Current, C. McConaghy, F.F. Becker, C. Andrews: Dielectrophoresis-based programmable fluidic processors, Lab Chip **4**(4), 299–309 (2004)

20.29 T. Nisisako, T. Tori, T. Higuchi: Droplet formation in a microchannel network, Lab Chip **2**, 24–26 (2002)

20.30 Y.C. Tan, J.S. Fisher, A.I. Lee, V. Cristini, A.P. Lee: Design of microfluidic channel geometries for the control of droplet volume, chemical concentration, and sorting, Lab Chip **4**(4), 292–298 (2004)

20.31 S.L. Anna, N. Bontoux, H.A. Stone: Formation of dispersions using *flow focusing* in microchannels, Appl. Phys. Lett. **82**(3), 364–366 (2003)

20.32 Y.C. Tan, V. Cristini, A.P. Lee: Monodispersed microfluidic droplet generation by shear focusing microfluidic device, Sens. Actuators B **114**, 350–356 (2006)

20.33 T. Kawakatsu, G. Trägårdh, C. Trägårdh, M. Nakajima, N. Oda, T. Yonemoto: The effect of the hydrophobicity of microchannels and components in water and oil phases on droplet formation in microchannel water-in-oil emulsification, Colloids Surf. **179**, 29–37 (2001)

20.34 B.J. Briscoe, C.J. Lawrence, W.G.P. Mietus: A review of immiscible fluidmixing, Adv. Colloid Interface Sci. **81**(1), 1–17 (1999)

20.35 T. Thorsen, R.W. Roberts, F.H. Arnold, S.R. Quake: Dynamic pattern formation in a vesicle-generating microfluidic device, Phys. Rev. Lett. **86**(18), 4163–4166 (2001)

20.36 V. Cristini, Y.C. Tan: Theory and numerical simulation of droplet dynamics in complex flows – A review, Lab Chip **4**(4), 257–264 (2004)

20.37 J.D. Tice, H. Song, A.D. Lyon, R.F. Ismagilov: Formation of droplets and mixing in multiphase microfluidics at low values of the Reynolds and the capillary numbers, Langmuir **19**(22), 9127–9133 (2003)

20.38 J.D. Tice, A.D. Lyon, R.F. Ismagilov: Effects of viscosity on droplet formation and mixing in microfluidic channels, Anal. Chim. Acta **507**, 73–77 (2003)

20.39 R. Dreyfus, P. Tabeling, H. Willaime: Ordered and disordered patterns in two-phase flows in microchannels, Phys. Rev. Lett. **90**(14), 144505 (2003)

20.40 B. Zheng, J.D. Tice, R.F. Ismagilov: Formation of droplets of alternating composition in microfluidic channels and applications to indexing of concentration in droplet-based assays, Anal. Chem. **76**(17), 4977–4982 (2004)

20.41 Q. Xu, M. Nakajima: The generation of highly monodisperse droplets through the breakup of hydrodynamically focused microthread in a microfluidic device, Appl. Phys. Lett. **85**(17), 3726–3728 (2004)

20.42 I. Kobayashi, M. Nakajima, K. Chun, Y. Kikuchi, H. Fujita: Silicon array of elongated through-holes for monodisperse emulsion droplets, AIChE Journal **48**(8), 1639–1644 (2002)

20.43 S. Sugiura, M. Nakajima, M. Seki: Prediction of droplet diameter for microchannel emulsification, Langmuir **18**, 3854–3859 (2002)

20.44 T. Kawakatsu, H. Komori, M. Nakajima, Y. Kikuchi, T. Yonemoto: Production of monodispersed oil-in-water emulsion using crossflow-type silicon microchannel plate, J. Chem. Eng. Jpn. **32**(2), 241–244 (1999)

20.45 T. Kawakatsu, G. Trägårdh, C. Trägårdh: Production of W/O/W emulsions and S/O/W pectin microcapsules by microchannel emulsification, Colloids Surf. **189**, 257–264 (2001)

20.46 S. Okushima, T. Nisisako, T. Torii, T. Higuchi: Controlled production of monodisperse double emulsions by two-step droplet breakup in microfluidic devices, Langmuir **20**, 9905–9908 (2004)

20.47 A.S. Utada, E. Lorenceau, D.R. Link, P.D. Kaplan, H.A. Stone, D.A. Weitz: Monodisperse double emulsions generated from a microcapillary device, Science **308**, 537–541 (2005)

20.48 K. Handique, M.A. Burns: Mathematical modeling of drop mixing in a slit-type microchannel, J. Micromech. Microeng. **11**(5), 548–554 (2001)

20.49 T. Nisisako, T. Torii, T. Higuchi: Novel microreactors for functional polymer beads, Chem. Eng. J. **101**, 23–29 (2004)

20.50 H. Song, M.R. Bringer, J.D. Tice, C.J. Gerdts, R.F. Ismagilov: Experimental test of scaling of mixing by chaotic advection in droplets moving through microfluidic channels, Appl. Phys. Lett. **83**(22), 4664–4666 (2003)

20.51 H. Song, J.D. Tice, R.F. Ismagilov: A microfluidic system for controlling reaction networks in time, Angew. Chem. Int. Ed. **42**(7), 768–772 (2003)

20.52 J.M. Kohler, T. Henkel, A. Grodrian, T. Kimer, M. Roth, K. Martin, J. Metze: Digital reaction technology by micro segmented flow – Components, concepts and applications, Chem. Eng. J. **101**(1–3), 201–216 (2004)

20.53 L.H. Hung, K.M. Choi, W.Y. Tseng, Y.C. Tan, K.J. Shea, A.P. Lee: Alternating droplet generation and controlled dynamic droplet fusion in microfluidic device for CdS nanoparticles synthesis, Lab Chip **6**, 1–6 (2006)

20.54 D.R. Link, S.L. Anna, D.A. Weitz, H.A. Stone: Geometrically mediated breakup of drops in microfluidic devices, Phys. Rev. Lett. **92**(5), 054503 (2004)

20.55 Y.C. Tan, Y.L. Ho, A.P. Lee: Microfluidic sorting of droplets by size, Microfluid. Nanofluid. **4**, 343–348 (2008)

20.56 G. Yi, S. Jeon, T. Thorsen, V.N. Manoharan, S.R. Quake, D.J. Pine, S. Yang: Generation of uniform photonic balls by template-assisted colloidal crystallization, Synth. Met. **139**, 803–806 (2003)

20.57 S. Sugiura, T. Oda, Y. Izumida, Y. Aoyagi, M. Satake, A. Ochiai, N. Ohkohchi, M. Nakajima: Size control of calcium alginate beads containing living cells using micro-nozzle array, Biomaterials **26**, 3327–3331 (2005)

20.58 L.-H. Hung, A.P. Lee: Microfluidic devices for the synthesis of nanoparticles and biomaterials, J. Med. Biol. Eng. **27**(1), 1–6 (2007)

20.59 E. Lorenceau, A.S. Utada, D.R. Link, G. Cristobal, M. Joanicot, D.A. Weitz: Generation of polymersomes from double emulsions, Langmuir **21**(20), 9183–9186 (2005)

20.60 R.C. Hayward, A.S. Utada, N. Dan, D.A. Weitz: Dewetting instability during the formation of polymersomes from block-copolymer-stabilized double emulsions, Langmuir **22**(10), 4457–4461 (2006)

20.61 G. Yi, T. Thorsen, V.N. Manoharan, M. Hwang, S. Jeon, D.J. Pine, S.R. Quake, S. Yang: Generation of uniform colloidal assemblies in soft microfluidic devices, Adv. Mater. **15**(15), 1300–1304 (2003)

20.62 G. Yi, V.N. Manoharan, S. Klein, K.R. Brzezinska, D.J. Pine, F.F. Lange, S. Yang: Monodisperse micrometer-scale spherical assemblies of polymer particles, Adv. Mater. **14**(16), 1137–1140 (2002)

20.63 J.R. Millman, K.H. Bhatt, B.G. Prevo, O.D. Velev: Anisotropic particle synthesis in dielectrophoretically controlled microdroplet reactors, Nature **4**, 98–102 (2005)

20.64 V. Srinivasan, V.K. Pamula, R.B. Fair: An integrated digital microfluidic lab-on-a-chip for clinical diagnostics on human physiological fluids, Lab Chip **4**(4), 310–315 (2004)

20.65 B. Zheng, J.D. Tice, R.F. Ismagilov: Formation of arrayed droplets by soft lithography and two phase fluid flow, and application in protein crystallization, Adv. Mater. **16**(15), 1365–1368 (2004)

20.66 M. He, J.S. Edgar, G.D. Jeffries, R.M. Lorenz, J.P. Shelby, D.T. Chiu: Selective encapsulation of single cells and subcellular organelles into picoliter- and femtoliter-volume droplets, Anal. Chem. **77**, 1539–1544 (2005)

20.67 S. Sugiura, T. Oda, Y. Aoyagi, R. Matsuo, T. Enomoto, K. Matsumoto, T. Nakamura, M. Satake, A. Ochiai, N. Ohkohchi, M. Nakajima: Microfabricated airflow nozzle for microencapsulation of living cells into 150 micrometer microcapsules, Biomed. Microdevices **9**, 91–99 (2007)

20.68 H. Shintaku, T. Kuwabara, S. Kawano, T. Suzuki, I. Kanno, H. Kotera: Micro cell encapsulation and its hydrogel-beads production using microfluidic device, Microsyst. Technol. **13**(8–10), 951–958 (2007)

20.69 S. Sakai, I. Hashimoto, K. Kawakami: Agarose-gelatin conjugate for adherent cell-enclosing capsules, Biotechnol. Lett. **29**(5), 731–735 (2007)

20.70 C.H. Yang, K.S. Huang, J.Y. Chang: Manufacturing monodisperse chitosan microparticles containing ampicillin using a microchannel chip, Biomed. Microdevices **9**(2), 253–259 (2007)

20.71 S. Abraham, E.H. Jeong, T. Arakawa, S. Shoji, K.C. Kim, I. Kim, J.S. Go: Microfluidics assisted synthesis of well-defined spherical polymeric microcapsules and their utilization as potential encapsulants, Lab Chip **6**(6), 752–756 (2006)

20.72 J.W. Kim, A.S. Utada, A. Fernandez-Nieves, Z.B. Hu, D.A. Weitz: Fabrication of monodisperse gel shells and functional microgels in microfluidic devices, Angew. Chem. Int. Ed. **46**(11), 1819–1822 (2007)

20.73 V. Srinivasan, V.K. Pamula, R.B. Fair: Droplet-based microfluidic lab-on-a-chip for glucose detection, Anal. Chim. Acta **507**, 145–150 (2004)

20.74 I. Shestopalov, J.D. Tice, R.F. Ismagilov: Multi-step synthesis of nanoparticles performed on millisecond time scale in a microfluidic droplet-based system, Lab Chip **4**, 316–321 (2004)

20.75 Z. Guttenberg, H. Müller, H. Habermüller, A. Geisbauer, J. Pipper, J. Felbel, M. Kielpinski, J. Scriba, A. Wixforth: Planar chip device for PCR and hybridization with surface acoustic wave pump, Lab Chip **5**, 308–317 (2004)

20.76 M.A. Burns, C.H. Mastrangelo, T.S. Sammarco, F.P. Man, J.R. Webster, B.N. Johnson, B. Foerster,

D. Jones, Y. Fields, A.R. Kaiser, D.T. Burke: Microfabricated structures for integrated DNA analysis, Proc. Natl. Acad. Sci. USA **93**, 5556–5561 (1996)

20.77 B. Zheng, L.S. Roach, R.F. Ismagilov: Screening of protein crystallization conditions on a microfluidic chip using nanoliter-size droplets, J. Am. Chem. Soc. **125**, 11170–11171 (2003)

20.78 B. Zheng, J.D. Tice, L.S. Roach, R.F. Ismagilov: A droplet-based, composite PDMS/glass capillary microfluidic system for evaluating protein crystallization conditions by microbatch and vapor-diffusion methods with on-chip x-ray diffraction, Angew. Chem. Int. Ed. **43**, 2508–2511 (2004)

20.79 P.S. Dittrich, M. Jahnz, P. Schwille: A new embedded process for compartmentalized cell-free protein expression and on-line detection in microfluidic devices, ChemBioChem. **6**, 811–814 (2005)

20.80 R.N. Beer, B.J. Hindson, E.K. Wheeler, S.B. Hall, K.A. Rose, I.M. Kennedy, B.W. Colston: On-chip, real-time, single-copy polymerase chain reaction in picoliter droplets, Anal. Chem. **79**, 8471–8475 (2007)

20.81 M. Srisa-Art, A.J. deMello, J.B. Edel: High-througput DNA droplet assays using picoliter reactor volumes, Anal. Chem. **79**(17), 6682–6689 (2007)

20.82 D. Dendukuri, K. Tsoi, T.A. Hatton, P.S. Doyle: Controlled synthesis of nonspherical microparticles using microfluidics, Langmuir **21**, 2113–2116 (2005)

20.83 A. Terray, J. Oakey, D.W.M. Marr: Microfluidic control using colloidal devices, Science **296**(7), 1841–1844 (2005)

20.84 B.R. Acharya, T. Krupenkin, S. Ramachandran, Z. Wang, C.C. Huang, J.A. Rogers: Tunable optical fiber devices based on broadband long-period gratings and pumped microfluidics, Appl. Phys. Lett. **83**(24), 4912–4914 (2003)